Advances in
Energy Systems and Technology

Volume 1

Advances in Energy Systems and Technology

Volume 1

PETER AUER

Upson Hall
Cornell University
Ithaca, New York

ACADEMIC PRESS New York San Francisco London 1978
A Subsidiary of Harcourt Brace Jovanovich, Publishers

ACADEMIC PRESS, INC.
111 Fifth Avenue, New York, New York 10003

United Kingdom Edition published by
ACADEMIC PRESS, INC. (LONDON) LTD.
24/28 Oval Road, London NW1 7DX

Library of Congress Catalog Card Number: 78-4795

ISBN 0-12-014901-X

PRINTED IN THE UNITED STATES OF AMERICA

Contents

GEOTHERMAL ENERGY
Vasel Roberts

CLEAN FUELS FROM COAL
Harry Perry

DISTRICT HEATING WITH COMBINED HEAT AND ELECTRIC POWER GENERATION
Richard H. Tourin

List of Contributors

Numbers in parentheses indicate the pages on which the authors' contributions begin.

Harry Perry (243), Resources for the Future, National Economic Research Associates, Washington, D.C. 20036

David Pimentel (125), College of Agriculture and Life Sciences, Comstock Hall, Cornell University, Ithaca, New York 14853

Vasel Roberts (175), Electric Power Research Institute, P.O. Box 10412, Palo Alto, California 94304

D. G. Shepherd (1), Sibley School of Mechanical and Aerospace Engineering, Upson Hall, Cornell University, Ithaca, New York 14853

Richard H. Tourin (327), Stone & Webster Engineering Corporation, One Penn Plaza, New York, New York 10001

Walter Vergara (125), Department of Agricultural Engineering, Riley-Robb Hall, Cornell University, Ithaca, New York 14853

Preface

The subject of energy is now highly topical. The very rapid growth in the number of published articles and books on the subject provides clear proof of this. Is there a need for more books on energy, one might well ask. Our answer to this is a firm yes, for it is our expectation that the serial publication we are embarking on will serve a unique purpose.

The chapters appearing in this volume, and in each subsequent volume, are intended to furnish detailed critical reviews of specific topics within the general field of energy. They address largely technological issues, or issues in a somewhat broader systems context, which in turn are closely related to technological issues. We intend to have each article provide a breadth of coverage greater than that generally found in review articles prepared for journal publication or a standard review series, yet less than what may be expected from a textbook devoted to the same subject. Thus, each of the articles appearing here should serve as a valuable reference work for an extended period of time.

The publication is addressed both to the serious student and research investigator engaged in some aspect of energy study, as well as to policy analysts and energy planners who seek a fuller understanding of the technical factors underlying energy developments. It may well serve as a reference text in university level courses.

The scope of this serial publication will be broad, encompassing recent developments in the technologies of energy supply and delivery, end use and conservation, as well as in methodologies for policy analysis. The present volume includes topics ranging from such renewable resources as wind and biomass, to the assessment of the potential of geothermal energy, the prospects of clean fuels from coal, and, finally, to the possibility of introducing modern district heating systems to the United States. Future volumes of this publication may concentrate to a greater degree on a common theme or select a group of topics that appear to be particularly timely.

Finally, it is our intent to attract an international audience of readers to the series, for the energy problems of concern are inherently international in nature, though they may be punctuated by important considerations that vary from one region of the globe to another.

Peter Auer

Wind Power

D. G. Shepherd

Sibley School of Mechanical and Aerospace Engineering
Cornell University
Ithaca, New York

ISBN 0-12-014901-X

I. INTRODUCTION AND BACKGROUND

The development of wind power has been characterized generally by periods of progress and longer periods of neglect, but nevertheless the employment of wind power has never been absent from the time of its inception many centuries ago in ancient Babylon and the Middle East, and possibly before that to a suggested genesis from the prayer mills of Tibet. It is convenient, if an oversimplification, to date the modern era from the time of World War I, when powered flight led to the burgeoning of the science of aerodynamics and its adaptation to the development of fluid machines in general. Application of propeller theory to wind turbines by Betz, Glauert, and others still furnishes the basis of modern analysis.

The 25 years from the end of World War II have seen increased, although sporadic, interest in the western countries, of which Denmark, Germany, France, Great Britain, and the United States may be perceived to have made major contributions.

In Denmark, following the earlier pioneering work of La Cour at Askov, particularly in the use of wind power for electricity generation, the concept of wind power for making a significant impact was energetically revived. The complete absence of fossil fuels and the vicissitudes of two world wars led to a not-too-common united effort of government, industry, and public utility to carry out experimental work and to finance, design, operate, and test a number of wind energy conversion systems (WECS), culminating in the Gedser Mill of 200 kW placed in continuous operation in 1958 (Juul, 1956, 1964). Danish activity died down in 1967 after the Gedser Mill suffered a mechanical breakdown after 10 years use, and an adverse economic survey led to the conclusion that wind power was not competitive and the mill was idle for 10 years. However, the current energy shortage has revived interest and recommendations have been made to the government for a five-year, $9,000,000 program in experiment and pilot operation (Hinrickson and Cawood, 1976; Danish Academy of Technical Sciences, 1975). The Gedser Mill itself was recommissioned in 1977 for further experimental work, particularly more detailed measurements of the rotor performance (Merriam, 1977–1978). A 2-MW wind turbine has been designed, built, and placed in operation at Tvind by a local community headed by a school group, with consultant help (see Section XII).

In Germany the tradition of Betz (see Section II) has been carried on with Hütter as the major contributor. His work covers many years and stands out, along with that of Juul in Denmark, as having a wide scope in all areas: analysis, design, construction, testing, operation, and economics (Hütter, 1964a, 1977). His name is associated with the Allgaier units produced commercially for many years and with the design of the ERDA–NASA 100-kW MOD-O test unit at Plum Brook, United States. He has also pioneered in the use of composite materials for rotor blades.

French work for the period in question includes that of Lacroix (1969) and Vadot (1957), plus the manufacture and testing of two large units. One, the plant at Nogent-Le-Roi, was rated at 800 kW at a wind speed of 16.7 m/sec (≈37 mph), and this appears to the third largest output wind turbine yet operated. However, the blade diameter was only 30.2 m (≈100 ft) owing to the high wind speed for the rated output. The second unit, at Saint-Remy-des-Lourdes, was rated at 132 kW at 13.2 m/sec (≈30 mph), with a blade diameter of 21.2 m (≈70 ft). Tests were also made in Algeria on the Andreau-type plant designed and manufactured in Britain (see later) and small output units for lighthouse duty, etc. An account of French work is given by Bonneville (1974).

In Britain, the years since World War II have been dominated by the work of E. W. Golding and his colleague and successor A. Stodhart at the Electrical Research Association. Golding is the author of the classic text on wind power, which although published in 1955 is still valid in most respects and after being out of print for many years has now been reissued with some revisions by Harris (Golding, 1976). The work is particularly valuable for its emphasis on aspects of the motive power itself, i.e., the characteristics, distribution, surveys, measurement, and general data relating to the wind and its behavior. The aerodynamic design of blades is treated lightly, but on the whole is an excellent introduction to all aspects of wind power. During the 1950s and 1960s, three 100-kW units were built and operated in the United Kingdom; two being conventional propeller types, one being in the Orkney Islands north of Scotland and the other in the Isle of Man, with the third being an Andreau type wherein hollow blades ejected air from the tips, with the resulting suction used to drive the turbine at ground level via a connecting duct. The latter unit was originally erected at St. Albans near London, and later removed to Algeria, where it was operated for a period. These machines were regarded as experimental and as models for larger units of possibly 1-MW output (Golding, 1955; *BEAMA J.*, 1955). Again the time was not right for economic development and so technical development likewise came to a halt.

In the United States, the period is highlighted first of all by a negative development, the almost complete disappearance of the rural windmill used for pumping water and for low-voltage dc lighting. These units, of 0.5–3 kW, were ubiquitous in the Middle West, where distances were too great to afford the distribution cost of public utility power to scattered farms. Their disappearance was due to the Rural Electrication Acts of the

1930s, but it is said that up to about 1930 there were still some 50,000 units in operation. Ironically, these well-engineered, reliable windmills are now a collector's item and those extant are being vigorously sought for overhaul and resale. Their basic design still appears to be the equal of anything yet proved in quantity production for small-output plants.

The second development is that of the Smith–Putnam 1250-kW wind turbine erected in 1941 on "Grandpa's Knob" in New Hampshire, which operated (intermittently and with one 2-year outage due to war shortages) until March 1945 (Putnam, 1948). This ambitious project, with an output far greater than hitherto attempted, still remains the second largest unit operated to this day and this was 35 years ago. The abandonment of the wind power project as a whole, although precipitated by mechanical failure, was due to economic reasons pertinent at the time, and not to basic unsoundness in either mechanical or aerodynamic design.

The third major event was the reports by Thomas (1945, 1946, 1949) for the Federal Power Commission on the possibilities and design features of wind power plants. These were comprehensive studies and still relevant in many respects, although no action resulted at the time.

Thus the quarter-century 1945–1970 was a period of activity but an uneven one, with interest stimulated by individuals or small groups and in nearly every case with promising developments arrested by virtue of adverse or nonproven economic merit. By the last half of the 1960s, activity was at a low point, judging by the lack of reported work during this period. The following quotation is pertinent:

> The recent resurgence of interest in wind power generation is due to a number of causes. Among these are the costs of fuels and their high rate of exhaustion in some countries, the need for alternative sources of energy in countries where the end of the exploitation of economic water power sources is in sight, the desire for independence of imported fuels and the urge to make fuller use of some of the under-developed areas of the world where a main supply of electricity would be out of the question in the early stages of development.

Thus Golding (1955)—in 1954. Prophetic words unheeded then, repeated by scores 20 years later. An additional reason for the resurgence in these times is that of concern for the environment and the search for "clean" sources of energy. Wind power plants have access to an inexhaustible source of energy and yield no chemical or thermal pollution. It would seem that their environmental impact is limited to visual aspects, to noise, and to interference with high-frequency broadcast transmission. An individual wind turbine can be esthetically satisfying but it is unlikely that rows of turbines on the passes and peaks of the White Mountains of New Hampshire or the Pennines of Northern England will lack fierce opposition from environmentally concerned groups or indeed from the public at large.

Given the motivation for the use of wind power quoted above, but on the other hand being aware of the disappointments of the recent past, are there any new reasons for reexamination at this time? It would seem that the interest of much of the general public is engendered by the environmental aspects, which fit the new awareness of the need for restraint in the spread of hard technology, while the not-inconsiderable support of many governmental institutions throughout the world is based primarily on the rapidly diminishing availability of the convenient fossil fuels and their equally rapidly increasing cost. With respect to technological reasons, the continuous development of new materials and manufacturing methods would be expected to allow greater freedom in design and lower production cost, while the advances in the aerodynamics of wings and blading, particularly as related to experience with helicopter rotors, together with greatly improved techniques in stress analysis applicable to vibration and flutter of rotor blades, should provide substantial help in areas posing particularly severe problems for wind turbines. All technical and economic concerns now have today's vastly more sophisticated computer techniques to help in obtaining optimal solutions for performance, reliability, and life, and, equally important, in load control and scheduling, It is the purpose of the following material to first provide a framework by which recent and current work may be related and assessed and then to summarize such work as a starting point for evaluation of the present status of wind energy systems, and for appraising some of the major problems for continuing development.

II. ENERGY, POWER, AND MOMENTUM CONSIDERATIONS

A. Energy Available

The power available in the wind is taken as the flux of kinetic energy through the active cross-sectional area intercepting the apparatus which is utilizing this energy for mechanical or electrical output. This flux is $\dot{m}V_\infty^2/2$ and with $\dot{m} = \rho A V_\infty$, it is expressed as $\rho A V_\infty^3/2$. The density cannot be controlled except within the limits of altitude siting, and usually other considerations prevail. However, it might be noted that the density of the standard atmosphere is nearly 16% lower at an altitude of 1500 m ($\approx$5000 ft) than it is at sea level. Using the sea-level value of density, 1.225 kg/m^3 (0.0765 lb/ft^3), the available power $\dot{W}$ in kW is given by $6.125 \times 10^{-4} A V_\infty^3$, where A is in m^2 and V_∞ in m/sec (or $\dot{W} = 5.05 \times 10^{-6} A V_\infty^3$, with A in ft^2 and V_∞ in mph). This relationship is the key to the whole problem of utilizing wind power, in the first place because of the very low energy density, and secondly, because of the powerful effect of the cube relationship. Evaluating the above expression for power with $V_\infty = 5$ m/sec ($\approx$11.2 mph), then an area of 13.15 m^2 ($\approx$142 ft^2) is required for 1 kW available power, or a circle of 4.09 m diameter ($\approx$13.4 ft). Average wind speeds of only 4–5 m/sec are typical for many regions of the world where wind power might be desirable and although diligent effort can sometimes find individual sites on a generally poor area which can be more favorable, economic application in such areas is likely to be minimal. The dependence of power on V_∞^3 is thus of paramount importance, because it implies that only a small variation of estimated wind behavior can be the difference between success and failure in economic terms. Thus siting is all-important and this requires long-term detailed data for wind behavior to ensure reliable system behavior. Such data are very meager and unlikely ever to be available in kind and degree to the desired extent and hence much effort is going toward developing generalized statistical relationships which may be used to estimate reliable probable values from a minimum of measurements.

This power relationship provides the motif for the design of wind turbines and the problems associated with its economic development. In the first place, the energy density is very low, comparable to that of solar radiation, thus requiring a large machine for appreciable output. It must also be noted that the power levels quoted are for available power and not that actually delivered by a wind power unit. A combination of an inherent physical limitation and the aeromechanical efficiency means that at best, only about 40% of the available power is likely to be utilized. The large size carries with it the concomitant need for low rotational speed, owing to stress restrictions, and this in turn connotes a speed increasing device for electrical generation. The cube law for wind velocity stresses the importance of siting and of setting a design condition optimal between failure to utilize fully the energy available and high first cost due to overrating of the unit. There is also the fact that there is a minimum windspeed for which usable power is delivered (cut-in speed) and a maximum value beyond which the unit must be shut down for reasons of structural safety (cut-out speed), also sometimes called "furling speed." The advantages are a free and inexhaustible energy supply and low maintenance, thus obviating problems of cost inflation.

It will be these characteristics that guide the following discussion. There is no doubt of the technical feasibility of the use of wind power—the problems lie in providing cost-effective systems.

B. Performance Boundaries

The starting point in wind turbine analysis lies in the control volume approach contained in the activator disk concept originated by Rankine and extended by the Froudes for marine propellers. It was codified by Betz (1926) for wind turbines and has received further treatment in recent years from others, as noted in following analyses. Figure 1 shows a stream tube with an actuator disk representing the wind turbine, the air being decelerated from approach speed V_∞ to velocity V at the disk and thence to V_e well downstream of the disk. Application of the momentum and energy relationships for simple axial flow having no tangential components, with the assumptions of continuous deceleration and discontinuous pressure decrease at the disk, with accompanying extraction of energy, leads to the relationship

$$V = \tfrac{1}{2}(V_\infty + V_e) \tag{2.1}$$

Thus half the total velocity decrease $\frac{1}{2}(V_\infty - V_e)$ occurs upstream of the disk and half downstream. This retardation is expressed in the form of an axial *induction*, *interference*, or *perturbation* factor a, such that $V = (1 - a)V_\infty$ or $a = 1 - (V/V_\infty)$. Hence $V_e = V_\infty(1 - 2a)$ and therefore a has a maximum value of $\frac{1}{2}$ when $V_e = 0$.

The power output can be expressed either as the product of the force or thrust on the disk $\rho AV(V - V_e)$ times the air velocity V at the disk, or as the total flux of kinetic energy, $\rho AV(V_\infty^2 - V_e^2)/2$, where ρAV is the mass flow of air in the stream tube. Using the factor a, this gives

$$\text{power } \dot{W} = (\tfrac{1}{2}\rho AV_\infty^3)4a(1 - a)^2 = 2\rho AV_\infty^3 a(1 - a)^2 \tag{2.2}$$

For a given disk and wind speed, the power is seen to have a maximum value at $a = \frac{1}{3}$, when the discharge velocity V_e is $\frac{1}{3}$ of V_∞.

The effectiveness of a wind turbine may be expressed in several ways but that known as the power coefficient C_p has proved to be the most useful. This is the ratio of the actual power developed by the rotor to the free stream energy flux through a disk of the same area as swept out by the rotor, i.e., $\rho AV_\infty(V_\infty^2/2) = \rho AV_\infty^3/2$.

$$C_p = \text{power}/\tfrac{1}{2}\rho AV_\infty^3 \tag{2.3}$$

Using (2.2), the value of C_p is then given by $4a(1 - a)^2$ and its maximum

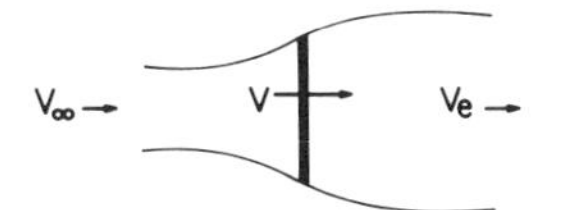

Fig. 1. Actuator disk diagram for a wind turbine.

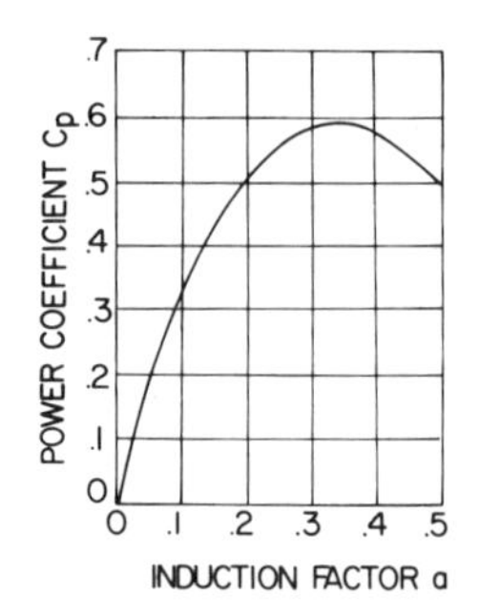

Fig. 2. Power coefficient versus induction factor.

value for $a = \frac{1}{3}$ is $\frac{16}{27} = 0.593 \approx 60\%$. This is sometimes called the Betz coefficient. It should be noted that occasionally the power coefficient of a wind turbine is used as the ratio of the actual power output to the ideal Betz value, i.e., $\dot{W}/(16/27\rho AV_{\infty}^3)/2$, but this is not common practice. C_p as a function of a is shown in Fig. 2, and it is seen that although there is some latitude for a in the region $0.2 < a < 0.5$, there is a rapid decline of C_p for $a < 0.2$. [Note that this definition of power coefficient is not the "efficiency" of the rotor, which would be expressed as the ratio of the power output to the actual kinetic energy flux based on the approach area A_{∞} of the stream tube. The ratio $A_{\infty}/A = 1 - a$, and thus this efficiency is given by $4a(1 - a)$, hence with $a = \frac{1}{3}$ for maximum C_p, the corresponding efficiency is then $\frac{8}{9} \approx 0.89$. The maximum efficiency of unit occurs at $a = \frac{1}{2}$, when $V_e = 0$ and this represents a limit as V_e cannot be less than zero.]

The question arises of possible staging of rotors and although it intuitively appears unrewarding in view of the cubic relationship of power to wind speed, the situation may arise in some forms of vertical axis units in which the air flows over a blade twice. Figure 3 shows an idealized arrangement for actuator disks in tandem, together with the nomenclature. The upstream velocity for the downstream unit is the downstream ve-

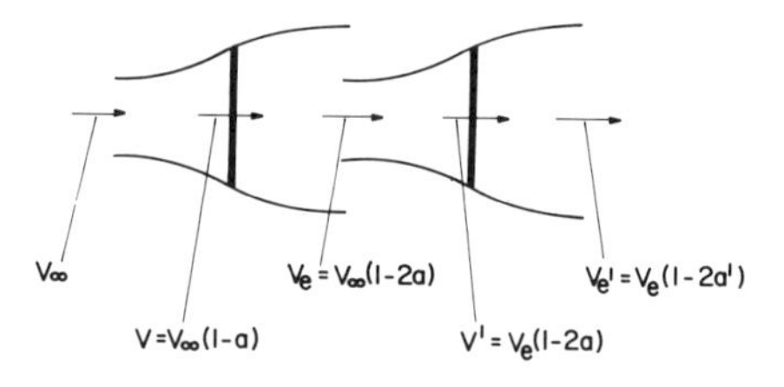

Fig. 3. Actuator disks in tandem.

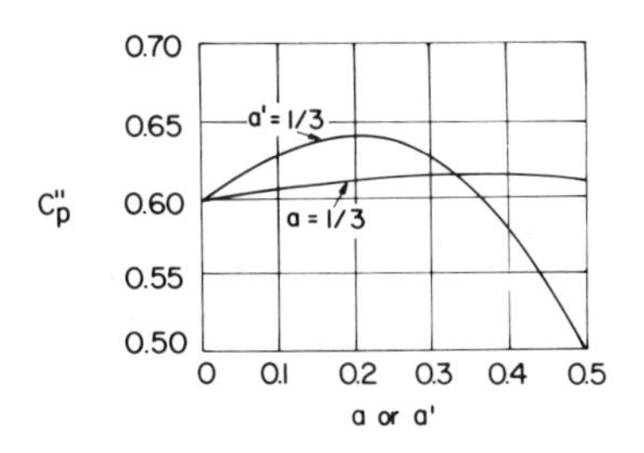

Fig. 4. Power coefficient for tandem disks versus induction factors.

locity of the upstream unit. Using the prime symbol to denote downstream values, then

$$\dot{W} = C_p \rho A V_\infty^3/2 \qquad \text{and} \qquad W' = C_p' \rho A V_\infty'^3/2$$

$$C_p = 4a(1-a)^2 \qquad \text{and} \qquad C_p' = 4a'(1-a')^2$$

The combined power coefficient C_p'' based on the upstream or free-stream wind speed V_∞, and substituting $V_\infty' = V_e = V_\infty(1 - 2a)$, is then given by

$$C_p'' = \frac{\dot{W} + \dot{W}'}{\rho A V_\infty^3/2} = C_p + C_p'(1 - 2a)^3$$

The maximum value of C_p'' occurs when C_p' is a maximum, i.e., when $a' = \frac{1}{3}$ and $C_p' = 0.593$. With this value of C_p', then the maximum value of C_p'' is 0.640 with $a = 0.2$ and $C_p = 0.512$. A plot is given as Fig. 4, which also shows that operating with C_p at a maximum and allowing C_p' to vary, the increase in C_p'' is negligible, for the reason that the wind velocity available to the second stage is very small.

III. TURBINE TYPES AND TERMINOLOGY

It is helpful to classify wind turbines into a number of types, provided that this is not carried out with the attempted rigor of biological taxonomy. The usefulness of a taxonomy lies in being able to categorize a group of machines as having similar features, functions, or means of operation. One possible classification is geometrical, being the orientation of the axis of the rotor, i.e., in the direction of the wind or normal to it. The latter type is also known as a *crosswind-axis* turbine. The shaft orientation is also used to differentiate the *horizontal-axis* turbine and the *vertical-axis* turbine, which have considerable overlap with the first classification, as axes in the direction of the wind must be horizontal, and vertical axes must be normal to it. However, the crosswind type can have a horizontal axis, and thus the terms are not exclusive. Axis classification implies a rotational motion, but a wind turbine can have a translational motion only, the "tracked-vehicle" concept, in which a force is generated by the wind on a surface attached to a carriage, which thus can be moved along a track, with electric power developed by a generator driven from the axles. Another major classification is based on operative force, whether of *lift* or *drag*. Vadot (1975) uses two main categories: machines in which the rotor blades move in the direction of the wind and those that move perpendicularly to the wind. He subdivides the first group into those with "simple drag" and those with "drag difference," and the second group as having lift, but it is difficult to agree with some of his classifications. It is not possible in the limited space here to mention all types of wind turbines and Eldridge (1976) gives a useful illustrated "taxonomy," with many diagrammatic representations.

About the only significant type of machine in the horizontal-axis class is the *propeller* type, characterized in history by the Dutch windmill, and in later years by the multivaned farm windmill, through the Danish prototypes presaging the modern form with two or three blades of exact aerodynamic shape and appearing almost indistinguishable from an airplane propeller, whence the name. It is also a lift type with blades moving perpendicularly to the wind, and has the following major characteristics:

1. It must face the wind, i.e., the plane of the rotor is perpendicular to the wind, and therefore the rotor has to be moved into the correct direction as the wind changes. For small machines, this can be accomplished automatically by a downstream tail vane, but in large units there must be a restraining mechanism to avoid overstressing due to sudden Coriolis and other forces, and a means of slowly moving the rotor into the required direction.

2. It may have a high blade-tip/wind speed ratio, yielding a relatively high rotational speed, very desirable for electrical generation.

3. Its aerodynamic losses are small, placing it among the most efficient types of wind turbine known.

4. It has to be mounted on a tower to allow for blade clearance, and this increases the first cost and difficulty of maintenance. The tower has to be strong enough to support both rotor and generator under all wind conditions, while in addition it acts as a source of undesirable vibrational effects on the rotor. There are some ingenious designs minimizing the tower problem, one such type being the Andreau turbine, which has hollow blades which expel air from the tips via centrifugal action, drawing this air through a duct near ground level. A turbine is placed in this duct near

entry, operating by the suction produced by the rotating blades. Thus tower structure is minimized and the generator is more conveniently inspected and serviced.

5. It may have few blades (two or three) or many, and this determines its speed and torque characteristics (see later).

The analysis and performance of the classic propeller wind turbine is given in Section IV. Another type of horizontal-axis turbine is the sail-wing rotor, so called because its "blades" are made of a flexible, sail-like material, with a spring-mounted trailing edge which allows a variable tension and shape depending on the force and direction of the wind, akin to that of a sailing boat. It is an interesting variant with several advantageous features, but which so far does not seem to have progressed beyond an experimental stage.

It has been said that it is impossible to invent a new type of wind turbine—they have all been thought of a long time ago. While not literally true, it is certainly difficult to devise a radically new design, and most innovations turn out to be variants of relatively few distinct species. This seems to be true of vertical-axis units, of which there are certainly very numerous examples. It is possible to use three classifications for these, the *lift* type, the *drag* type, and the one here named the *turbo* type, although there may be others which have escaped attention.

The lift type uses blades of airfoil profile as in the propeller design, but with their span arranged in a generally longitudinal direction along the axis of rotation rather than being perpendicular to it. The blades may be straight or curved and be fixed or capable of variable incidence, see Fig. 5. This type is at present characterized by the *Darrieus* rotor or "egg beater," Fig. 5(b), as it is receiving considerable attention lately. Lift may also be utilized for translational motion, as there are two significant designs using the tracked-vehicle concept. One consists of straight airfoil blades mounted vertically on a carriage and acting similarly to an airplane wing, and the other of a rotating cylinder on a carriage, employing the

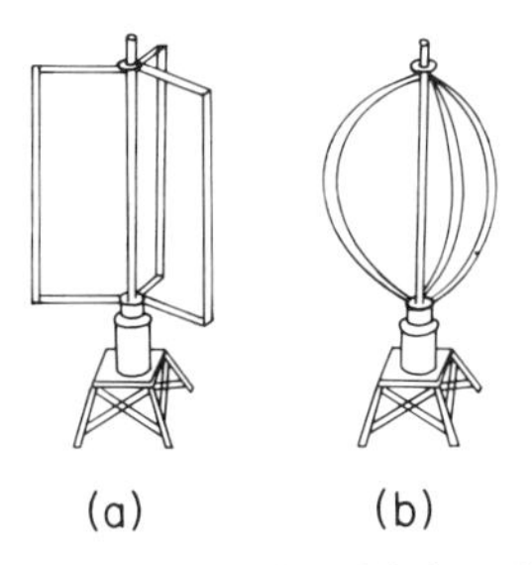

Fig. 5. Darrieus rotor. (a) Straight blades. (b) Curved blades.

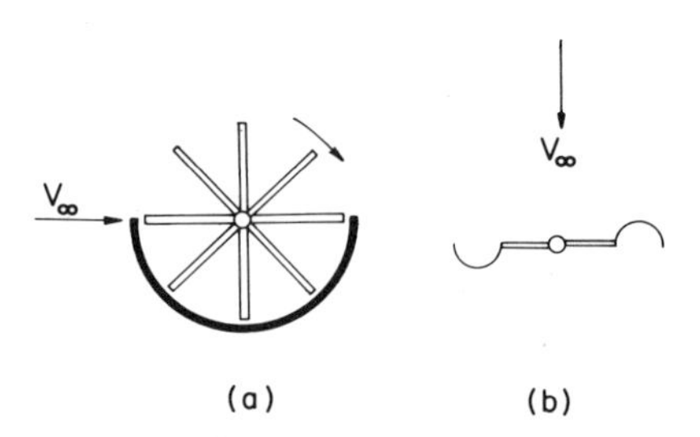

Fig. 6. Drag-type turbines. (a) Paddle wheel. (b) Cup anemometer.

Magnus effect exemplified by the Flettner rotor used in earlier years for ship propulsion.

The drag type appears with a variety of straight or curved blading. One example is the simple paddle-wheel rotor, Fig. 6(a), with the wind acting on one half, with the other half shielded (Vadot's simple drag type), and another (Fig. 6(b)) is the familiar cup anemometer utilizing differential drag on a hemisphere as it revolves about the axis (Vadot's drag difference type). Drag types are capable of myriad variations, with vanes which are movable as they revolve to provide variable drag with angular position. The turbo type, Fig. 7, has the features of industrial turbines in which the fluid, water, steam, or gas can be thought of as producing power by a momentum change in flowing through a bounded passage formed by adjacent blades, rather than by a force acting on a surface. One example is the well-known Savonius rotor with central gap, Fig. 7(a), and another is modeled after the Banki water wheel, Fig. 7(b). (Vadot classes these as drag difference types, but their performance is vitiated rather than improved by increase of drag.)

The characteristics of vertical-axis turbines are as follows:

1. Nearly all types, certainly all those in which the blades or vanes have a circular motion around the axis, develop power in winds from any direction, thus not requiring tail vanes or positioning mechanism and

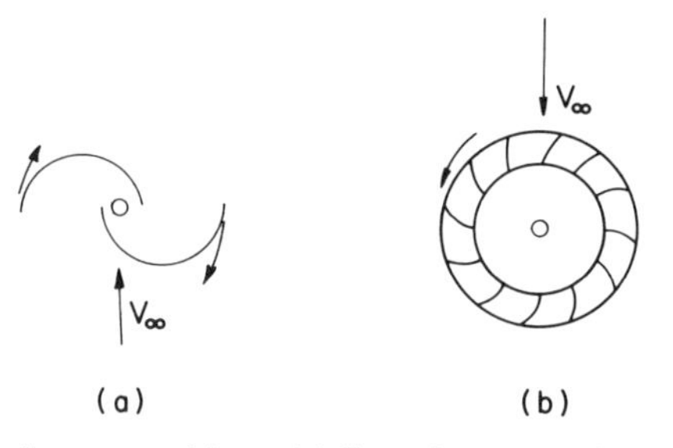

Fig. 7. Turbo-type turbines. (a) Savonius rotor. (b) Banki wheel.

being able to take immediate advantage of wind veers. For this reason they are often called *panemones,* sometimes pananemones ("all winds").

2. High towers are not necessary and support may be afforded by a bottom bearing with guy wires from the top. The attached load, such as a generator, is at or near ground level, with obvious advantage.

3. The blade-tip-speed/wind-velocity ratio depends on whether the unit is a lift or drag type. It may be as high as that of the propeller type for the former, but the drag type necessarily has a value less than unity, since the vane cannot outrun the fluid when traveling in the same direction. (This is one reason for classifying the Savonius rotor as a turbo type, because it can have this ratio a little greater than unity.)

The Darrieus rotor, which is taking a prominent part in today's development, is discussed in more detail later, together with the characteristics of some of the other more prominent vertical-axis types.

The terminology of wind power is varied, with the terms windmill, wind generator, wind machine, etc. being used along with various modifying adjectives. The Wind Energy Conversion Systems Workshop of 1973 (Savino, 1973) endeavored to establish nomenclature and definitions, but these are not yet universally followed. Their recommendations are generally used here, with the term wind energy conversion system (WECS) used for wind-powered machines, with "wind turbine" for the motive power unit itself (with rotor used on occasion). An acronym which has general use is VAWT (vertical-axis wind turbine), which will also be used herein, and also HAWT is used for horizontal-axis wind turbines. These are convenient terminologies, but are apt to be used for specific designs which the proponent apparently considers the only possible type for either a horizontal-axis or vertical-axis turbine, i.e., the propeller type or the Darrieus type, hence misunderstanding can arise.

IV. AERODYNAMICS OF WIND TURBINES

A. Flow through a Rotor

The simple linear momentum theory of Section II is an idealized concept and contains assumptions which are either not true or which we have no real reason to expect to be true. Little if any experimental work has been done on wind turbines or even models thereof in order to visualize the actual flow field, and much of the general behavior has been deduced from propeller and helicopter technology. This is mainly because of the restricted interest and financial support for wind power studies until very recently, with such support as has been available going to solve more immediate problems. Even with interest and support, prototype experiment is very difficult owing to the size of unit and the unsteadiness of the wind over sufficiently long periods. Model experiments are possible but are handicapped by the effect of wind tunnel walls (blockage) on all but very small models, because the wake is expanding rather than contracting as with propellers. On a generalized basis, helicopter experience has been used to shed light on turbine behavior. Apart from the earlier analyses of Glauert (1935), who systematized the development, taking into account rotational flows in actuator disk theory, contributions have been in later years by Rosenbrock (1952), Wilson *et al.* (1976), Wilson and Lissaman (1974), Stoddard (1977), Hütter (1977), Lapin (1975), and Rohrbach (1976).

The axial-momentum theory neglects any rotational component ω of the flow after passing through the actuator disk. This is obviously a simplification and its effect must be examined. This was done by Glauert (1935) following the original analysis for propellers of Joukowski and has been recapitulated by Wilson and Lissaman (1974). The addition of a vortex component must reduce the torque and power given to the disk because it is given to the fluid in passing through the disk. It is thus discontinuous at the disk and also is opposite to the disk rotation. It can be shown that for constant angular momentum in the wake flow (irrotational vortex, ωr^2 = constant) then

$$a = \tfrac{1}{2}a_\omega\{1 - [a_\omega^2(1 - a)/4\lambda^2(a_\omega - a)]\} \tag{4.1}$$

where a_ω is the wake axial induction factor with wake angular momentum, i.e., $a_\omega = 1 - V_e/V_\infty$; λ is the tip speed ratio, $\Omega R/V_\infty = U/V_\infty$, with Ω the rotor angular velocity, R the rotor radius, and U the rotor tip speed equal to ΩR.

Figure 8 presents a plot of the relationship (4.1) from Wilson and Lissaman, and it is seen that for the ideal actuator disk with $\lambda = \infty$, then

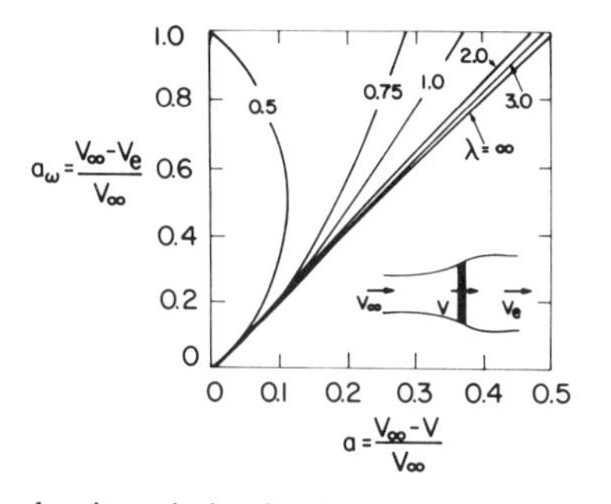

Fig. 8. Effect of tip-speed ratio on induction factors. (From Wilson and Lissaman, 1974.)

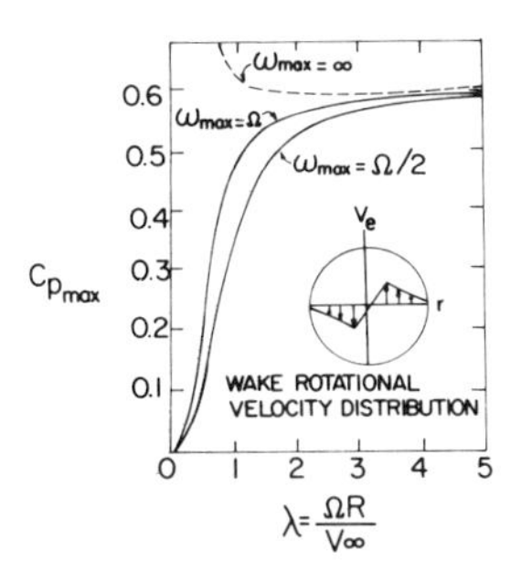

Fig. 9. Maximum power coefficient versus tip-speed ratio for a rotor with Rankine vortex wake. (From Wilson and Lissaman, 1974.)

$a_\omega = 2a$ exactly. For $\lambda \gtrsim 2$ then a_ω is little different from $2a$, but rapidly departs from this value for $\lambda < 1$. The second term within the bracket in (4.1) is usually small and if a_ω is replaced therein by its approximate value from the axial analysis, $2a$, then

$$a \approx \tfrac{1}{2}a_\omega\{1 - [a(1 - a)/\lambda^2]\} \tag{4.2}$$

The power coefficient with a rotational component is found to be

$$C_p = a_\omega^2(1 - a)^2/(a_\omega - a) \tag{4.3}$$

and with $a_\omega \simeq 2a$ this reduces to

$$C_p = 4a(1 = a)^2 \tag{4.4}$$

again the exact relationship for the axial theory. The assumption of the free vortex condition, ωr^2 = constant, across the wake becomes untenable near and at the axis and Wilson and Lissaman replace it with a Rankine vortex having a finite value for $\omega_{\max}$. This is shown in Fig. 9 as $C_{p_{max}}$ vs $\Omega R/V_\infty$, with values of $\omega_{\max} = \infty$, Ω, and $\Omega/2$. This gives an indication that $C_{p\max} \approx \frac{16}{27}$ is approached quite closely for $\Omega R/V_\infty$ greater than about 2, but decreases rapidly for low-speed rotors. For a given power, low values of λ imply high torque and the previous analysis shows that this also yields lower values of C_p. This is the effect of higher wake rotational energy due to higher torque. But high-torque turbines obviously have better starting characteristics than those with low torque, i.e., they can overcome starting resistance and produce useful power at lower values of wind velocity. Thus a basically more efficient WECS will yield reduced energy at minimum wind speeds. This parameter $\lambda = U/V_\infty$ is a very important one in wind turbine performance, as is its inverse $V_\infty/\Omega R$ in propeller performance. It is a dimensionless kinematic parameter and is analogous to the U/C_0 of the steam and gas turbine in which C_0 is the ideal "nozzle spouting velocity" or the velocity attained by an ideal expansion to maximum kinetic energy in the nozzle. It contains the three most important variables of turbine design and analysis; the wind speed, the rotor size, and the rotor rpm. It is used as a parameter against which the effect of aerodynamic variables as well as overall performance parameters such as power output, C_p, etc. may be expressed.

The actuator disk analysis is an idealized concept and efforts continue to be made to refine it to provide more exact answers. It does seem important that if possible there should be a firm standard representing maximum performance capability against which to judge actual wind turbine behavior. The simple axial disk analysis leads to the result that the induction factor a cannot be greater than 0.5 as this yields zero downstream velocity. However increasing thrust coefficient values are obtained for $a > 0.5$. The anomaly has been discussed by earlier writers, and Rosenbrock (1952) considers the effects of a thrust exerted by air which does not pass through the disk. It does no work but modifies the flow, so that the air speed at the disk is no longer the mean of its initial and final speed. Entering such a force F in the momentum and energy equations in the actuator disk model yields a modifying term $u = aV_\infty - (a^2V_\infty^2 - 2F/\rho A)^{1/2}$. u is zero when $F = 0$ and the usual relations result. When $F > 0$, the term is positive and Rosenbrock argues that $F = 0$ for $0 < a < 0.5$, the ideal regime. For $a > 0.5$, $V_e = 0$ and introducing the force $F > 0$ for $0.5 < a < 1$ allows for physical explanation of the actual behavior, with $C_p = 1 - a$ for $V_e = 0$. Thus C_p increases linearly from zero at $a = 1$ to 0.5 at $a = 0.5$, reaches a maximum of $\frac{16}{27}$ at $a = \frac{1}{3}$ and then decreases to zero when $a = 0$. He concludes that the assumption of the force F removes the barrier of no solution for $a > 0.5$ and justifies the accepted maximum value of C_p of $\frac{16}{27}$. It is pointed out that an unsteady slipstream would invalidate a major assumption of the simple theory and hence could modify its conclusions.

Following on the early work of Glauert (1935) and others on flow regimes in propellers (and to some extent with wind turbines), Wilson and Lissaman (1974), Wilson *et al.* (1976) and Stoddard (1977) have enlarged autogyro and helicopter rotor experience. It is possible by using the momentum theory alone to analyze several flow patterns for a propeller-like rotor and to gain an understanding of general behavior under a variety of circumstances. There are three overall flow states for a rotor which can be distinguished, each for a range of the induction factor a. These are shown diagrammatically in Fig. 10, which is based on Wilson and Lissaman (1974), and Wilson *et al.* (1976) with additions from Stoddard (1977). For $a < 0$ (i.e., negative values), the rotor is acting as a propeller accelerating the flow, with thrust opposing flow and requiring power to be

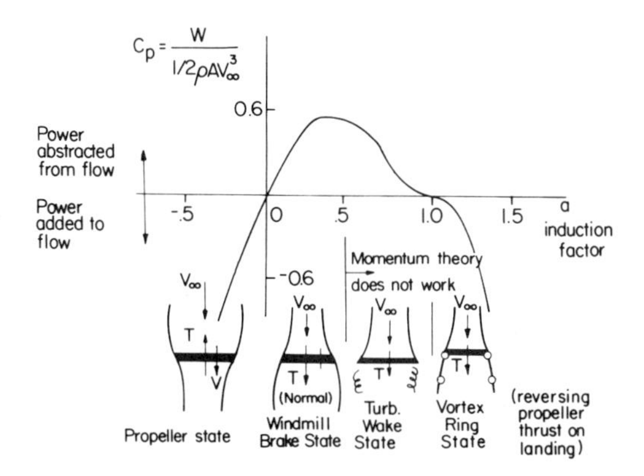

Fig. 10. Rotor flow states. (From Wilson and Lissaman, 1974; Wilson *et al.*, 1976; Stoddard, 1977.)

added. The thrust and power coefficients C_T and C_p are then negative. The second regime has two subdivisions, one for $0 < a < 0.5$, and one for $0.5 < a < 1.0$. Both are for decelerating flow, with C_T and C_p positive, and hence power is extracted from the flow. This is then the "windmill" state, with the former subdivision termed the (normal) "windmill brake state," with C_T a maximum at $a = 0.5$ and C_p a maximum at $a = \frac{1}{3}$. At $a > 0.5$, the wake velocity is zero, streamlines cease to exist, and the regime is incompatible with the simple momentum assumptions. For $0.5 < a < 1.0$, the simple theory would require flow reversal in the far wakes and zero velocity somewhere between disk and infinity. It is therefore necessary to look to observed behavior, since the rotor continues to operate in this anomalous state. It is called the "turbulent wake state," characterized by large recirculating flows and high turbulence. Values of C_T increase in this state, which is likened to that of a solid disk perpendicular to flow, with analogous downstream reverse flow and turbulence, and a drag coefficient of 1.2 and greater, which is actually observed. For $a > 1$, the rotor enters the propeller brake state, with power being added to the flow to create downwind thrust, corresponding to reversing propeller thrust on landing. For values of a not greatly over unity, the flow regime is called the vortex ring state, experienced by helicopter rotors during part-power descent. Airplane propellers pass through this state to that of their brake state with reversing thrust, $a \gg 1$.

Wind turbines will then normally operate in the region of $0 < a < 0.5$, but it is pointed out by Wilson and Lissaman that when a rotor operates at tip-speed ratios appreciably different from the design value, then the tips may by driven into the propeller brake state. Also, they note that it might be possible to use the confused flow of the propeller state to dump energy when it is necessary to prevent rotor overspeed due to high winds or reduced shaft-torque loads. Hütter (1964, 1977) discusses this problem of the rotor flow states and reasons that the existence of small values of $\zeta = V_e/V_\infty$ means a considerable expansion of the stream tube at the rotor tips and an expanding wake (contrary to the contracting wake of a propeller in the normal state), resulting in significantly changed induced velocities. Viscosity effects prevent any discontinuity at the rotor, and he reasons that the Betz coefficient of $\frac{16}{27}$ is not a definite upper limit, so that ζ should be shifted to smaller values than so far have been assumed to be best. Thus increases in values of the theoretical maximum C_p from 4 to 13% could be attainable. He also points out that a tighter grouping of WECS is made possible by such design because the shorter axial distances to complete wake development allow quicker recovery of the free wind velocity via turbulent exchange. Rohrbach (1976) likewise points out that the vortex ring flow condition and a vortex interference condition with the tip vortex of one blade passing close to the vortex of the following blade both may occur during start-up. Hence there seems reason to investigate these transient states more deeply and to recognize their possible development in practice.

B. Aerodynamic Analysis of Rotors—Propeller Type

The global approach via the actuator-disk model supplies the basic analysis, but it is necessary to introduce aerodynamic theory to design a rotor to yield a known performance in terms of force, torque, and power, or alternatively to determine these elements for a given rotor. The origin of such analysis lie in nineteenth century work on marine propellers and it was greatly extended by the advent of aeronautics and the application of wing theory to propeller design, which in turn serve as the starting point for wind turbines. It is not the purpose here to examine these analyses in any detail, but again the basic approach is recapitulated to establish the method as it applies to wind turbines and to focus on their particular problems. The model used is that of the "propeller" type of turbine, quite obviously in view of the history of the analysis, but it is capable of application to any lift type of rotor with some modification, including the Darrieus VAWT of today's interest. Thus the terminology and geometry used is that of the horizontal-axis airfoil-bladed rotor.

Figure 11(a) shows a diagrammatic front view of such a rotor blade, with a blade profile shown at radius r, while Fig. 11(b) shows an enlarged profile section with the relevant velocities and forces. The wind velocity V_∞ is modified at the rotor by the axial induction factor a and the rotor velocity Ωr is modified by a rotational induction factor a' defined as

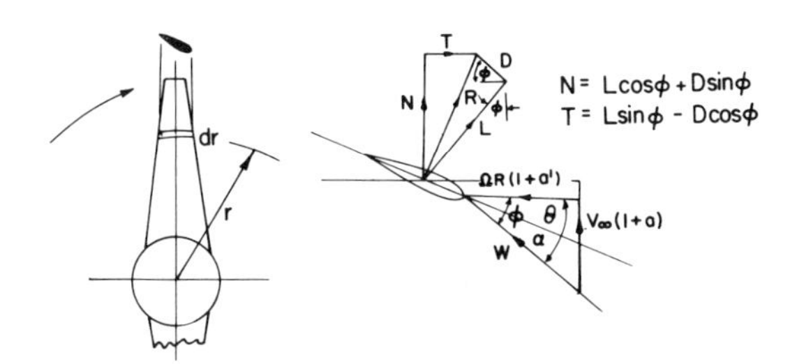

Fig. 11. Airfoil blade geometry. (a) Front view. (b) Profile section.

$a' = r/2\Omega$. The blade angle is given by θ, and ϕ is the angle between the plane of rotation and the relative velocity W. The air flow gives rise to a normal force F_N and a tangential force F_T, whose resultant can be resolved into components of lift L and drag D, parallel to and perpendicular to the relative wind velocity W, respectively. From the geometry we have

$$F_N = L \cos \phi + D \sin \phi \tag{4.5}$$

$$F_T = L \sin \phi - D \cos \phi \tag{4.6}$$

The thrust dT on the number B of blade elements dr of chord c is given by

$$dT = Bc\ \tfrac{1}{2}\rho W^2 C_N\, dr \tag{4.7}$$

where C_N is the normal force coefficient. Applying the momentum equation to a control volume formed by the annulus containing dr from V_∞ to V_e, this force may be written as

$$dT = d\dot{m}(V_\infty - V_e)$$

and with $d\dot{m} = \rho(2\pi r\ dr)V$ and using the momentum relationships of Section II,

$$dT = \rho(2\pi r\ dr)(1 - a)V_\infty(2aV_\infty) = 4\pi r\rho V_\infty^2 a(1 - a)\ dr \tag{4.8}$$

Equating (4.7) and (4.8) and substituting $W \sin \phi = V_\infty(1 - a)$ yields

$$a/(1 - a) = \sigma C_N/(4 \sin^2 \phi), \qquad \text{where} \qquad \sigma = Bc/2\pi r \tag{4.9}$$

Considering the drag force to be negligible, then $C_N = C_L \cos \phi$, and

$$a/(1 - a) = \sigma C_L \cos \phi/(4 \sin^2\phi) \tag{4.10}$$

A similar analysis can be made for the torque dT from the force F_T and this is equated with the angular momentum of the air. Using the approximation $a_\omega = 2a$, as before, results in

$$a'/(1 + a') = \sigma C_L/(4 \cos \phi) \tag{4.11}$$

From Eqs. (4.5)–(4.11), together with previous expressions for torque and power, the elements of rotor blade design can be established. This is the basis of the *blade element* or *strip* theory, which assumes that each elementary strip or annulus behaves independently of every other strip. Thus Glauert develops the analysis to show the optimum variation of the angle ϕ along the blade and the number and shape of the blades assuming a constant value of C_L along the blade. This gives rise to the relationship of solidity σC_L and velocity ratio $\Omega R/V_\infty$, where R is the blade-tip radius, and the characteristic form of fast-running wind turbines resembling a propeller with rather wide blades and of slow-running turbines with a large number of blades with rather large blade angles. Wilson *et al.* (1976) demonstrate an iteration procedure for these basic relationships by assuming values of a and a', then calculating ϕ, α, and the force coefficients from airfoil data and substituting back in 4.10 and 4.11 until agreement with the assumed values is obtained.

The next step in rotor analysis is the introduction of vortex theory, that is, the discontinuity introduced by the rotor into the undisturbed wind is represented by a vortex sheet or tube behind the rotor. This has been used in the development of propeller theory for many years but its application to turbines would seem to require replacement of the rigid wake vortex of a propeller by more flexible models. Useful summary presentations are made by Wilson *et al.* (1976), Wilson and Lissaman (1974) and by Rohrbach (1976), Rohrbach and Worobel (1975), who discuss various approaches and the development of programs applicable to the design of wind turbines.

Wilson and Lissaman (1974) give a lucid presentation of the vortex model and discuss its relevance to the simple blade element theory of noninteracting annular elements. The basic picture of the vortex flow model is shown in Fig. 12, with a system of bound vortices along the rotor arms which produce the lift on the blade and free vortices at the rotor tips which produce the induced velocity components. This yields a vortex tube of increasing radius downstream together with a central vortex to fulfill the Helmholtz vortex continuity laws. A high tip-speed ratio λ or a large number of blades result in a "tighter" helix, passing in the limit to a

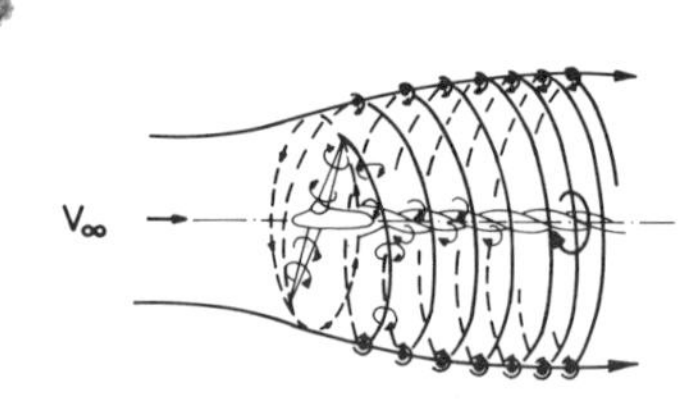

Fig. 12. Vortex flow model for a propeller turbine. (From Wilson and Lissaman, 1974.)

continuous bounding vortex system for an infinite number of blades. The expanding nature of the wake vortex leads to some relief from difficulties which occur with the contracting wake of the propeller, e.g., less reduction in circulation at the rotor tips for a finite number of blades (Hütter, 1961) and other effects dealt with shortly.

Rohrbach (1976) and Rohrbach and Worobel (1975) describe several vortex theories applied to wind turbines, the "Goldstein propeller" method, the "prescribed wake" program, the "rotor wake geometry" program, and the "skewed wake" program. These are summarized very briefly below. The first named is a well-established technique for propellers and is based on a constant-diameter wake. Comparison of results with the Hütter design indicates that it is probable that the expansion has a secondary influence on blade loadings. Comparison of the Goldstein design method gives good general agreement in predicting the effects of variation in geometric and aerodynamic parameters in performance, as substantiated by the limited experimental data available.

The restriction of the rigid-wake geometry of the Goldstein method to improve design accuracy can be removed by introducing arbitrary wake geometries. This is done through the use of finite filament vortex theory using the Biot–Savart law relating induced velocities to the filament circulation strength. This general method is of long standing and of recent years has been applied to helicopters and lift fans, but requires numerical solutions instead of the closed form of the Goldstein method. The difficulties in determining the wake shape or location of the expanding path of the wake vortices have been handled by two methods adopted from helicopter and propeller techniques. One uses parameters derived from visualization techniques, the other on analytically defined wakes. Rohrbach's prescribed wake method follows the former and his rotor wake geometry follows the latter. The skewed wake determines rotor performance when the freestream wind velocity is at an angle to the rotor axis. The wake shape is of constant diameter and is located by use of the Biot–Savart equation, following which the axial-induced velocities can be computed. A momentum balance is achieved by iteration methods and the induced velocities calculated.

The prescribed wake program for propellers utilizes a smoke technique for obtaining wake shape geometry and is the basis for experimental work for turbines. The rotor wake geometry program is another iterative technique based on the distribution of circulation and an arbitrary estimated wake shape. The Biot–Savart law allows induced velocities to the calculated and integrated over a small time interval to obtain a new wake shape. This is used for a prescribed wake analysis as above, giving a new circulation distribution. Iteration is continued until wake geometry and

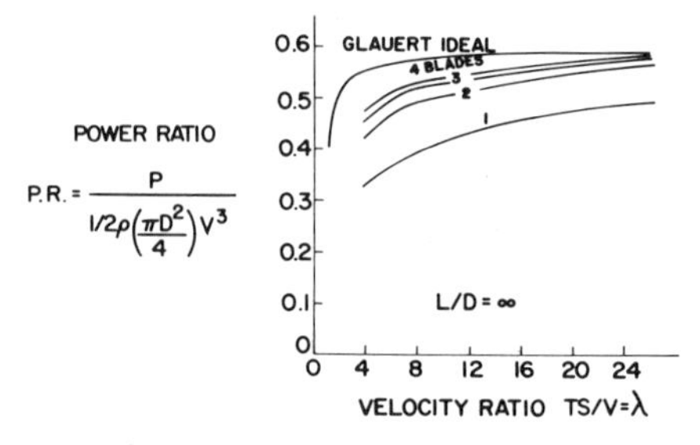

Fig. 13. Effect of number of blades on power coefficient. (From Rohrbach and Worobel, 1976.)

circulation distribution are compatible. Rohrbach gives a description of a check case of a two-bladed rotor with $\lambda = 10$ to determine the tip vortex geometry, but the iteration with the prescribed wake technique had not been completed at the time of the report. However, it was concluded that the influence on generated power of tip vortex distortions relative to the undistorted helical geometry was expected to be small because (1) the tip vortices being behind the rotor tend to make such distortions small relative to the distance of the vortex from the blades; (2) distortions are in a direction away from the turbine (i.e., expanding wake) and hence blade tip–vortex interference is small; and (3) again for an expanding wake, tip vortices do not pass directly under a blade and this eliminates the strong velocity gradient near the tip associated with opposite directions of the axial induced velocity component on opposite sides of a near-tip vortex.

Rohrbach (1976; Rohrbach and Worobel, 1975) goes on to discuss the aerodynamics of optimum wind turbines as calculated by the Goldstein method. It was seen earlier that the power coefficient C_p and the velocity ratio λ were primary parameters for design and performance and can be given in terms of the rotor geometry, that is, the number of blades and their airfoil section as expressed by lift/drag ratio L/D. Some of the results are shown in Figs. 13 and 14. Figure 13 shows the effect of number of

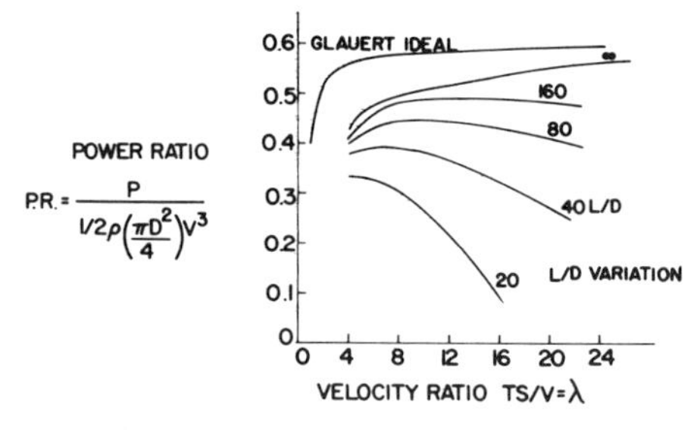

Fig. 14. Effect of lift/drag ratio on power coefficient. (From Rohrbach and Worobel, 1976.)

blades N for $L/D = \infty$ (no drag), with C_p increasing with N and λ, with significant but not major difference for the former. Figure 14 shows the strong adverse effect of decreasing L/D ratio, particularly the decreasing C_p with λ for low values of L/D.

The effect of blade width, which is a vital factor in structural design and cost, is given in terms of an activity factor (AF), which is a blade area–weighted solidity parameter in terms of radial distribution of chord width, i.e.,

$$\mathrm{AF} = \frac{10^5}{16} \int_{\mathrm{hub}}^{\mathrm{tip}} (b/D)(r/R)^3 \, d(r/R)$$

where b is the section chord, D is blade outside diameter, r is the local radius, and R is the rotor tip radius. Figure 15 shows AF vs λ for a two-bladed rotor with values of C_L from 0.5 to 1.5 and for a rotor with C_L of unity for n from 1 to 4. Solidity decreases as λ increases and structural problems may arise as the blade narrows and Rohrbach suggests a minimum value of AF of 15, as shown in the figure. Because high rotor speed implies reduced step-up gear ratio for ac electrical generation, high λ and low solidity turbines are favored. Figures 13–15 show very clearly the interaction of design variables for optimum performance. On the basis of the data shown, together with a consideration of airfoils of known performance having high L/D values consistent with practical blade geometries and performance levels, Rohrbach suggests that a combination of C_L of 1.0, an AF limit of 15, and a two-blade rotor can be considered a very reasonable configuration for optimum performance combined with structural integrity and cost. There does remain, however, a trade-off between design velocity ratio λ and L/D. This is summarized in Fig. 16, showing C_p vs λ for an optimum two-bladed turbine, with performance limits of minimum λ of 6 and AF of 15 for values of C_L from 0.5 to 1.5 at high values of λ. Figure 17 shows four blade planforms within this range of optimum turbines. It would appear that although minimum cost and

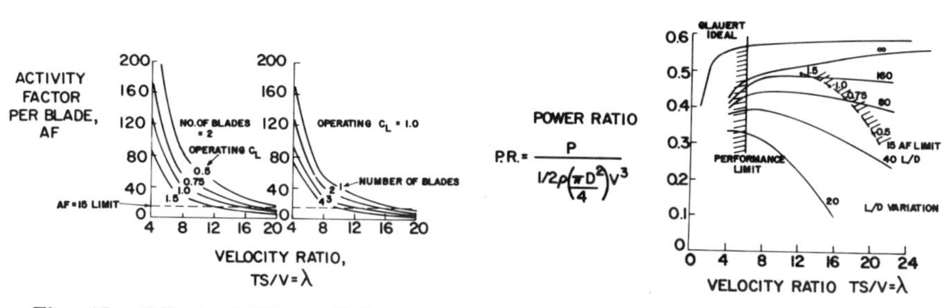

Fig. 15. Effect of lift coefficient and number of blades on activity factor. (From Rohrbach and Worobel, 1975.)

Fig. 16. Effect of lift coefficient and activity factor on power coefficient. (From Rohrbach and Worobel, 1975.)

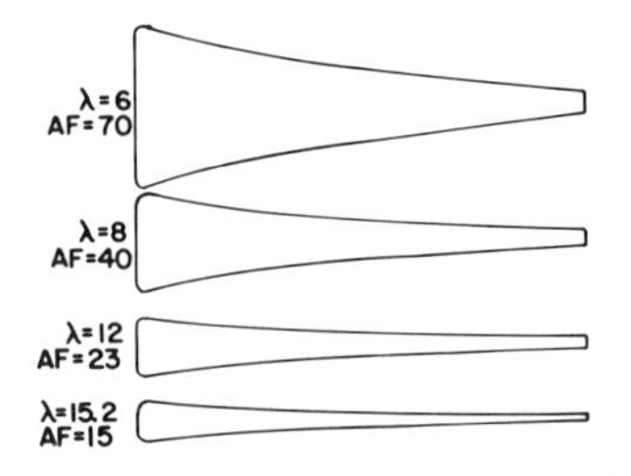

Fig. 17. Optimum blade planforms for optimum turbines. (From Rohrbach and Worobel, 1976.)

structural effectiveness may not coincide with optimum aerodynamics performance, the performance compromise is not likely to be great (2–3% at low λ, $\approx$10% at high λ).

C. Aerodynamic Analysis of Rotors—Cross-Wind Axis Type

Significant attention to the analysis of cross-wind types has been quite recent, and is limited mostly to the Darrieus rotor, with some work on the Savonius type and the tracked-vehicle wing or Magnus type.

1. Darrieus-Type Rotors

The obvious difference between such rotors and the wind-axis type is that there is a continuously varying sinusoidal aerodynamic condition as a blade rotates from "upwind" or "windward" to "downwind" or "leeward." The previous general aerodynamic approach via wake momentum analysis and blade element theory can be made, but there are differences in analysis according to the assumptions made and the degree of complexity of solution. A quasisteady condition is usually assumed and effects of the upwind wake turbulence on the blade in downwind position are neglected. Figure 18 shows a diagrammatic plan view of a Darrieus with nomenclature used in the succeeding discussion.

Templin's (1974) analysis is about the earliest, as the work of the NRC of Canada beginning in the mid 1960s heralded the present interest in the Darrieus concept (originally introduced by him in 1925). It is simple and straightforward and results in good agreement within certain limits with the meager available overall performance test data. It treats the flow as a single streamtube with the induced velocity constant across the rotor, which allows a closed solution, but which limits its use to lightly loaded blades and circumstances in which there is no significant variation of wind velocity across the flow area. The analysis includes aerodynamic drag of the blades and it is concluded that performance is sensitive to drag, particularly at high values of λ.

Other analyses (e.g., Ashley, 1977; Muraca *et al.*, 1975; Shankar, 1976; Strickland, 1975, 1977; Holme, 1977; Lissaman, 1977) adopt the multiple-streamtube concept, for which an iteration procedure is required, as the ratio of wind speed V at the rotor to the freestream velocity V_∞ is not constant and must be obtained by matching of momentum and blade element relationships. A preliminary analysis by Wilson and Lissaman (1974) using a model with straight blades parallel to the axis, inviscid flow, and a lift coefficient $C_L = 2\pi \sin \alpha$ demonstrates very simply some of the essential behavior of the Darrieus rotor. Thus the local induction factor $a = 1 - V/V_\infty$ is given by $a = \sigma\lambda|\sin\theta|$, where σ = solidity $= Bc/2R$, with B equal to the number of blades, c the chord, R the blade radius (span), and $\lambda = \Omega R/V_\infty$. θ is the blade angular position, Fig. 18, defined with $\theta = 0$ at 3 o'clock for wind from the 12 o'clock direction and counterclockwise rotor rotation.

The time-averaged power coefficient is given by

$$C_p = 2\pi\sigma\lambda[\tfrac{1}{2} - (8/3\pi)\sigma\lambda + \tfrac{3}{8}\sigma^2\lambda^2] = 2\pi a[\tfrac{1}{2} - (8/3\pi)a + \tfrac{3}{8}a^2]$$

This yields a maximum value of $C_p = 0.554$ when $a_{\max} = 0.401$, indicating a slightly lower value than the 0.593 for the simple actuator disc model. Also, the maximum angle of attack occurs near $\theta = \pi/2$ and $\tan\alpha = V/U = (V/V_\infty)/(U/V_\infty) = (1 - a)/\lambda = 1/\lambda - Bc/2R$. Thus if $\alpha_{\max}$ is the stalling angle $\approx 14°$, then the starting value of $\lambda \approx 4/(1 + 2Bc/R)$. This means that for typical values of B, c, and R, the minimum starting value of λ will be the order of 2–3. This first approximation is borne out in practice

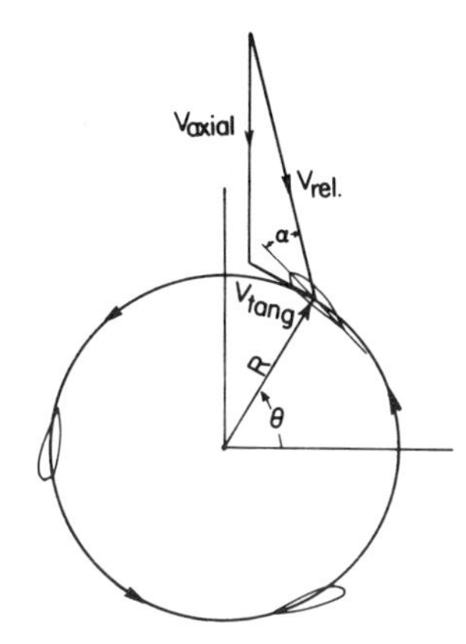

Fig. 18. Geometry of Darrieus rotor.

and the Darrieus rotor needs to have a starting device of some sort. If drag is considered for this model, it can be shown that the reduction in power coefficient is given approximately by

$$\Delta C_p = -C_p(Bc/2R)\lambda^3 = -C_D a\lambda^2.$$

In a later report of Wilson *et al.* (1976), an analysis of a curved rotor is carried out by introducing the local angle between the blade tangent and the axis of rotation and integrating for a given geometry. A circular arc rotor shape yields $C_p = 0.536$ at $a_{\max} = 0.461$. In this same work, the analysis is carried forward by taking into account the fact that the assumption used to $C_L = 2\pi \sin \alpha$ implies a linear relation between the circulation and the component of the relative velocity of the wind perpendicular to the blade. This is not valid for a real airfoil resulting in the necessity for a numerical calculation in place of a closed term solution.

A paper by Ashley (1977) extends the single-stream tube analysis (constant induced velocity) without blade stall to include unsteady flow effects, as possibly useful as a compromise between the simple steady-flow approximations and complex vortex theories. Thus his quasi-steady analysis takes into account the effect due to a time-averaged lift component developed on a surface exposed to a sinusoidally varying gust, and proceeds by the general integral procedure of Templin discussed previously. There are some complexities to the mathematics in the form of elliptic integrals, but these may be evaluated using quadrature. The major result is that the unsteadiness of the flow produces rather a large influence on the power output as against a steady flow analysis. Thus even at a very low value of 0.04 for the reduced frequency $\Omega c/2\Omega R \equiv c/2R$, the "solidity per blade," which Wilson uses as a criterion for quasi-steady flow, the reduction in C_p is about 15%, with greater reductions as $c/2R$ increases. Ashley recognizes the approximations in his analysis and suggests that only careful experimental work can determine the quantitative validity of his results.

The analyses of Muraca *et al.* (1975) and Strickland (1977) are multiple streamline approaches essentially similar to those of Wilson *et al.* The first has a computer program in the body of the report and one for the latter is available in Strickland (1975). Holme (1977) has conducted the most elaborate analysis for two-dimensional inviscid flow through a straight-bladed VAWT. He takes into account bound vorticity as well as the wake vorticity and uses the Biot–Savart relationship for obtaining the induced velocities from both vortex sheets. His results show that the velocity induction effect is such that half the retardation takes place within the turbine itself and hence that blade incidence and aerodynamic loading are much higher on the windward side than on the leeward side.

Other analyses do not show these unsymmetrical conditions. A numerical calculation shows that for a value of $\sigma = 0.2$, a maximum C_p of 0.545 at $\lambda \approx 4$ is obtained. The angle of incidence was found to vary from $-13.5°$ to $+8.5°$. He gives an expression for correction of C_p for viscous effects, with $C_{p_{max}}$ reduced to 0.494 for a blade drag coefficient of 0.01. He also shows that the effect of drag is considerably increased for lower values of solidity, particularly for higher blade-tip/wind-velocity ratios as do other analyses.

A diagram for nonviscous flow with straight blades from three analyses (Wilson *et al.*, 1976) is shown in Fig. 19 for C_p vs $\lambda\sigma$. This is a comparison of the single-streamtube method of Templin (1974), the multiple stream-tube method of Wilson *et al.* (1976), and the more complex method of Holme (1977). Although the values for the last two are very close, perhaps it is interesting that all three of these analyses are so generally in agreement. The Holme method, although complex and time consuming, contains information which may well be of considerable value for structural analyses and behavior of the rotor in limiting circumstances.

Lissaman (1977) presents in summary a general performance theory for crosswind-axis turbines based on that published with Wilson *et al.* (1976; Wilson and Lissaman, 1974). He points out that although there is a periodic induced flow in an inertial reference frame in a propeller-type turbine due to rotation of a finite number of blades, the induced flow fields relative to the blades are constant and if the product λb is large, then the downstream wake unsteadiness is low. He emphasizes the importance of a parameter he calls the *chordal ratio C*, where $C = c/R$, chord/radius of

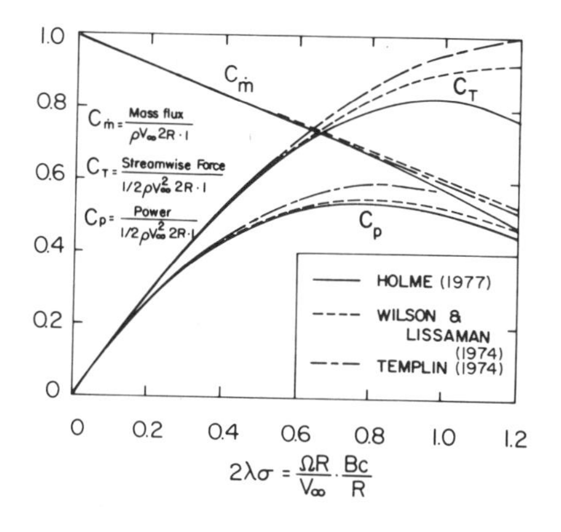

Fig. 19. Comparison of performance coefficients for three streamtube methods of analysis. (From Wilson *et al.*, 1976.)

rotation. This distinguishes types of crosswind rotor, examples being one of low C, like the Darrieus rotor, and one of high C, like the Savonius rotor. With low C, it is possible to assume a quasisteady condition and even allowing the induction factor to vary in the crosswind direction i.e., $a = a(\theta) = \frac{1}{2}C\lambda|\cos\theta| = a^*|\cos\theta|$, yields the ideal C_p of 0.554 mentioned above.

For high values of chordal ratio C, for which the Savonius rotor is modeled as an example, the situation is very different and the analysis becomes more complex and requires more assumptions of greater significance. (The Savonius model used is that of the "S rotor" without gap.) Three potential functions are used to represent the perturbations due to (1) the normal flow perpendicular to a flat plate, (2) the flow parallel to a cambered plate, and (3) the flow due to a flat plate rotating about its midpoint, making up the elements of flow in an S rotor. Space is not available here to pursue this analysis and also, as Lissaman comments in relation to a full analysis of the unsteady flow, the operational value of the Savonius turbine is debatable. It will suffice to say that although quantitatively uncertain, the main results have considerable interest qualitatively and picture the performance surprisingly well. For example, we have that the mean torque is equal to the static moment, the same result that would have been obtained if the unsteady forms were ignored and the static moment integrated over a full cycle: that $L/D \approx 2$–4 and as it is only the drag term that is giving useful work, there is a large penalty in crosswind force for power extraction: that in terms of reduced parameters $\lambda^* = \lambda/(1 - a)$, $C_L^* = C_L/(1 - a)$, etc., then the torque is a function only of camber, the lift only of tip-speed ratio, and drag is coupled to both. Viscous effects are difficult to incorporate in any simple fashion but it appears that boundary layer effects themselves must be quite insufficient to account for the difference between the inviscid model and actual test results. Hence the discrepancy is attributed to separation, which does not appear surprising in view of the geometry, although in nonsteady flow it would be expected to be less severe than for the steady condition. Wilson *et al.* (1976) present pictures from a film of smoke-flow visualization of a Savonius rotor (attributed to Edgewood Arsenal) and discuss the phenomena. It should be noted, however, that the rotor has a central gap between semicircular blades of approximately one-third of a diameter and it would appear that flow through this gap might change the pattern significantly from that of a continuous S vane. Lissaman suggests that a rotor with a chordal ratio C of about 0.5 might be worth studying as a model combining some of the good features of both the Darrieus ($C \approx 0.07$) and the Savonius ($C \approx 1.0$) types. He also notes that two contrarotating high-C crosswind types placed side by side would yield a net crosswind force of zero and might provide increased power due to greater mass flow and wind acceleration.

2. *Tracked-Vehicle Concept Analysis*

Another VAWT type, the tracked-vehicle airfoil concept, has received significant attention by two investigators. One work is that of Powe and his co-workers, summarized in Powe (1977; and very similarly in Powe *et al.*, 1974), which presents an approximate performance comparison with the propeller type. The other is that of Lapin (1976), who also reports a parallel investigation of the Madaras concept utilizing the Flettner rotor and makes a comparison of these two systems.

Lapin presents his aerodynamic analysis in considerable detail following the general lines of the momentum and blade element approaches of propeller-type theory for the airfoil type and using a speed ratio $\alpha = \omega R/V \equiv$ rotor spin velocity/wind velocity at the location of the rotor, corresponding to the equivalent of angle of attack, for the rotating Flettner rotor type. The analysis is made for aspect ratios of 4, 6, and 9, using the optimum incidence for maximum torque at each angular position chosen for calculation around the track. Wing data for C_L and C_D (including induced drag) are available and reliable for known airfoils but data for rotors are scarce and not reliable to the extent comparable with those for wings. Lapin uses what is available but emphasizes that more experimental values are needed. Most of the calculations are for a circular track but it is apparent that a linear track perpendicular to the wind will yield significantly greater output. Thus a flattened oval or race course track is better for a wind predominantly from one direction, as the propulsive effect is maintained at its greatest value over a longer fraction of the circuit. The loss of power due to rotors in tandem, i.e., differences of performance in upstream positions to downstream positions, is dealt with in detail. The analysis is based along the lines of the tandem actuator disk theory as outlined in Section II,B but the relationships are somewhat difficult to handle directly. However, a computational procedure is given and the procedure carried out semigraphically. Performance data are given for this loss of power and also for the case of no loss, i.e., assuming that mixing between upwind and downwind positions restores the free-stream condition.

Lapin takes into account the additional drag forces due to the unconventional system, which are significant. These are rolling friction forces due to (1) the weight of the rotor, cart, etc. and to (2) the cross-track force. Rolling friction coefficients from railroad freight car technology are used. The cross-track force is computed in terms of the aerodynamic force expressions, and thus varies with position. The maximum cross-track force is important for calculating structural stresses and the tipping moment. There is also cart drag, which is calculated with a drag coefficient equal to the aerodynamic equivalent flat plate area, taken as invariant with flow angle, and integrated around the circuit for varying angular position. He gives examples of overall power coefficients and power losses for a variety of system parameters. Lapin mentions the disadvantageous effect of the tracked-vehicle system having to operate in the lower wind speeds near the ground due to the boundary layer, but does not specifically include any factor for this in his data.

Powe (1977; Powe *et al.*, 1974) starts off with a momentum analysis of the control volume for an airfoil-type tracked-vehicle system, taking into account a velocity variation due to ground effect and indicates that theoretically this type of turbine would give about 19% more power than a propeller-type turbine with each unit having the midpoint of its overall blade span H at a height $H/2$ above ground. For systems with the midpoint of the blade elevated above this minimum height, the tracked-vehicle airfoil concept yields about 26% more power than the corresponding propeller type. This is his basis for an investigation of this type of WECS.

The analysis considers only a flattened oval (race course) type of track, with no power contribution being taken for the turns. In the two papers of Powe (1977; Powe *et al.*, 1974), the aerodynamic analysis is minimal and the problem of the value of the interference factor a and the actual value of the flow velocity of the disk is not discussed. Several airfoils from the NACA series were investigated, with the ubiquitous 0012 type being selected for the detailed calculations, using corrections for finite aspect ratio. The system resistances are categorized as the two rolling resistances due to the vertical and horizontal forces on the track, a resistance due to track curvature, and an aerodynamic resistance. A simple expression for rolling resistance was taken from railroad literature in terms of weight, with the weight being taken to include the aerodynamic forces of lift and drag which yield a moment giving rise to a vertical force acting on the track. Aerodynamic resistance of the cars was broken down into a value for the leading car as that of a streamlined locomotive, with the remainder as that for streamlined railroad cars, and a third value for unstreamlined cars being used for the turns.

The airfoil spacing was considered, with the evaluation being made with the vertical axes of adjacent airfoils being spaced one chord apart. This was based on a consideration of minimum spacing to minimize land area and track length, together with the fact that the airfoil axes have to be at least one chord apart to rotate in the turns. The effect of incomplete recovery was accounted for in one particular calculation by taking the downwind value of wind speed as 75% of the freestream values. The effect of variation of wind speed and direction in finding an actual energy output over a period of time was evaluated by using data from an Air Force Base in Montana, as detailed values were available over a 25-year period.

A computer program was compiled with the system parameters of geometry and weights, airfoil characteristics, and wind spectrum as input. Output is given in overall terms as energy per month, and as energy per month per unit area, with area as both total swept area including turns and as per unit blade area. A fourth output figure is energy per month per unit weight of the system. Other output data computed but not quoted are carriage velocities, angles of attack, etc., pertinent to analyzing component performance and operating conditions.

The performance of these tracked vehicle systems of both Lapin and Powe *et al.* will be discussed later.

V. WIND ENERGY CONCENTRATORS AND AUGMENTORS

With one of the prime features of freestream wind energy being its low energy density, considerable attention has been given to forms of turbines which operate with this available energy concentrated into a smaller cross-sectional flow area. A similar state of affairs occurs with direct solar energy, where parabolic and other geometries are used to overcome the low specific energy capacity of flat-plate collectors. The idea is not new (it is apparent in the slanted entry of primitive Persian paddle-wheel windmills) but has received increasing attention in the immediate past few years. The energy available per unit of freestream area of course remains the same, but as the moving element of the system, the rotor itself, is the most critical and probably the most expensive component, any decrease in size for a given power or increase of power for a given size owing to the cube law of power and wind speed can be a major factor in economic viability.

First it is useful to mention a basic aerodynamic aspect of augmentation pointed out by Lissaman (1977). It was shown in Section II that the air velocity V at the actuator disk is given by $(V_\infty + V_e)/2$, but if it is assumed that there is also another axial perturbation ΔV, induced by bound vorticity which does not affect the far wake, then $V = \frac{1}{2}(V_\infty + V_e) + \Delta V$. Thus if ΔV can be made to occur, as by a special geometry, than C_p can be greater than $\frac{16}{27}$. However, the ΔV created must not develop significant velocity perturbations in the wake, otherwise the power output will be reduced. One such method is to surround the actuator system with a vortex ring, which will create a ΔV without shedding vorticity.

Wind concentration or augmentation can be categorized into two main classes, one utilizing a translation acceleration of the ambient freestream velocity, naturally or artificially induced, and the other utilizing a rotational (vortical) motion, again either naturally or artificially induced. The first group may be subdivided into two, one essentially occurring with acceleration by area change, i.e., the Venturi effect, the other activated by flow around an obstacle, e.g., acceleration at the edge of a flat plate or around the surface of a cylinder. Loth (1976), who with his co-workers has instigated several types of concentrators, distinguishes these basic patterns as three main groups, which he calls the Venturi type, the obstruction type, and the vortex type.

Concentration by reduction of area and hence of pressure with increase of velocity according to the Bernoulli principle is probably a very old one, as the concept is visually observed by natural water flow in gorges and canyons and experienced by air flow in mountain passes or between any two steeply inclined surfaces with a narrow base. In more recent times, about forty years ago, Vezzani (1950) put forward an ambitious scheme combining energy storage with wind power, some parts of which are now receiving attention in a number of quarters. The project envisaged a number of ducts clustered around a central duct at the top of a hill, each duct facing directly into the wind from a given quarter, with the air passing through a turbine whose discharge was a duct of increasing area acting similarly to the draft tube of a hydraulic turbine. A feature of present-day prominence is that of pumped storage using power from the wind turbine to raise water from a low-level reservoir to a higher one when excess power is generated, and making hydropower available in times of wind shortage.

The ducted turbine concept goes by other names such as "shrouded turbine" or "diffuser-augmented turbine." Recent work has been carried on by Igra (1976a) and co-workers in Israel and by Oman *et al.* (1977; Oman, 1977) in the United States. The earlier work was that of Igra *et al.*, and their results have been confirmed by the latter group, and the work extended in some directions over a range of variables with attempts at optimization. The principle of operation is to increase the mass of air flowing through the turbine, compared to that of a conventional turbine, by recovering the exhaust kinetic energy with a diffuser and producing a much lower pressure immediately downstream of the turbine. The diffuser must be short (large divergence angle) in order to lower cost and to minimize structural problems, but must have a good efficiency (avoiding separation.)

Two methods have been suggested for obtaining such short, large-angle diffusers of high performance, those of boundary layer control and ring-wing shroud sections. The former uses the relatively high-energy external air to energize the diffuser boundary layer through tangential slots, the shroud being divided into several segments, as shown in Fig. 20. The ring-wing concept uses diffuser shroud segments of airfoil section, with the inner side being the high-velocity low-pressure surface or upper surface of an airplane wing, and such profiles can have a high lift-to-drag ratio. A very important feature of these types of diffuser is to provide a static pressure *lower* than atmospheric at the diffuser discharge plane, normally not a characteristic of subsonic flow from a duct into the surrounding atmosphere. This is a consequence of the turbulent momentum interaction between the inside and outside air flows. This phenomenon was first reported by the Israeli group and confirmed by Oman, and it results in the major augmentation effect of the system, a phenomenon which had not been previously recognized, with the result that for many years ducted turbines were thought not to be capable of giving a satisfactory performance. It is also important that the effect is considered to be mainly due to the ideal flow pattern and not to viscous action, as the test program was conducted with small models and with low Reynolds numbers, hence extension of scale might show disappointing results if viscous effects were predominant.

Turning first to the boundary layer control diffuser, much work has been carried out by Oman *et al.* (1977; Oman, 1977). The diffusing cone is made in segments with annular gaps, as shown in Fig. 20, and a considerable number of configurations were tested in an open-jet wind tunnel, using a wire-mesh screen to simulate work extraction by a turbine. This

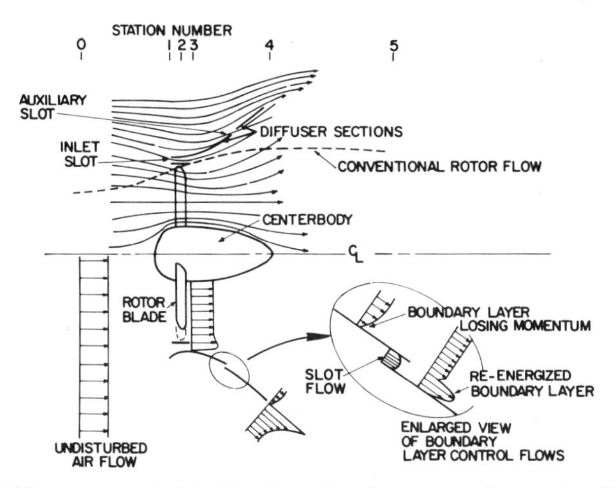

Fig. 20. Diffuser-augmented turbine, boundary-layer control geometry. (From Oman *et al.*, 1977; Oman, 1977.)

provides a ready means of changing the pressure drop, but does not give any residual swirl which might occur with a turbine. (However, the effect of this is likely to be small except possibly at conditions considerably off design.) Over 100 different diffuser combinations were tested, the range of geometries being 40°, 60°, 80°, and 90° included angle, area ratios of 1.28–4.94, and disk loadings from 0.37 to 0.93, with the last named defined as the total pressure drop Δp_t across the screen divided by the dynamic head at screen inlet, q_i. The major performance parameter is the augmentation ratio r, defined as the ratio of the power dissipated by the screen (equivalent to actual power extracted by a turbine) divided by the ideal available power from an unducted turbine of the same cross-sectional area in a freestream, i.e., $r = \Delta p_t V_i / \frac{16}{27} q_\infty V_\infty$. Values of r up to about 2 were obtained and the optimum area ratio was just under 3. For the important base pressure reduction effect discussed above, static pressure coefficients at discharge up to about 0.60 were obtained, with each configuration yielding a maximum value at a particular area ratio. This is a critical parameter for optimization. The number, size, and arrangement of diffuser segments and the characteristics of the main (inner) slot and auxiliary slots all have a bearing on the overall performance and also on the final cost balance, with possible tradeoffs between necessary turbine size, diffuser first cost and output, but it is not possible to make a sufficiently succinct summary here. Two additional features of the ducted turbine can be significant in total energy output in a given location: (1) the system can utilize lower wind speeds, i.e., have a lower cut-in point, and (2) there is only a very slight degradation of performance in a shear flow, and the ground effect due to the geometry was actually favorable, even with the diffuser exit edge on the ground.

The ring-wing concept of high-lift annular airfoil sections replacing the normal diffuser of linear taper operates to give a lift force acting towards the centerline and creates a reaction force on the internal flow that tends to direct it radially outward as well as axially downstream (see Lissaman, 1977, discussed above), i.e., the circulation induced around the lifting airfoil section results in an increased flow rate and the turbine power is thus increased. Figure 21 (from Igra, 1976a) shows the general configuration, in this case of a high-lift airfoil main shroud, with a high-lift axisymmetric flap. The flap position is critical and optimum performance is obtained when the radial gap between flap and shroud is 4% of the shroud length and there is no axial gap. Igra's latest work in a large tunnel shows that the shroud performance is maintained for a yaw angle with respect to the wind up to 25°, Figure 22, showing slightly better performance immediately before the stall point at about 25°. This is because the ring-wing airfoil has a lift-curve slope about double that of a linear wing, and there-

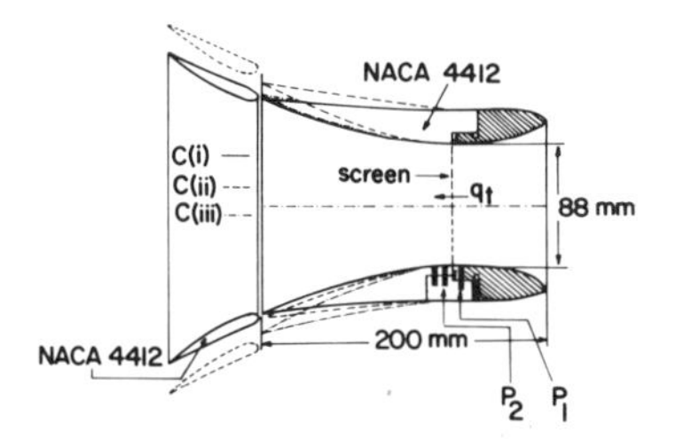

Fig. 21. Diffuser-augmented turbine, ring-wing geometry. (From Igra, 1976a. Reprinted by permission from the American Institute of Aeronautics and Astronautics, AIAA paper 76-181, Fig. 7.)

fore increase of yaw angle increases lift until the stall angle is reached. A shroud with flap has a better performance at zero yaw, but has a much smaller range of stall yaw angle. The augmentation ratio r increases over a wide range of loading coefficient (turbine work) although it does reach a maximum, with the optimum value of loading coefficient being reduced with increase in shroud exit area. Igra (1974, 1976b) has shown that a wind turbine for use with the shroud concept, based on conventional axial-flow turbine design as used in gas or steam turbines, can give satisfactory performance over a range of wind speeds without variable pitch.

Oman *et al.* (1977) discusses the theory of the use of high-lift airfoils as ring wings for diffuser augmentation and gives test results of three different airfoil sections, the NACA 4412 section (as used by Igra), a high-lift Williams airfoil A, and a high-lift Liebeck laminar rooftop section. Standard methods of calculating velocity and pressure distributions around an airfoil by conformal transformation are available, but the presence of an

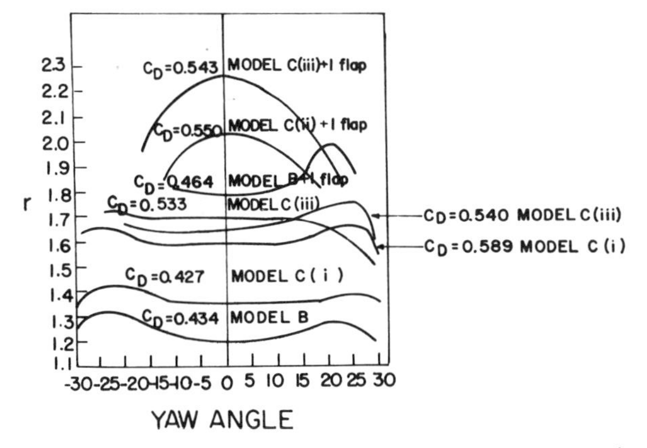

Fig. 22. Effect of yaw angle on ring-wing diffuser turbine augmentation. (From Igra, 1976a. Reprinted by permission of the American Institute of Aeronautics and Astronautics, AIAA paper 76-181, Fig. 18.)

energy absorbing or dissipating device (turbine or screen) requires reformulation of the analysis. It is reported that this is in hand but results are not available to date (Dec. 1977). Tests showed that augmentation increased with disk loading but no maximum was reached as per Igra. The Liebeck airfoil was disappointing, but the other two gave good results, and the choice lies as much in cost effectiveness as in absolute performance.

A concentrator of the Venturi type, but with the augmentation and diffusion generated in a different manner, is that of "tipvane induced diffusion," as introduced by van Holten (1974a, 1977) and co-workers (van Bussel, 1977). The ring-wing diffuser previously discussed develops lift on the stationary shroud, with the reaction on the air deflecting the flow outward and providing rapid diffusion downstream. In van Holten's model, this effect is produced by tip vanes attached to the ends of the turbines, i.e., the stationary shroud is replaced by rotating shroud segments, Fig. 23.

Van Holten first shows that the time-averaged radial force and induced velocity due to discrete tip vanes is the same as for the model with a continuous ring-wing shroud and thus the resulting diffusion is the same. If the induced velocity increment is ΔV, then using the results and nomenclature of Section II, the augmented mass flow rate is $\rho A[\frac{1}{2}(V_\infty + V_w) + \Delta V]$ and the power absorbed by the turbine is equal to the decrease of kinetic energy between freestream and for wake, i.e.,

$$P/A = \tfrac{1}{2}\rho[\tfrac{1}{2}(V_\infty + V_w) + \Delta V](V_\infty^2 - V_w^2) \tag{5.1}$$

Setting $dP/dV_\infty = 0$, then

$$(V_u/V_\infty)_{\text{opt}} = -\tfrac{1}{3}(1 + 2\,\Delta V/V_\infty) + [\tfrac{1}{9}(1 + 2\,\Delta V/V_\infty)^2 + \tfrac{1}{3}]^{1/2} \tag{5.2}$$

Substituting this into Eq. (5.1) yields an expression for the maximum power, but for a simple answer this may be linearized for values of $\Delta V/V_\infty \ll 1$ thus,

$$(V_w/V_\infty)_{\text{opt}} \approx \tfrac{1}{3}(1 - \Delta V/V_\infty)$$

and the optimum specific power becomes

$$P_{\text{opt}}/A = \tfrac{16}{27}(\tfrac{1}{2}\rho\, V_\infty)^2(1 + \tfrac{3}{2}\,\Delta V/V_\infty) \qquad \text{for small } \Delta V/V_\infty \tag{5.3}$$

or

$$P_{\text{opt}}/P_0 = 1 + \tfrac{3}{2}\,\Delta V/V_\infty \tag{5.4}$$

where P_0 is the ideal output of a simple turbine of the same area without

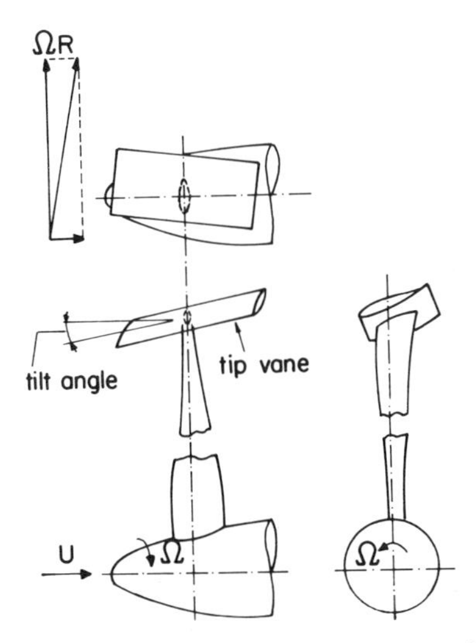

Fig. 23. Tip vane-induced diffusion augmentor. (From van Holten, 1974a, 1977.)

augmentation. The gain in power $\Delta P/P$ is $\frac{3}{2}\,\Delta V/V_\infty$, which is greater then the gain from increased mass flow alone. This is because for the optimum condition

$$(V_w/V_\infty)_{\text{opt}} = \tfrac{1}{3} - \tfrac{1}{3}\,\Delta V/V_\infty \tag{5.5}$$

i.e., less than the value of one-third of the simple axial momentum analysis and there is an added diffusional effect, as observed for ring-wing shrouds. The next step is to obtain a relation between $\Delta V/V_\infty$ and the magnitude and distribution of the applied radial forces, in order to estimate a power output from the geometry of the shroud or tip vanes. This requires a calculation of two coefficients α and β depending upon the distribution of vorticity connected with the shroud or tip vanes. Van Holten gives this in a separate report (1974b), as it is rather complicated. In order to check the validity of the linearized model analysis and the values of the two coefficients α and β, van Holten used a driven turbine with tip vanes, i.e. as a propeller, thus requiring only a change of signs in the relationships, as nothing is affected in the theory. It is shown that the diffusion ratio D defined by

$$D = \left(\frac{\pi/4\, D_s^2}{\pi R^2}\right)^{1/2} \tag{5.6}$$

where R is the turbine radius and D_s is the diameter of the slipstream far downstream, is a function of nondimensional terms for the radial forces K_r and the axial force or thrust T, thus

$$D = D\left(\frac{K_r}{\rho V_\infty^2 R}, \frac{T}{\rho \pi R^2 V_\infty^2}\right). \tag{5.7}$$

For the special case of the static propeller condition, D becomes a function of the ratio of these two parameters only, and it can be shown that $D = D(K_r\beta/T)$. It is difficult to measure the radial forces and velocity increment directly so that a special case of the propeller was used to demonstrate the validity of the analysis, but this was still in the normal working range of the propeller. With the arrangement shown in Fig. 23, K_r and T are generated by the same aerodynamic surfaces, so that they can be related by the tilt angle γ, i.e., $D = D(\gamma)$. The experimental approach was then to visualize and photograph the flow and measure D_s for a given geometry. The test values of D as a function of γ are given by van Holten and show excellent correlation with the theory.

Now most of the above summary analysis is valid for shrouds and tip vanes and hence the crux of van Holten's design is the relative size and cost of the two concepts. It is simple to show for a first-order approximation that for a given diffusion ratio, the ratio of total wetted surface areas A_t and A_s of tip vanes and shroud, respectively, is equal to the square of the ratio of freestream wind speed to turbine tip speed, i.e.,

$$A_t/A_s \approx (V_\infty/U)^2 = 1/\lambda^2 \tag{5.8}$$

If $\lambda = 7$, then $A_t \approx 0.002A_s$, a very striking potential reduction of area, hence of material and cost.

However, because the tip vanes are rotating, their drag decreases the ideal power gain and both indirect drag and viscous drag have to be considered. The induced drag is zero for the time-averaged flow field, as no vorticity is shed from the cylindrical ring of circumferential vorticity, but with a finite number of vanes, a fluctuating component is added which does represent a loss of kinetic energy. Van Holten's analysis indicates that given certain conditions including a vane planform having an approximate parabolic lift distribution and a tilt angle (Fig. 23) of the vanes to align them with the local undisturbed flow (≡ without tip vanes), the fractional power loss P_i/P_0 due to induced drag can be kept quite small. It is inversely proportional to the number of vanes, and directly proportional to λ, which then must be considered carefully.

The power loss P_d due to viscous drag is given by

$$P_d/A = C_p \tfrac{1}{2}\rho U^3 \sigma_t \qquad \text{or} \qquad P_d/P_0 = \tfrac{27}{16} C_p \lambda^3 \sigma_t \tag{5.9}$$

where σ_t is the tip-vane solidity ratio. This loss can be considerable if C_D is appreciable, especially at high tip speeds. Hence not only must C_D be kept low, but high-lift airfoils are necessary and Liebeck airfoils are being investigated for use as tip vanes. The calculated overall performance using the above relationships, comparing the power output with that of a conventional turbine of 80% efficiency, is shown in Fig. 24. The maxima occur at high values of λ and are dependent on solidity ratio, but values of power augmentation of about 2 appear possible at tip speed ratios of about 6.

It is reported by van Bussel (1977) that research on the practical feasibility of this tip vane concept is being carried out as part of the Dutch National Program in wind power, in cooperation with the United States Energy Research and Development Administration who will be responsible for empirical development by testing an existing propeller-type turbine modified with tip vanes. The Dutch program envisages a 4-m model to be developed in 1978.

The second class of wind concentrators utilizing acceleration in translational flow covers those in which freestream air is slowed down in front of an abstacle and hence is speeded up around the sides of the obstruction. This may occur with a flat surface normal to flow, such as a plate, or a rounded surface, like a cylinder. In nature it occurs continually over rough terrain or as an effect of human artifacts, particularly buildings. The airplane wing is perhaps the outstanding example of the use of a solid device to produce a differential pressure and velocity pattern by simple flow around a surface. Loth (1975, 1977) and co-workers (Walters, 1975a) have reported investigations on this type of concentrator and a

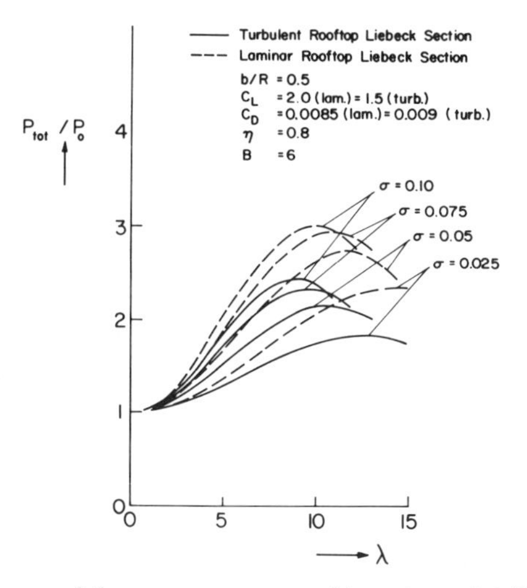

Fig. 24. Performance of tip-vane augmentor with various airfoils. (From van Holten, 1974a, 1977.)

power concentration ratio R is defined as the ratio of the local wind kinetic energy available for harnessing to the energy available in the freestream flow without a concentrator. If V_{av} is the average local inviscid velocity at the rotor inlet and V_∞ is the undisturbed freestream velocity then $R = (V_{av}/V_\infty)^3$. The local pressure coefficient C_{pr} is given by $(p - p_\infty)/(\frac{1}{2}\rho V_\infty^2)$, where p is the local static pressure, and hence for inviscid flow $p - p_\infty = \frac{1}{2}\rho(V_\infty^2 - V^2)$ and $C_{pr} = 1 - (V/V_\infty)^2$. Thus R can be expressed as $R \approx (1 - C_{pr})^{3/2}$. For flow around a circular cylinder, the maximum theoretical value of the velocity is $2V_\infty$, giving $C_{pr} = -3$, and this occurs at the surface of the cylinder at the ends of the diameter perpendicular to flow. However, the boundary layer separates just before the point of maximum velocity and the finite size of any work extraction device limits the actual attainable value of R to about 3 or less. Similarly to the simple actuator disk momentum analysis, the total available power cannot be extracted because this would require zero exit velocity and hence no through flow, so that with respect to the local velocity, the Betz limit still applies, although the overall effect has been enhanced by the increased inlet velocity (cf. the vortex concentrator discussed later).

Loth describes several devices which utilize this type of augmentation, the basic method utilizing two Savonius rotors placed one on each side of a fixed cylinder. The cylinder can be replaced by a stationary flat plate or a combination of the two with the cylinder stationary and the flat plate coupled to and moving with the two Savonius rotors as wind direction changes, Fig. 25. A model of this sort was tested in a wind tunnel with a

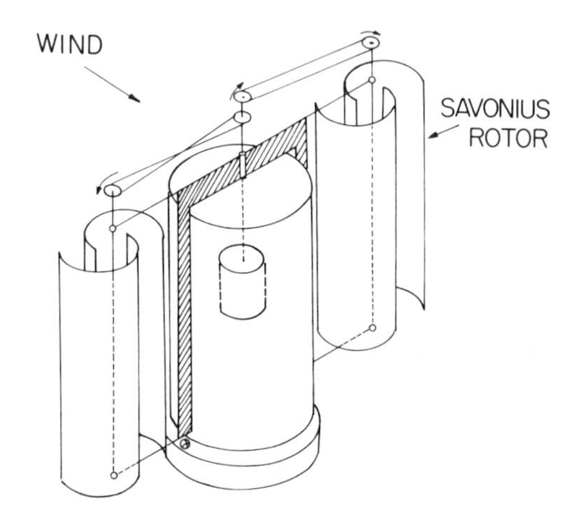

Fig. 25. Obstruction type augmentor—cylinder with Savonius rotors. (From Loth, 1977.)

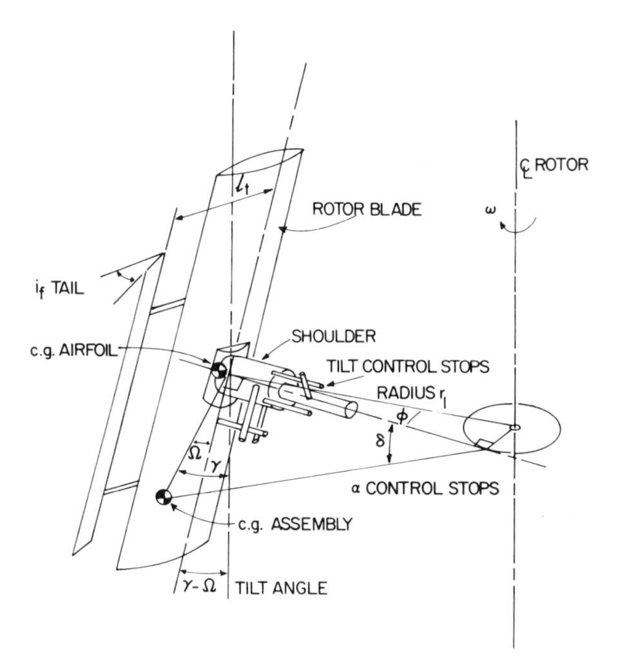

Fig. 26. Wind energy concentrator—Darrieus type. (From Loth, 1977.)

belt-drive system coupling the two rotors to eliminate rpm fluctuations of the individual rotors. Using one cylinder of diameter D and two pairs of rotors of differing diameters d, the results showed values of $R \approx 2.1$ for an area ratio $A_r = (D + 2d)/2d \approx 2.7$ and $R \approx 2.35$ for $A_r \approx 3.45$ with these average values covering a wide range of tunnel velocities.

Loth (1977) also describes a type of Darrieus rotor, Fig. 26, which he places in the energy concentrator category or at least as a type not subject to the theoretical Betz limit coefficient. He pictures it as a porous cylindrical obstruction to the wind which, especially for high-solidity rotors, will increase the wind speed for the blades operating near the maximum width, due to the blockage caused by the inner blades similar to the flow around a cylinder. To take full advantage of this flow regime requires vanes capable of varying angle of attack. A 1.5-m rotor has been constructed to test an automatic angle of attack actuator, a self-starting feature, and automatic feathering. The control is based on a double pivot system, one in the radial arm direction and one in the airfoil spanwise direction, and it is the action of a stabilator tail, the center of gravity of both airfoil and assembly, and a pivot with stops that provide the desired performance. Test results are not available at this time.

The vortex-flow type of concentrator makes use of the low-pressure high-velocity area developed with decreasing radius in a free vortex, in which the tangential velocity varies inversely with radius, and hence with no torque or friction loss, the pressure varies according to the Bernoulli equation. The free or potential vortex is a very common feature in nature, seen in multitudinous ways, e.g., the drainpipe vortex, the whirlpool, the dust devil, the automobile exhaust, the tornado, the hurricane, and even on the cosmic scale. Vortex motion has been studied intensively by aerodynamicists, as it is a major tool in analysis of fluid behavior, and this behavior has been utilized for wind turbine use. Three major investigations have been undertaken and significant analyses published in the last few years, two of them using flow regimes naturally occurring around particular shapes of surface (Loth, 1975, 1977; Walters, 1975; Sforza, 1977) and one by designing for a particular vortical arrangement (Yen, 1977).

The ideas of Loth and Sforza are occasioned by the vortices associated with airfoil profiles providing lift, the former by utilizing the trailing vortex at the tip of a wing section and the latter using the vortex formed by flow separation from sharp, highly swept-back delta-type planforms.

Loth points out that the vortex concentrator allows all the rotational kinetic energy to be extracted without reducing the mass flow through the rotor. Using tangential velocity V_θ as an average value, then the pressure coefficient $C_{pr} = 1 - (V_\theta^2 + U_\infty^2)/U_\infty^2 = -V_\theta^2/U_\infty^2$. The power available at rotor entry with the vortex concentrator is $\dot{m}(V_\theta^2 + U_\infty^2)/2 = \rho A U_\infty(V_\theta^2 + U_\infty^2)/2$, while without it, it would be $\rho A U_\infty(U_\infty^2)/2$. Thus the concentration ratio is

$$R = \frac{(V_\theta^2 + U_\infty^2)/2}{U_\infty^2/2} = 1 + \frac{V_\theta^2}{U_\infty^2} = 1 - C_{pr} \tag{5.10}$$

Figure 27 (from Loth, 1975) shows the trailing vortex system from a wing. The bound vortex providing the lift of the wing continues as trailing

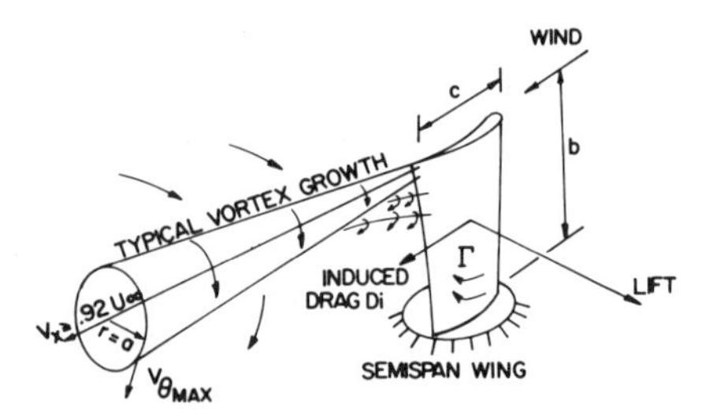

Fig. 27. Vortex-type concentrator—trailing edge vortex arrangement. (From Loth, 1975.)

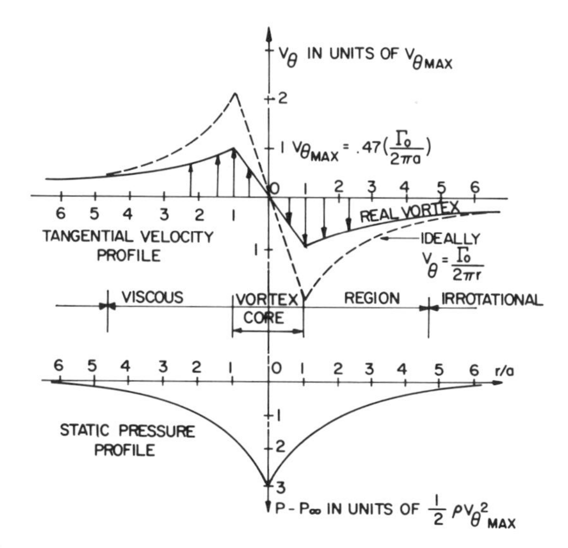

Fig. 28. Vortex-type concentrator—trailing edge vortex fluid pattern. (From Loth, 1975.)

vortices from the wing tips and the low pressure in these vortices is proportional to the square of the wing lift coefficient. A downwash velocity is generated which leads to an induced drag D_i and for ideal flow, the pressure drag due to the low pressure equals the rotational kinetic energy of the vortex. However, the vortex is not ideal toward the axis because the very high velocities there lead to considerable viscous shear, and the actual velocity distribution is more like that shown in Fig. 28 (from Loth, 1975). The vortex core is delineated by the radius of maximum velocity a inside of which the rotation is that of a solid body, and outside of which it is more like an irrotational vortex, although viscous effects are significant out to about a radius of $4.5a$. The maximum velocity at radius a is only about one-half of the ideal value.

Loth 1977 analyses the augmentation by a vortex in the following manner. The rotational energy entering a turbine of diameter d placed coaxially with the vortex as a fraction f of total induced drag D_i is given by

$$f = \frac{2\pi}{D_i} \int_0^{d/2} \rho \, \frac{V_\theta^2}{2} \, r \, dr \tag{5.11}$$

This value for one chord length downstream of a wing tip is shown in Fig. 29. Then the concentration ratio is given by $R = 1 + fZ$, where Z is

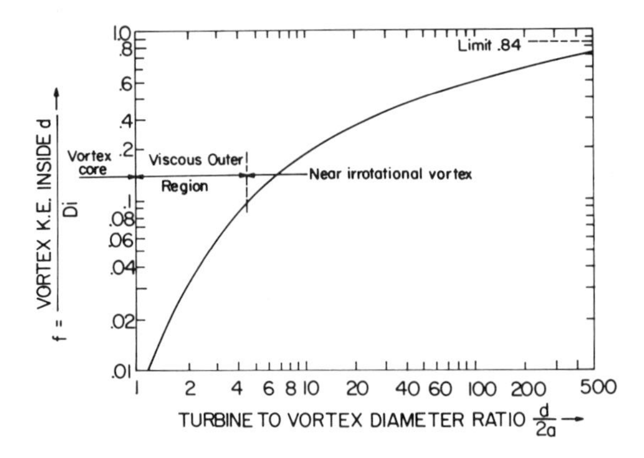

Fig. 29. Vortex-type concentrator—rotational energy fraction. (From Loth, 1975.)

the ratio of the induced drag to the freestream kinetic energy available to the turbine inlet, i.e.,

$$Z = D_i/\tfrac{1}{2}\rho V_\infty^2 \tfrac{1}{4}p\, d^2 \tag{5.12}$$

and with $D_i = C_{D_i} \frac{1}{2}\rho V_\infty^2\, bc$, where $C_{D_i} = C_L^2/\pi A_R e$, b is the wing semi-span, C is the chord, A_R is the aspect ratio $= 2b/c$, and e is a spanwise loading efficiency,

$$Z = (2/e)(cC_L/\pi d)^2 \tag{5.13}$$

Thus low e and high C_L are required. The total area of concentrator and turbine is given by

$$A_{\text{tot}} = \frac{\frac{1}{4}\pi\, d^2 + bc}{(\pi/4)\, d^2} = 1 + \frac{2A_R}{\pi(d/c)^2} \tag{5.14}$$

This is shown in Fig. 30 for a uniformly loaded high-lift semispan wing, with the vortex core radius d/c found experimentally to be 0.03. High C_L implies high A_R and hence A_{tot} increases rapidly, so a compromise must be made. If $R_{\max}$ is taken when $d = 3a$, then

$$R_{\max} = 1 + \tfrac{2}{3}(V_{\theta\max}/V_\infty)^2 \approx 1 + C_L^2/2e \tag{5.15}$$

A model concentrator with a Liebeck airfoil set vertically in the ground, having $C_{L_{\max}} = 4$, height of 2.5 times the chord length, and $A_R = 5$, should operate with $R = 3$, $A_{\text{tot}} = 10$ and turbine diameter/chord, $d/c = 0.6$.

Sforza investigated the use of the vortex flow field generated by flow separation from sharp-edged highly swept-back delta-type planforms, as shown diagrammatically in Fig. 31. The vortex field formed by delta planforms has a considerable literature but there is a dearth of detailed mea-

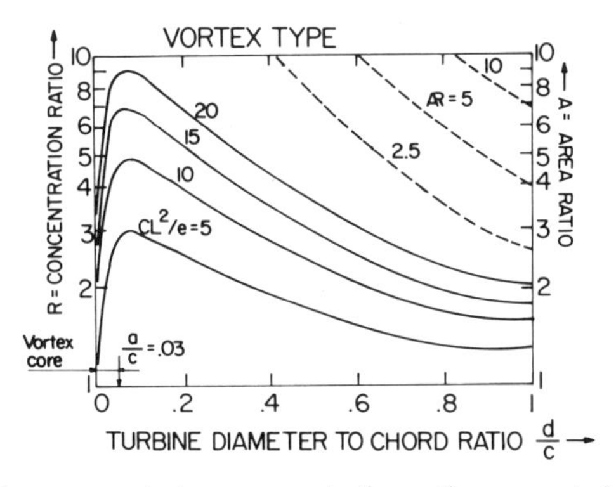

Fig. 30. Vortex-type concentrator—concentration ratio versus turbine diameter/chord ratio. (From Loth, 1975.)

surements of the velocities. However, what is available shows that the axial components are appreciably larger than the freestream flow and that the tangential component is of the same order. There is thus a considerable power augmentation effect, due to both mass flow rate and rotational energy. Initial wind tunnel tests were made by Sforza to demonstrate the concept, using a flat plate delta wing with sharp leading edges and total apex angle of 15° set at an angle of attack of 20.4° with the freestream flow. Such an arrangement gives a three- to fourfold increase over unaugmented operation and this can be increased by adding a rear flap and by a ramplike centerbody. The tests also showed the advantage of an augmentor is being able to operate at much lower wind speeds, in this case of the order of 25% of that for a conventional turbine.

Following these initial tests, detailed work on many configurations was carried out. It was found that the velocity field is well ordered although not truly axisymmetric and that the vortex center moves upward and

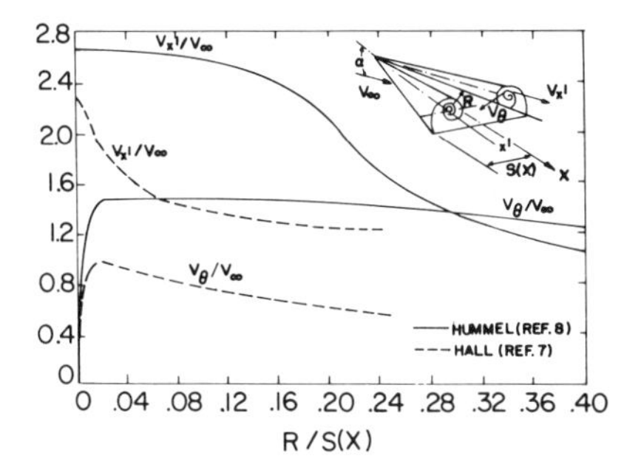

Fig. 31. Vortex-type concentrator—swept-back delta wing–type arrangement. (From Sforza, 1977.)

outward with increasing angle of attack. Thus simple change of angle of attack, or together with flap deflection, may be used as a power and speed control. Furthermore, the vortex formation process smoothes out nonuniformities in the approach flow, a useful feature in regulation of output and for structural considerations. Patents have been applied for Sforza's concept.

The third type of vortex concentrator is called the "tornado type" by its progenitor Yen (1977), for which patents have also been applied for. Figure 32 shows one such type of wind power system, consisting of an open-top stationary, cylindrical tower with adjustable vanes, at the bottom of which is mounted a vertical-axis turbine of much smaller diameter, located in a convergent duct drawing air from an annular opening next to the ground. The vanes, which would be open to windward and closed to leeward, direct air tangentially to create a vortex, generating a high-velocity low-pressure flow at the turbine discharge. The turbine air and the vortex air exhaust through the top of the tower. In this manner a large pressure difference is created across the turbine to yield a high-density power output. This would seem to be similar in principle to the Venturi-type augmentation in providing a pressure lower than atmospheric at the system discharge, but in this case by means of an axisymmetric vortex rather than the toroidal one of the ring-wing concept. The pressure drop across the turbine is much larger with the Yen concept than with the ring-wing design. The total quantity of air is many times that of the simple turbine of equivalent area, but the collecting surface of the tower and all is also much larger and the direct connection between total frontal area and increased power output is not immediately apparent.

Yen develops two "scaling laws," which are in effect relationships governing power output and power coefficient, deduced from applying the conservation equations of mass, momentum, and energy first to the turbine and then to the tower and turbine as a whole. The flow in the tower is regarded as a central vortex of radius r_c with a nearly constant pressure p_c

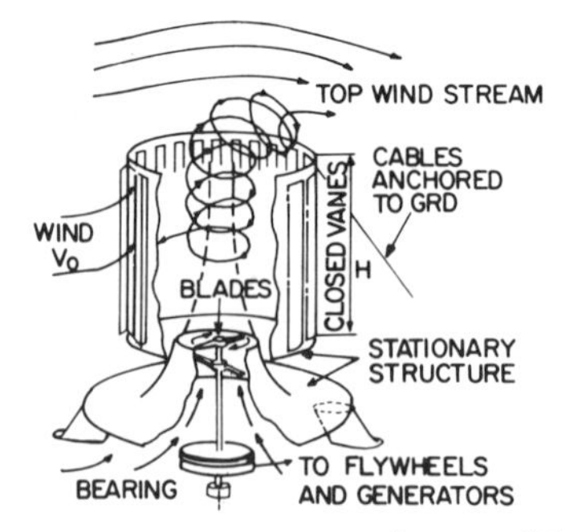

Fig. 32. Vortex-type concentrator—tornado type. (From Yen, 1977.)

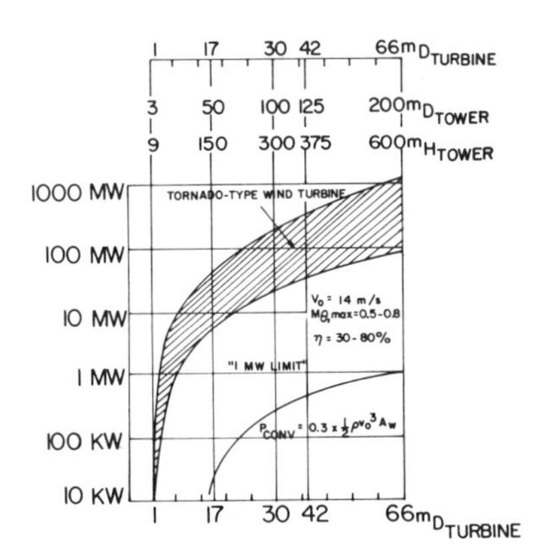

Fig. 33. Tornado-type concentrator—performance. (From Yen, 1977.)

and nearly constant axial velocity $V_{z,c}$, surrounded by a potential vortex reaching to the wall where there is a maximum tangential velocity $V_{\theta,\max}$. The turbine radius is taken as greater than r_c. The turbine output is given by

$$\dot{W} = \eta_t \, \Delta_{p_t} V_i A_t \tag{5.16}$$

where subscript t indicates the turbine and V_i is the turbine inlet velocity. Δ_{p_t} is the turbine pressure drop and from considerations of momentum and isentropic flow theory for the turbine, expressions for an optimized value of Δ_{p_t} are found. These give Scaling law I. The second scaling law results in an expression for the power coefficient as

$$C_p = \frac{\dot{W}}{\frac{1}{2}\rho V_\infty^3 HD} = \eta \frac{A_t V_t}{HDV_\infty} \left(\frac{M_{\theta,\max}}{M_\infty}\right)^2 \tag{5.17}$$

where H is the tower height, D is the tower diameter, and M indicates Mach number. Some small-scale tests have been carried out in order to validate these scaling flows and Yen reports that Scaling law I underestimates, while Scaling law II agrees with the data. A system efficiency η of 50–70% and an augmentation factor of 6 were obtained. Predicted output of large systems using the above analyses are shown in Fig. 33 together with a comparison with a conventional turbine.

VI. WIND CHARACTERISTICS AND SITING

A. Wind Data, Velocity, and Power Duration, Rated Values

In addition to its low energy density, the characteristics of wind power are its variability, its unpredictability except statistically over long periods of time, and its untransportability—it has to be used at a fixed location. Because of the cubic relationship for power/speed, it is of the utmost importance that the wind characteristics at a potential site are known—once *in situ* it would be disastrous to have to move a system of any size. In spite of what appeared to be very careful consideration of the site, the 1250-kW Putnam wind turbine on Grandpa's Knob in the United States experienced an average wind speed of only 16.7 mph as against the expected value of 24 mph. This represents only about $\frac{1}{3}$ of the expected power output, a fact which crippled its economic viability. Unfortunately, to obtain wind data of any reliability at a given location has to be a lengthy business, a period of a year being seemingly a minimum. Even then there are "windy years," although statistically over a long period of time it is reasonable to expect that the average velocity in any one year will be within limits of ±15–20% of the long-term mean, with perhaps 90% probability of being within ±10%. While there is a considerable amount of wind data available from meterorological stations, much of this is often of marginal value, because the WECS is not wanted at the particular location, or other factors rule out the possibility, and because the data may not be in a usable form for a truly realistic appraisal. However, data have become increasingly available in recent years, due to the rising complexity of global culture being more dependent on weather prediction and especially due to the ubiquity of the airplane throughout the whole world.

The starting point for an appreciation of wind characteristics, surveys, sites, etc. is Golding (1976). While this is mostly a quarter century old, much of it is still highly pertinent, and it is mostly this section of the text which has been updated in the recent reprinting. New parameters of more utility in statistical analyses are coming into use today but Golding's discussion still provides the basic understanding of the problem.

Another overview of wind characteristics and applicable data and techniques is given in *Initial Wind Energy Data Assessment Study* (Changery, 1975), which reports presentations and discussion by numerous meteorologists and engineers concerned with wind power at a NSF/NOAA Meeting in 1974. Sweden has initiated a comprehensive meteorological program to obtain data for site evaluation and for studies of optimal grouping and performance characteristics of WECS on a local, regional, or national scale (Ljungström, 1975). A useful discussion of available sources and methods of presentation of data for United States locations is given in the *GEC Wind Energy Mission Analysis* (1977a). Figure 34 shows a wind energy density map of the United States (Blackwell and Feltz, 1975).

Data are most likely to be nonexistent or extremely sparse in many areas of particular interest for wind power, e.g., mountainous and wilder-

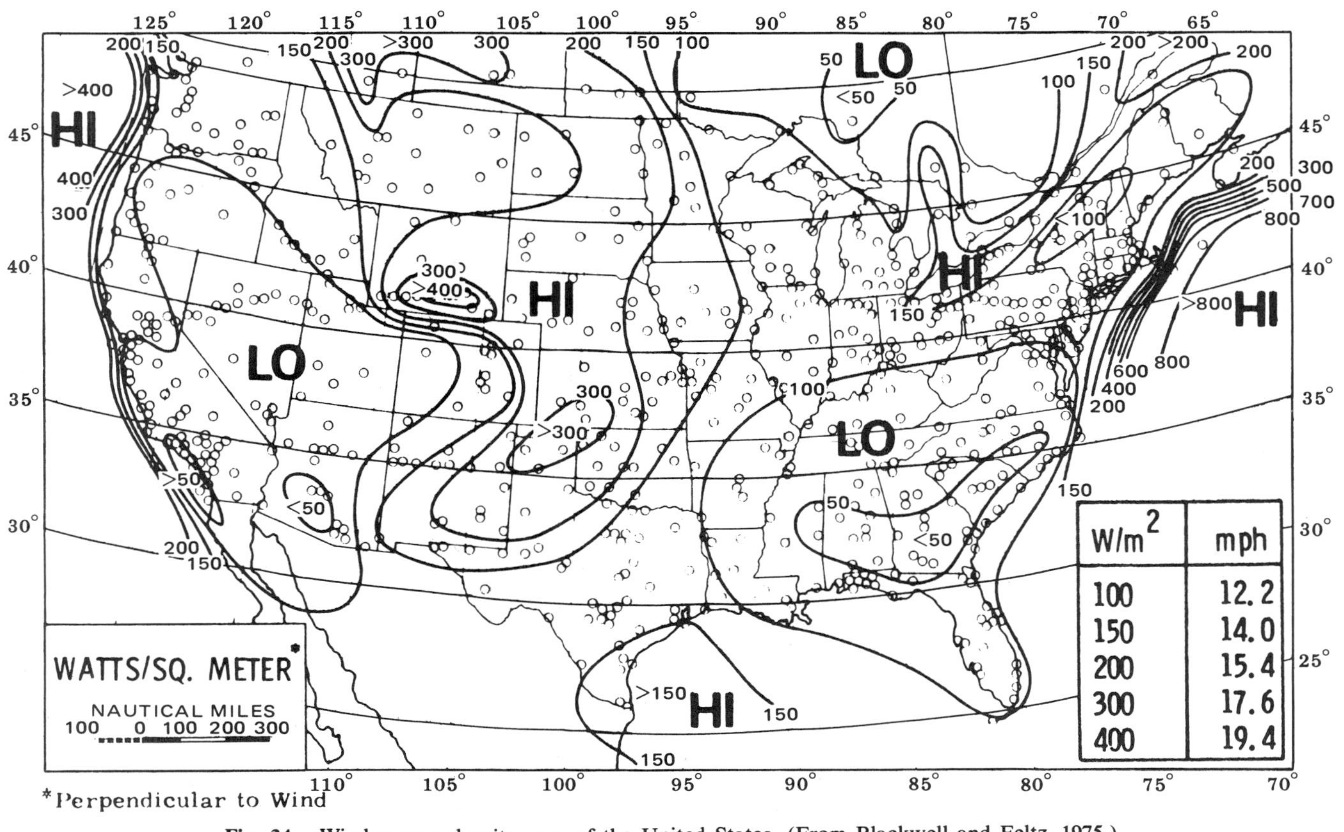

*Perpendicular to Wind

Fig. 34. Wind energy density map of the United States. (From Blackwell and Feltz, 1975.)

ness areas or regions of sparse population where meteorological records are unimportant or impractical. Because the specific site is so important and the period required for detailed information is so long, increasing attention has been paid in recent years to statistical analysis applied to prediction from a minimum of input. Only accumulated experience will prove the accuracy of such methods, but to accomplish the widespread use of wind power necessitates some short cuts in site selection.

The single most useful variable is the average wind speed. For other than special locations, the average wind speed over much of the world's terrain is between about 3.5 and 5.5 m/sec (7.8–12.2 mph), a range which just brackets the "gentle breeze" classification on the Beaufort scale! This is not to deny the fact that there are many sites having greater speeds, but these have to be searched for. They are usually found on or near coastlines and in high or mountainous terrain. Because of the cube law again, the average velocity is not enough for a valid economic design and it is necessary to have information on the wind speed distribution, that is, the annual duration of wind speeds of a given value, which requires measurement of hourly wind speeds throughout the year. An instructive way to use these data is to plot a *velocity–duration* curve, i.e., the range of wind speed against the number of hours per year that the speed equals or exceeds a particular value. Figure 35 (Golding 1976) shows such curves for a number of widely separated sites throughout the world. There is a similarity of shape and this can be put on a comparative basis by normalizing through division by the average wind speed, Fig. 36 (Lapin, 1976). Thus it is possible to construct an approximate velocity–duration curve from a value of mean wind speed only, if detailed data are lacking. Other ways of showing this distribution data are by a frequency curve, with duration in hours per years or fractionally as ordinate against

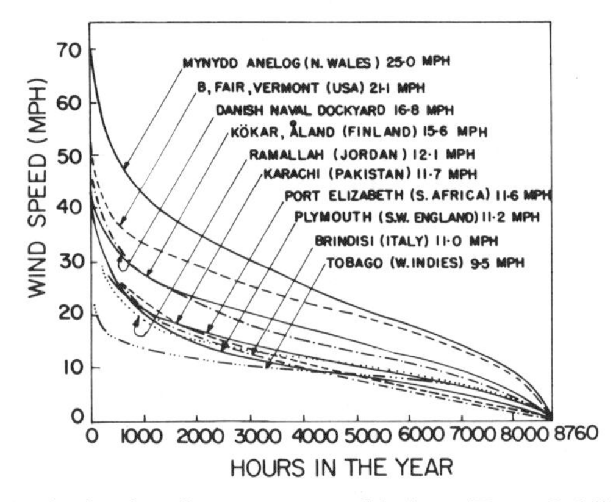

Fig. 35. Velocity-duration curve—world sites. (From Golding, 1976.)

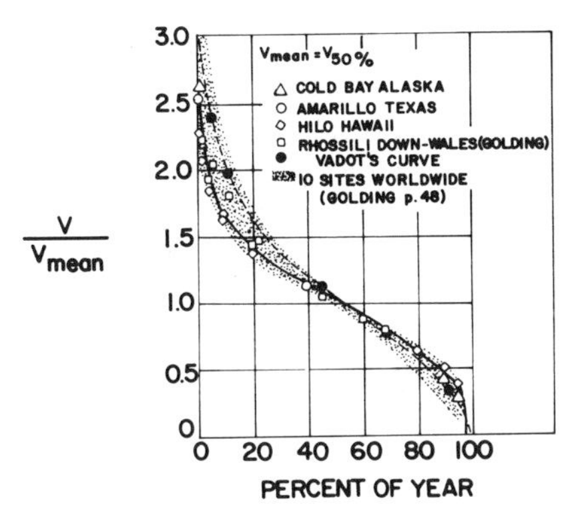

Fig. 36. Velocity-duration curves in terms of mean velocity. (From Lapin, 1976.)

wind speed, or as probability density versus speed or probability function (Smith, 1977). Figure 37 shows such a probability plot, also demonstrating that there is a considerable difference seasonally and this of course can be important. In the Northern Hemisphere, the maximum speeds occur in late winter or early spring, with minima in the summer. In the United States this pattern closely resembles the annual temperature curves, with maximum wind coinciding with minimum temperature, probably because maximum wind occurs with maximum north–south temperature gradients and hence maximum occurrence of frontal and wave cyclone activity (Garate, 1977a).

There is also diurnal variation and this is less certain and more subject to error than seasonal changes. In the United States, maxima generally occur just after midday and minima between midnight and 6:00 AM for

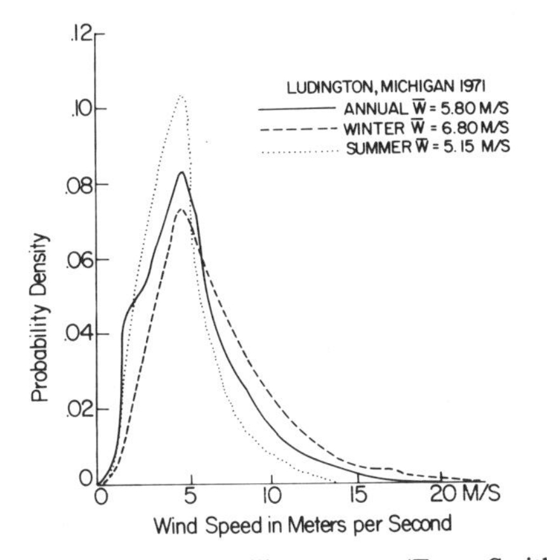

Fig. 37. Wind speed probability curves. (From Smith, 1977.)

locations which have low or moderate average wind speeds, but this is reversed for many locations having higher average speeds. The reason is that the low-speed locations are usually at lower altitudes and the measurements taken relatively close to the ground, while the high speeds are at higher altitudes and/or the measurements taken well above the ground, e.g., on a high tower or on a mountain top. These diurnal effects occur as a result of momentum exchange due to thermal gradients and the ground boundary layer (Garate, 1977a). Wind direction is relatively unimportant except for a few windpower systems, e.g., the tracked-vehicle and ducted augmentor types, because turbines can either inherently accept wind from all directions (VAWTs) or can move into the proper direction for maximum output (HAWTs). However, for large turbines with necessarily low rotational speed, the frequency of wind change can effect the nominal output estimated by wind speed only. There are some special locations, such as in valleys or situations which tend to tunnel the wind, when a fixed direction might be economically justified.

It is of course the turbine energy output which is important and this brings in the cube law, so that discrepancies between expected and actual wind speeds are greatly magnified in their effect. A power–duration or probability density curve can be constructed by cubing the ordinate of the corresponding velocity curve. Figure 38 shows a velocity-duration curve and a power-duration curve, the latter being commonly of monotonic slope even if the former is *S* shaped. There are other features of this diagram that are useful in depicting performance in convenient fashion. These are the cut-in, cut-out, and rated speeds. The cut-in speed, V_{ci}, is the lowest wind speed at which useful power is being delivered, lower

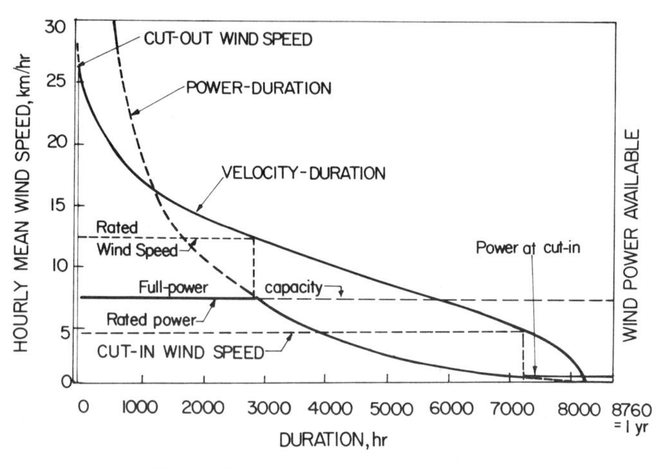

Fig. 38. Velocity- and power-duration curves.

values being capable of only just balancing mechanical and aerodynamic losses or for which the rotor vanes are stalled. For most types of turbine this is around 3–4 m/sec (7–9 mph) and is a matter of some importance in rotor design. The cut-out speed V_{co} (also sometimes called the "furling" speed from the background of Dutch-type windmill "sails") is that speed at which the wind may cause structural damage with the turbine in normal operating attitude, and measures have to be taken to "spoil" the wind by altering the blade pitch, plane of rotation, or introducing some screening or countering mechanism to ensure safety or to shut down completely. The rated speed V_R is that windspeed at which the system is delivering the maximum output $\dot{W}_R$ for which it is designed, and some power-shedding measure must be operative to prevent overloading. The rated speed is a most important parameter for economical operation and will be discussed shortly.

The power output is constant for all higher speeds than the rated speed, and the area under the power curve between the cut-in and rated power points represents the total annual energy output, the kilowatt hours, for example. The annual plant load factor or capacity factor is the ratio of this energy output to that which the system would deliver if it operated continuously at the rated windspeed. The latter is represented in Fig. 38 by the surrounding broken-line rectangle. With the power output $\dot{W} \propto AV_\infty^3$, then the rotor swept area $A \propto \dot{W}_R/V_\infty^3$ and for a propeller-type rotor, the diameter is proportional to $(\dot{W}_R/V_\infty^3)^{1/2}$. Wind speed, system power, and system energy at this limiting value of wind speed are known variously as the "rated" values, sometimes as "shaft capacity" values, and until the corresponding value of rated windspeed is known, a power output of so many kilowatts has limited meaning. The term "design" speed or power needs careful interpretation at times, as it is occasionally used in place of the term "rated," but generally it is reserved for values at the most efficient or minimum energy cost-operating conditions.

The determination of rated power is the starting point for the economic balance sheet of WECS. Consideration of Fig. 38 will show that for a given average wind speed, the lower the value of rated wind speed, the higher will be the specific output, but the rotor will have to be larger for a given output power. There is then an optimum value of the rated condition for the minimum annual cost of energy. For any system, the total investment cost has two parts, one which is a fixed amount and one which depends on the power rating and hence on the cube of the rated wind speed. The greater the proportion of cost dependent on rating, the lower is the optimum value of V_R/V_m. Two of the most costly items of a WECS are the rotor itself and the generator, and the ratio of their costs will affect the optimum rating. Various "energy pattern factors" have been suggested to compare the energy potential of different locations, but Golding (1976) concludes that their main use is "to show how fallacious it is to calculate available energy from long-period mean windspeeds. Such calculations are only useful when based upon the velocity–duration curve for the site."

Hewson and his co-workers have made extensive studies of wind power potential for the state of Oregon, and Hewson (1975) gives an interesting qualitative account of wind distribution problems and sitings in a general report on wind power. He shows that for two sites only about 25 km apart, Columbia Lightship at the mouth of the river and Astoria, just inland, there is a very pronounced difference in wind speed, which translated to power would be tremendous, Fig. 39. He also shows that for the Columbia Lightship again and a site in the Columbia gorge, Cascade Locks, there is, at the former, greater power in the higher speed wind but less in the lighter winds, i.e., the power–duration curves have very different slopes, Fig. 40. Another feature for the Lightship winds is that the power in 1973 was very much higher than the long-term annual average, Fig. 41. Such phenomena point up the difficulties in estimating available wind energy.

A recent analysis by Smith (1977) demonstrates the importance of rated values on cost and site selection. He defines an energy ratio R between the actual energy produced W_e and the energy produced if the system operated continuously at rated power $\dot{W}_R$. Thus

$$R = \frac{W_e}{W_R} = CD^2T \int_{V_{ci}}^{V_{co}} \eta P(V_\infty) V_\infty^3 \, dV_\infty / CD^2 \eta_R V_R^3 T \tag{6.1}$$

where CD^2 is a parameter representing a constant C (including air density) and a reference length D associated with the wind "capture" area or

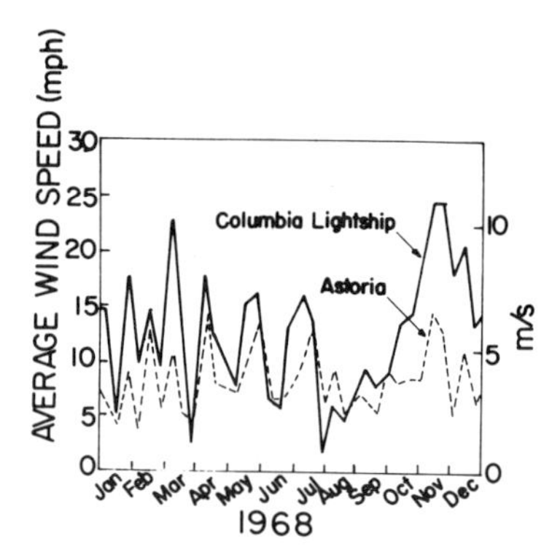

Fig. 39. Comparison of wind speed data. (From Hewson, 1975.)

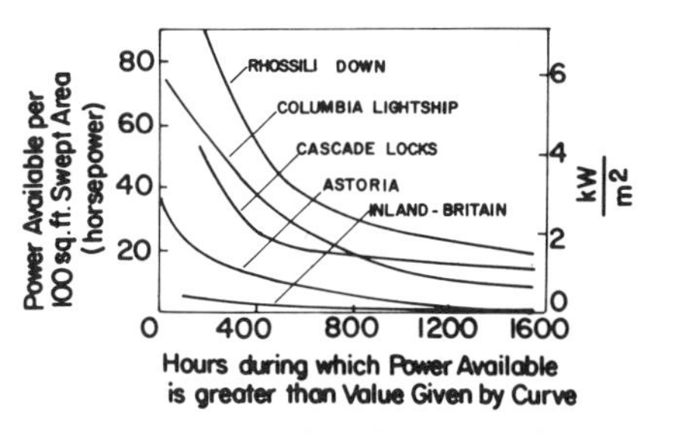

Fig. 40. Power-duration curves for five sites. (From Hewson, 1975.)

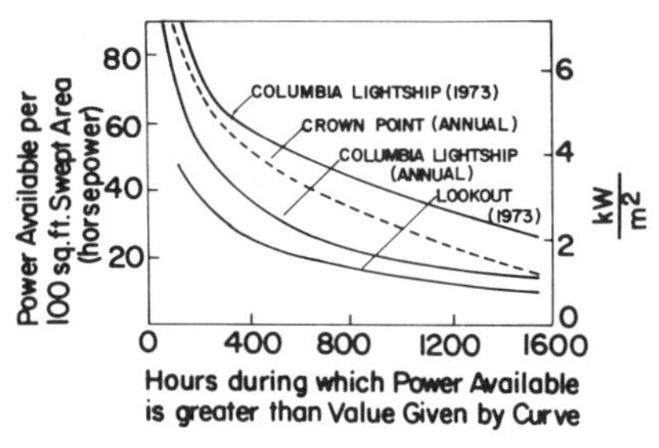

Fig. 41. Power-duration curves for various sites in Oregon. (From Hewson, 1975.)

blade–wind "interaction area," T is time (say, one year), η is the system efficiency, $P(V_\infty)$ is the wind density probability distribution, and subscript R indicates rated condition, with other symbols as previously defined. R and η can be established at any station from rotor coupling to busbar and is open to choice for a particular end result. $P(V_\infty)$ and η can be dependent on both V_∞ and azimuthal angle of the rotor if this is a factor and η is also dependent on the vertical wind profile. A simplification can be made without great error if η is taken equal to η_R and then, referring to a speed-duration curve as shown in Fig. 37, R can be obtained as

$$R = \int_{V_{ci}}^{V_{co}} P(V_\infty)\left(\frac{V_\infty}{V_R}\right)^3 dV_\infty = \int_{V_{ci}}^{V_R} P(V_\infty)\left(\frac{V_\infty}{V_R}\right)^3 dV_\infty + \int_{V_R}^{V_{co}} P(V_\infty)\, dV_\infty \tag{6.2}$$

Smith gives curves showing R as a function of V_R/V_m, indicating that the effect of V_{ci} is not large. The effect of V_{co} is considered negligible and ignored. Then function R can be used to establish two parameters. Thus in the first place,

$$W_e = R(\dot{W}_R T) \qquad \text{or} \qquad \dot{W}_R T/W_e = 1/R \tag{6.3}$$

Secondly, $\dot{W}_R$ can also be expressed as $\dot{W}_R = \frac{1}{2}\rho\pi D^2 \eta_R V_R^3$ and hence

$$\frac{CD^2 T \eta_R V_m^3}{W_e} = \frac{1}{R(V_R/V_m)^3} \tag{6.4}$$

where the average wind speed V_m is introduced for normalizing. Equations (6.3) and (6.4) define a power parameter and an area parameter respectively. These are plotted in Fig. 42 for the wind probability densities given in Fig. 37, showing lines of constant V_R/V_m. V_R can then be selected for minimum cost for a desired average annual energy, as it does

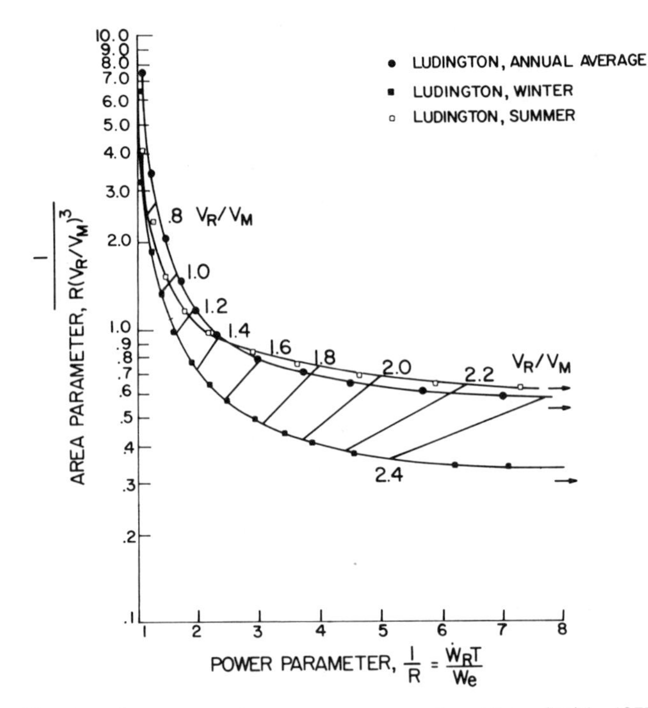

Fig. 42. Area parameter vs power parameter. (From Smith, 1977.)

not depend on W_e, V_m, η, or C. For some value of V_R, the values of D and $\dot{W}_R$ are then determined. Smith states (and this is found elsewhere in the literature) that a value of V_R/V_m of about 2 is currently recommended, although this is high by a number of actual designs, which tend to range from 1.6 to 1.8. Figure 42 shows very clearly that as V_R increases so must $\dot{W}_R$, and as it decreases, the rotor size, as D, must increase. The selected value of $\dot{W}_e$ affects the cost minimization curves for values of V_R, as this determines the size of $\dot{W}_R$ and D, which in turn affect the costs of the system. $\dot{W}_e$ may be known for a particular application, usually for small units, but for utility power it is likely that the maximum energy for the given site is required. It would seem apparent that a site with high values of V_m would be chosen over one with a lower value, but the wind distribution affects cost via the values of $\dot{W}_e$ and D. Thus it is possible that a lower V_m site with better wind distribution would lower the annual energy cost, although Smith states that for selected wind systems to date, this has not been apparent. However, it is clear that wind distribution characteristics are essential in fixing a rated power $\dot{W}_R$ and for selection between sites of about the same average wind speed.

B. Statistical Methods

Statistical parameters to express wind distribution speeds are very useful, and considerable work has been carried on in recent years, e.g., Justus and co-workers (1976), Wentink (1976), and Hennessey (1977). Widger (1976) developed a method of calculating wind power values from average wind speeds only, which was simple and appeared to have considerable value if generally valid. His results were based on records from six New England stations spread over four states using a linear relation between average and fastest wind speeds and normalizing the usual skewed distribution by a square root transformation. From comparing computed results for power output based on these theoretical values with the values calculated from the actual distributions, he concluded that his generalized tables and graph gave results within ±20%. However, Baker and Hennessey (1977) applied his method to four sites in Oregon, with the results that differentials of up to 60% were demonstrated. It would appear that wind distributions which are considerably skewed cannot be handled with a simple one-parameter distribution. These results are quoted here to emphasize the importance of wind distributions and the need for detailed analysis of a large number of wind records before generalized expressions can be used with confidence.

Justus *et al.* (1976) discuss the use of distribution parameters and consider a two-parameter distribution a practical alternative to complete precision with its accompanying complexity and cumbersome nature (e.g., the bivariate Gaussian distribution which requires five parameters to be specified). The Weibull and log-normal were selected for study because (1) the possible problem of applicability at low speeds is minimized by requiring data only between V_{ci} and V_{co}, the ready determination of parameters from observed data by least square techniques, and (2) both distributions have been used successfully in other analyses of wind phenomena (buildings and air pollution).

The Weibull distribution is expressed as

$$P(V)\,dV = (k/c)(V/c)^{k-1}e^{-(V/c)^k}\,dV \tag{6.5}$$

where c is the scale factor and k the shape factor. The former is closely related to mean speed V_m, since $V_m = -c\Gamma(1 + 1/k)$, where Γ is the ordinary gamma function of mathematics. For $1.4 < k < 3$, then $1.1 < c/V_R < 1.3$. k is inversely related to the variance σ^2 of wind speeds about the mean wind speed, with

$$\sigma^2 = c^2\{\Gamma(1 + 2/k) - [\Gamma(1 + 1/k)]^2\} \tag{6.6}$$

Values of c and k are determined from the cumulative probability data for site wind speeds within certain speed intervals. From Eq. (6.5), the cumulative probability of finding a speed less than V_x is given by

$$P(V \leq V_x) = 1 - e^{-(V_x/c)_k} \tag{6.7}$$

or rearranged to

$$\ln\{\ln[1 - P(V \leq V_x)]\} = k \ln c - k \ln V_x \tag{6.8}$$

k and c can be determined by least squares fit of $y = a + bx$, where y is the left-hand side of Eq. 6.8, $a = k \ln c$, $b = -k$, and $x = \ln V_x$. Least-square fits of Weibull and log-normal distributions showed that both were adequate representations, with the former being slightly better. Justus therefore prefers the Weibull distribution and used it to carry out a nationwide assessment of potential output from WECS (Justus *et al.*, 1976).

However, Wentink (1976) concluded in a study of the potentialities of Alaskan wind power that "the average power for a given windmill system can be predicted well from knowledge of only the average wind speed, at least in Alaska wind regimes." He made detailed analyses and formulations of measured and synthetic wind spectra. He gives five expressions, which he examined as representing the wind duration values designated as percent of total time when the wind speed V equals or exceeds that value designated by $\% \geq V$. These are

$$F1: \quad \% \geq V = e^{-KV}[A + BV + CV^2 + DV^3]$$

where K, A, B, C, and D are empirical constants.

$$F2: \quad \% \geq V = 100(k/a^4)^{e^{-ax}}[6 + 6ax + 3(ax)^2 + (ax)^3]$$

where a and k are constants, and $x = V/\hat{V}$, where $\hat{V}$ is another constant discussed below.

$$F3: \quad \% \geq V = 100e^{-(V/C)_k}$$

This is the Weibull Function as used previously here.

$$F4: \quad \% \geq V = 100\, e^{-a(V/V_m)^2}$$

where a is a constant and $V_m = V_{\text{mean}}$

$$F5: \quad \% \geq V = 100 \frac{k}{a^4} \sum_{m=1}^{\infty} \frac{e^{-\max}}{m^4} [6 + 6 \max + 3(\max)^2 + (\max)^3]$$

where m is integral 1, 2, etc., to, in practice, from 3 to 12 since convergence is usually rapid. $F1$ is empirical, of like form to $F2$ and $F5$, but with K, A, B, C, and D not having constraints. In $F2$, $\hat{V}$ can be taken as V_{median}, when a and k are fixed. $F4$ is $F2$ if $k = 2$ and $V = V_m$, and is the Rayleigh function.

$F5$ was deduced from the Planck frequency function, and was found no more useful than $F1$ or $F3$ and not easily used in curve fitting. The duration models $F1$ through $F4$ were then compared by fitting them to more than 107 cases of real wind data, with the result that the empirical $F1$ fitted the data best with 43% of the cases examined, the Weibull $F3$ with 30% and another 16% equally well by $F1$ and $F3$. For the Weibull $F3$, the fitted k ranged from 1.3 to 2.6 in individual cases, with the fitted mean $k = 2.08 \pm 0.15$ for all cases. Also in this range, the theoretical $V_m/c = 0.864 \pm 0.003$. These results covered the entire wind speed data range, and power estimates usually require only the speeds between cut in and cut out.

Now the Weibull distribution, Eq. (6.5), when integrated to give the velocity duration relationship, yields

$$\int_V^\infty P(V)\, dV = e^{-(V/C)^k} \tag{6.9}$$

with C related to V_m via a gamma function of k. Wentink found that for Alaskan coastal winds, k was very close to 2 and the product of C and the gamma function of $k \approx 0.78$. Thus

$$\% \geq V = 100be^{-0.78(V/V_m)^2} \tag{6.10}$$

where b is almost but not quite unity. This is the one-parameter Rayleigh-type function, requiring a knowledge only of average wind speed V_m, and hence leads to Wentink's optimism regarding a possible "universal" parameter. This carried over to the power outputs for three small wind turbines (3–6 kW), using measured wind data and the manufacturers power versus wind speed curves ("quasimeasured values") compared to the predicted values from the distribution expression. One machine, an Electro 6 kW, was tested and analyzed in detail. Using the Weibull distribution function, the percentage error when comparing the quasimeasured and predicted values of $\dot{W}_m$ (relative to the quasimeasured values) ranged from 0 to 7.4% for 44 points, with 2.6% mean error, for speeds of $11.4 < V < 26.4$ mph. Two additional lower velocity points, V_m of 10.7 and 8.9 mph, were in much greater error but still resulted in an overall average error of 3.7%. Hennessey (1977) discusses parameters for wind distributions and in particular the use of the Weibull model, which he considers a good one for facilitating the computation of the mean and the standard deviation of the total wind power density, the usable wind power density, and the density when the WECS is operating.

Kirschbaum *et al.* (1976) made a study of an offshore site along the New Jersey coast, using incomplete data for four approximately equal periods during one year, yielding five full weeks of data in each season, spring, summer, fall, and winter, a total of 3306 hours in all (i.e., about 38%). Three figures are reproduced here to show the results of applying a gamma distribution to such measured data. Figure 43(a) shows a frequency–speed diagram for the raw data and a smooth curve obtained by a gamma function from the mean wind speed (15.84 mph) and the standard deviation (5.45 mph). It was pointed out that the hourly average speeds were estimated by eye from strip chart grids with abscissa divisions every 2 mph and this, together with the propensity of many people to favor even over odd numbers, may explain the tendency to favor even speeds for the raw data points. Figure 43(b) shows the sum total for all four periods and 43(c) shows wind duration as percent of total time.

Several other studies of wind distribution have been made in varying degrees of depth. The selection discussed here has been made in an attempt to illuminate the nature of the problem, the methods being used to obtain simple but reliable generalizations from a minimum of data, and to acknowledge the efforts of some of those most active in furthering the state of the art. The problem has been put on a sound analytic basis and, at least for the time being, the Weibull distribution gives some reasonable answers. But much remains to do, and one aspect which appears interesting is a closer look at the particular wind speed distributions and site characteristics that do not fall into such general patterns as have been put forward.

C. Effect of Height

There are two other wind characteristics which are important, the variation of speed with altitude above ground level and the nature and effects of turbulence.

The effect of height above ground level on wind speed (i.e., the planetary boundary layer) has been a subject of considerable study for many years, much of it from the meteorological point of view, naturally, but it is certainly important for wind power. The immediate concern is, of course, to estimate variation of power with height due to rotor size and to tower height itself as a possible design criterion. But also, there is a great need to convert measured values of wind speed recorded at various heights to a standard height or at least the same height for all values. This last-named factor represents a considerable problem in the use of wind data because very few regular measuring stations have been installed with wind power in mind, and it is rare to find really long-term data in which either the specific location or height or measuring interval has been unchanged during the period. In earlier days, many measurements were taken at coastguard stations, lighthouses, lightships, etc. and installation

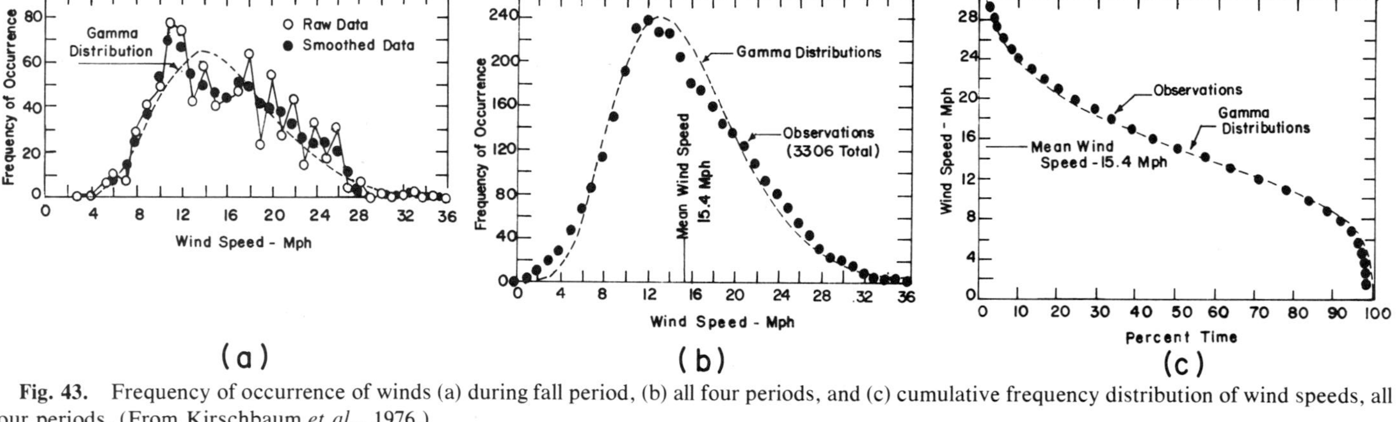

Fig. 43. Frequency of occurrence of winds (a) during fall period, (b) all four periods, and (c) cumulative frequency distribution of wind speeds, all four periods. (From Kirschbaum *et al.*, 1976.)

was dictated largely by convenience. With the exception of the Putnam 1250-kW WECS, wind power investigations were concerned with rather low hub heights, so that the nominally "standard" height of 10 m for anemometers was commonly used for all power units. Currently, the information required is largely for units upwards of 1 MW, with rotor radii of possibly 30 m and higher. Also, with the larger units there is a difference in the wind speed over the swept area from 6 to 12 o'clock, so to speak, and this requires evaluation in performance estimation and in structural design generally, perhaps particularly for the effect of cyclically varying stresses. It should be mentioned at this point that many of the measured speed data referred to earlier with respect to distribution characteristics may have had to be corrected by means of height relationship to establish standard conditions.

It is not to be expected that there can be any precise answer to the problem because there are too many variables, i.e., the general level of the terrain—a plain or a hill, the "roughness" factor—small scale or large scale, thermal equilibrium, frequency of directional change, etc. Golding (1955, 1976) provides useful background on this height question, particularly with respect to hill siting, but gives few precise figures. Published relationships are usually in the form of a logarithmic or power law and a simple one mentioned by Golding is $V_h = kh^{0.17}$ where V_h is the wind speed at the hub or height level h meters above a flat, open ground and where k is a proportionality constant.

Reed (1976) has made several studies of the height effect and points out that the problem is rendered somewhat easier by the fact that the winds of interest are moderate to strong, which promotes neutral stability, in distinction to pollution studies in which the effects are most troublesome at very low speeds. He states that rough terrain causes ground frictional effects to extend to higher levels and that stable air inhibits vertical circulation and frictional drag influences. With a usable value of wind speed turbulent mixing creates an adiabatic temperature lapse rate, a neutral stability. Thus some complexities can be neglected and an expression involving only height and roughness length may be formulated thus,

$$\frac{V_z}{V_a} = \frac{\ln Z/Z_0}{\ln Z_a/Z_0} \tag{6.11}$$

where Z is the height for which the velocity V_z is required knowing the reference velocity V_a at height Z_a, with Z_0 as a measure of terrain roughness. The roughness length is not an actual dimension, but an equivalent parameter which has been deduced from actual wind profiles. Level ground is approximated by $Z_0 = 0.03$ m and rough terrain by $Z_0 = 1.0$ m. Reed shows curves of height Z vs V_z/V_a for the two roughness lengths, demonstrating that for $V_z/V_a > 1$ (i.e., above the reference level of 10 m), the values for level ground ($Z_0 = 0.03$) are expressed with negligible error by the simple power relationship $V_z/V_a = (Z/Z_a)^{1/7}$. For rough terrain ($Z_0 = 1.0$), he considers that an exponent of 0.4 gives a useful simple expression, although it is in error by about 5% at 30 m. Reed also concludes that for level terrain, the error caused by using the velocity at hub height or midpoint is negligible compared with the exact value for integrating over the turbine diameter or height when using the $\frac{1}{7}$ power law.

Justus and Mikhail (1976) have enlarged the use of expressions for the height variation of wind speed to include height variations of wind distribution functions as well. They follow on the developments of Reed that (1) the power-law expression

$$V_z/V_a = (Z_z/Z_a)^n \tag{6.12}$$

is valid, with the exponent n being a function of V_a at Z_a, and (2) V_z and V_a can also be related by

$$V_z = \alpha V_a^\beta \tag{6.13}$$

where α and β are both functions of Z_a and Z_z. These two expressions can be simultaneously valid if

$$\alpha = (Z_z/Z_a)^a \tag{6.14}$$

and

$$\beta = 1 + b \ln Z_z/Z_a \tag{6.15}$$

with a and b depending on the reference height Z_a. Also, exponent n must have the form $n = a + b \ln V_a$.

Now Justus and Mikhail show that the Weibull distribution can be introduced and is given by the cumulative probability function $P(V)$ as [cf. Eq. (6.7)]

$$P(V_a) = 1 - e^{-(V_a/C_a)^{k_a}} \tag{6.16}$$

where C_a and k_a are scale factor and shape factor, respectively, at Z_a. Substituting the power law $V_z/V_a = (Z_z/Z_a)^n$ into this relationship, then for the power law to hold, $P(V_z)$ must be in the form

$$P(V_z) = 1 - \exp[-(V_z/\alpha)^{k_a/\beta}C_a^{-k_a}] = 1 - \exp[-(V_z/C_z)^{k_z}] \tag{6.17}$$

provided that $C_z = \alpha C_a^\beta$ and $k_z = k_a/\beta$. Thus the instantaneous wind and the Weibull scale factor both vary with height in like manner, and hence

$$C_z/C_a = (Z_z/Z_a)^n$$

with $n = a + b \ln C_a$. Then using data analyzed in Justus *et al.* (1976),

Justus shows that the observed behavior of C and k can be used to evaluate a, b, α, β and n, at least for heights up to 100 m. The results yield

$$\frac{k_z}{k_a} = \frac{1 - 0.0881 \ln(Z_a/10)}{1 - 0.0881 \ln(Z_z/10)} \tag{6.18}$$

and

$$C_z/C_a = (Z_z/Z_a)^n \tag{6.19}$$

where

$$n = \frac{0.37 - 0.0881 \ln C_a}{1 - 0.0881 \ln(Z_a/10)} \tag{6.20}$$

and heights are in meters.

For the corresponding problem of known wind speed V_a at height Z_a, wind speed V_z at height z is given by

$$V_z/V_a = (Z/Z_a)^n \tag{6-21}$$

with

$$n = \frac{(0.37 - 0.0881 \ln V_a)}{1 - 0.0881 \ln(Z_a/10)} \tag{6.22}$$

In terms of the wind speed law (6.13), then

$$\alpha = (Z_z/Z_a)^{0.37/[1-0.0881 \ln(Z_a/10)]} \tag{6.23}$$

and

$$\beta = \frac{1 - 0.0881 \ln(Z_z/Z_a)}{1 - 0.0881 \ln(Z_a/10)} \tag{6.24}$$

Because of the close correspondence of mean speed V_m and Weibull scale factor C, Eq. (6.21) can be used for mean values of V_z by substituting $V_{mz}/V_{ma} = (Z_z/Z_a)^n$, with n given by (6.22) substituting V_{ma} for V_a.

These relationships yield higher values of power-law exponent than in the purely empirical values of Golding (0.17) and Reed ($\frac{1}{7}$), e.g., a typical average velocity of 5 m/sec ($\approx 11\frac{1}{4}$ mph) requires $n \approx 0.225$.

Corotis (1976) has made a detailed statistical and probabilistic study of seven sites (five typical midwest, one high-tower midwest, and one Rocky Mountain) with the objective of establishing minimum data collection requirements to establish confidence limits on the wind characteristics of a potential site. He concludes that single-year velocity records appear acceptable for velocity statistics but are marginal for power estimates without additional information. He found much larger variations in the summer than in the winter, with late morning hours in both cases being fairly representative of the daily average. Sites with higher mean speeds have very little diurnal cycle, with maximum cycle deviation from the mean being up to 35% for low-duration sites in the midwest and 10% elsewhere. He points out that the decreasing diurnal effect with increasing height above ground is not accurately modeled in logarithmic or power-law formulas. Velocity probabilities exhibited reasonable fit with Chi-2 and Weibull distributions. Many other results of a probability analysis with respect to site evaluation are given, such as persistence of wind in terms of velocity and duration, etc.

D. Effect of Gusts and Turbulence

The problem of gusts has always been a major one for wind turbine designers, both for their effect on power output and control equipment, and more particularly, on transient loading phenomena and structural strength. Golding (1976) devotes considerable attention to gust patterns, but today's approach is built around statistical methods and computer simulations. There is a considerable amount of literature on gust phenomena in general because of its importance to aircraft, buildings, and other structures. As can be expected, the problem is a complex one and most work has been limited to one-dimensional analyses, with the hope that these will lead the way to a better understanding for an attack on the real three-dimensional case. It is pointed out by Base (1975) that, as has been realized for aircraft and buildings, it is not so much the effect of a single gust that is important, but the effect of several gusts in succession. Hence a simulation utilizing a continuous unsteady process seems very desirable.

In a recent paper Frost (1977a) discusses three methods of introducing turbulence into analyses of rotor response to wind fluctuations: the spectral approach, the discrete gust approach, and the turbulence simulation technique. He applies these to the response of a rotor in terms of a rotational fluctuation $\omega'(t)$ about an imposed constant angular rotation $\bar{\omega}$, the rotor being a simple, unloaded, constant lift/drag propeller type. The unloaded rotor implies one that is freely rotating and hence no account is taken of the generator load, so results do not apply directly to design but do illustrate design techniques and fundamental behavior of rotors in turbulent winds. Frost's three categories of techniques have been applied to aircraft and structures and he gives several references to such recent work in the area.

His spectral approach requires that a frequency response function $H(n)$ be known for the system. The power spectra of the system response, x, (which here is an angular velocity), is then determined from

$$\phi_x(n) = |H(n)|^2 \phi_W(n) \tag{6.25}$$

where $\phi_W(n)$ is the one-dimensional power spectral density for the gust component of interest in the turbulent field in question. From previous work (Frost, 1977b), $H(n)$ is computed by linearizing the moment equation about the hub axis. This linearized equation for the fluctuation in angular rotation $\omega'(t)$ is

$$M_1 I_1 (d\omega'/dt) - RI_2\omega'(t) = W'(t) \tag{6.26}$$

where $M_1 I_1$ and I_2 are dimensionless functions given in the paper but not quoted here. They are functions of the cross-sectional area, radius, chord length, and angular speed of the rotor, C_L and C_p, and λ. The first two terms in Eq. (6.24) are due to inertia force and aerodynamic force, respectively, and the relative values of these determine the sensitivity of the rotor to the turbulence. The power density spectrum for $\omega'(t)$ from Eqs. (6.25) and (6.26) is given by

$$\phi_{\omega'}(n) = \{RI_2^2[1 + (M_2 I_1 n/I_2)^2]\}^{-1} \phi_{W'}(n) \tag{6.27}$$

Models for the spectra of the turbulence components of the boundary layer to obtain $\phi_{W'}(n)$ are taken from the literature. Frost carries through the analysis and gives an example of results for the NASA 100-kW Mod-0 wind turbine, using five different construction materials for the rotor, varying from a light fiberglass-type material, through wood, concrete, and aluminum to stainless steel. The first named is the most responsive, having a root-mean-square fluctuation of 20.2% of the mean rotational speed, whereas the last named has only 4.2%, a direct consequence of the rotor inertia.

The discrete-gust analysis treates the turbulence of a series of gusts having various geometrical forms and has the advantage that the system response can be analyzed directly rather in a statistical manner. Its major disadvantage has already been referred to, namely that the critical system response is to the energy contained in a series of gusts, rather than the intensity of a single one and this is difficult to simulate easily, because as Frost points out, each gust is calculated as creating a transient from an equilibrium state and this restricts the application to well-damped response modes.

Again from previous work, Frost (1977b) gives a response of angular rotation to a gust input $W'(t)$ by

$$M_3 \frac{d\omega(t)}{dt} = C_L W(t) I_1'(t) - C_D \omega(t) I_2'(t) \tag{6.28}$$

where $I_1'(t)$ and $I_2'(t)$ are now functions of time. (Note that the discrete-gust analysis can handle nonlinear as well as linear systems.) This equation can be solved for a gust input, and Frost first gives one in trigonometric form and compares the response of the system computed with that of the linearized equation (6.26). There is relatively large disagreement at the end of the gust period, but Frost continues with the linearized approach because it permits closed form solutions and facilitates computations.

He defines a gust factor for determining the gust intensity V_i from the mean wind speed V_m as the ratio of the wind speed averaged over the period τ to the wind speed at 10-min periods. (Longer periods show little variation from that for 10 min.) The gust factor F_G is a function of V_n at 60 ft, the actual height Z, and the duration τ. Frost (1977b) shows some results as a plot (Fig. 44) of $\hat{F} = F(V_m, Z, \tau) - 1.0$ versus Z, with τ as parameter. $\hat{F} = V_i(Z,\tau)/Z(2,\tau)$, where $V_i(Z,\tau)$ is the difference between the wind speed average over τ and the mean wind speed, i.e., $V_i(Z,\tau) = V(Z,\tau) - V_m(Z, 10)$. He takes a gust shape based on experimental data, which is rather complex and not given here. The application of this technique is to first select the frequency or period for which the rotor is most responsive (using the spectral method to do this). These frequencies are then inverted to give τ and gust factors selected from Frost (1977b). Then $V_i = F_G V_m$. Frost applies this method to the five rotors as above for $V_m = 8$ m/sec (18 mph), showing that the response time of the heavier rotors is such that they react only with V_m, whereas

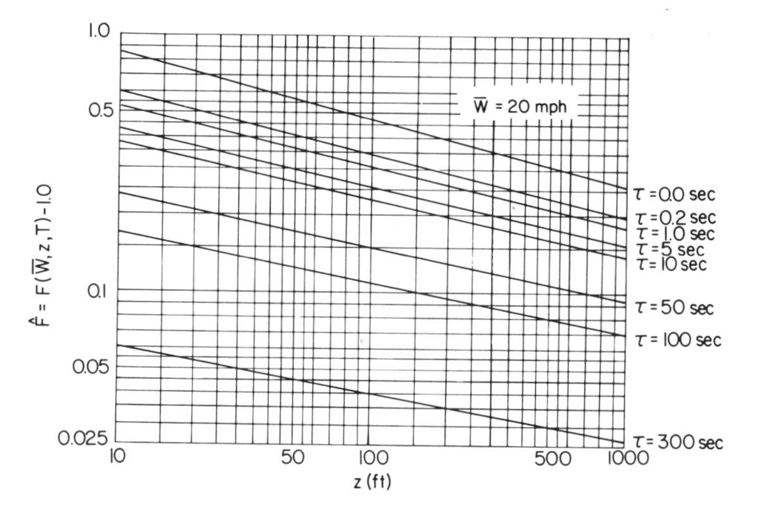

Fig. 44. Gust factors for a mean wind speed of 20 mph. (From Frost, 1977a.)

the spectral method showed that continuous turbulence appear to superimpose gusts and thus result in greater rotor fluctuations. However a direct comparison is difficult because of not knowing the probability of the gust response.

The third method described by Frost is that of turbulence simulation by generation of a random signal which has the statistical properties of the wind, this signal serving as the input function for the linear response equation (6.26) or the nonlinear one of (6.28). The problem in accurate simulation lies in the complexity of the filter used in computing the random signal. There appear to be three such systems developed at present: (1) a linear filter with a single Gaussian input, giving an output of the same power spectral density function as the turbulence prototype, and a Gaussian probability distribution of the simulated velocity fluctuations; (2) a nonlinear system with a more realistic probability and correct power spectral distribution; and (3) a linear system with a Gaussian probability distribution but with a turbulence simulation which takes into account the strong coherence of vertically separated layers, characteristic of atmospheric turbulence near the ground. This last is one developed by Frost who recommends future work to combine the nonlinear characteristics of (2) with the coherence matching of (3). Frost applied model (1) to three of the rotors categorized previously, with the result that the fiberglass rotor reached a quasi-steady-state almost immediately, the concrete rotor in 8 minutes and for the very heavy stainless steel rotor, the computation was not carried out far enough to establish such a steady state. The lightweight rotor followed the wind fluctuations very closely, while the heavy rotor was almost insensitive.

A comparison of the maximum fluctuations in angular speed given by the three models results in no clear-cut decision as to which is more appropriate for engineering application. The spectral model provides a quick method of estimating the rms response to continuous turbulence and allows for the possibility of estimating its probability of occurrence, but does not provide time histories nor allow for nonlinear effects. The discrete gust model does allow a time history but only to a sudden discrete gust and appears to predict low magnitudes of the maximum rotor response, possibly due to the noncontinuous nature of the input and loss of resonance behavior. It does, however, allow for nonlinear effects. The turbulence simulation model provides the most comprehensive analysis, allowing the time history to be evaluated, the standard deviation, and the number of times a given value of ω' is exceeded. It does not appear to provide the same maximum gust prediction as the spectral models, but to do this fairly, the same spectrum should be used for both models. The turbulence model requires a greater computational effort than the other two and provides a nonlinear response. Nonlinear solutions can also be obtained from a discrete gust model, but not with a simulation of continuous turbulence. Frost suggests the spectral analysis for a preliminary analysis and turbulence simulation for a final engineering design analysis. He emphasizes that analyses described by him are for longitudinal fluctuations only, neglecting lateral and vertical fluctuations, which are necessary for taking into account the very important factor of directional variations in the wind. There remains much to be done, but he considers that it is at present possible to extend the design procedure to give adequate information with respect to structural integrity, control, and performance of a wind turbine system.

Base (1975) employs what he names "pseudo-turbulent" flow models, with an array of three-dimensional vortices positioned randomly in space to represent the eddy structure of turbulence. The size and density of the vortices are arranged so that the irregular motion is reproduced. The vortices are convected along at a mean uniform flow velocity similar to Taylor's "frozen pattern" model, ensuring continuity but not the Helmholtz vorticity relation. To begin the calculation, the vortices are randomly positioned in four "boxes," whose dimensions are about three times the length scale of the turbulence being simulated. Rectangularly distributed pseudorandom numbers are used to calculate the initial coordinates of the vortices, with the first set of coordinates calculated so that half of the vortices are upstream of the origin of the coordinate system. With all the vortices positioned randomly, with a sign for each set to make the complete vortex system as stable as possible, the program is ready for running. It is assumed that the velocity at a fixed point in space is equal to the sum of the contributions from all vortices. The diagrams in Fig. 45 illustrate the method used to obtain a continuous velocity signal. Part (a) of the diagram shows the vortices set up in boxes and (b) shows them convected at a later time. The total time of a vortex model cycle is equal

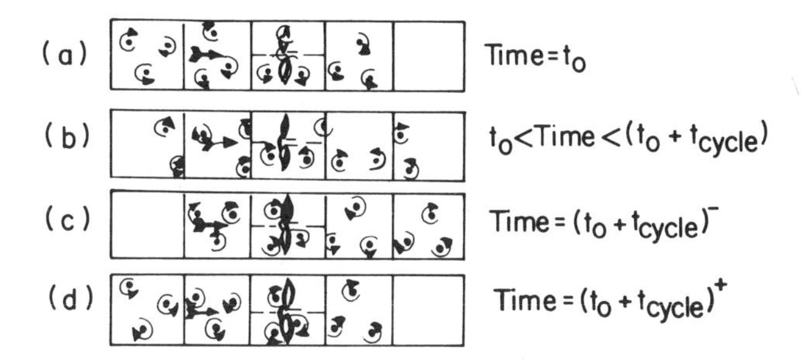

Fig. 45. Turbulence model with vortex "boxes." (From Base, 1975.)

to the time a vortex takes to travel one box length if it travels at the speed of the mean convection velocity. The program is scaled so that, as shown in (c), the vortices move about one box length from the original positions. About one-quarter of the total number of vortices, those furthest downstream, which have little influence at the measuring points, are removed and replaced by the same number of new ones, with new random positions and new signs, upstream of the measuring points where again they contribute little influence. The model that commences another cycle as shown in (a). When the changeover occurs at the end of each cycle, the velocity at the measuring points remains virtually unchanged since the changes are being made outside the bounds of the eddy. Hence the computer run can be made to provide the necessary sample length for statistical significance. It is possible to obtain conversion graphs from runs to enable real flow statistical data to be related to the vortex model parameters and Base gives tabular values of such a comparison for one run having a mean velocity of 12.8 m/sec, showing good agreement. He then develops expressions for the fluctuating drag force on a rotor via the momentum and blade element theories outlined here in Sections II and IV, resulting in an expression for the fluctuating circulation,

$$\Gamma' = \pi c(u' \cos \phi + V'_\theta \sin \phi) \tag{6.29}$$

where u' is the fluctuating axial and V'_θ is the fluctuating tangential component of velocity, c is the blade chord, and ϕ is defined as in Fig. 4.4. Then the fluctuating drag effect is given by

$$dD'/dr = \rho \pi U_\infty nc(u' \cos \phi + V'_\theta \sin \phi) \cos \phi \tag{6.30}$$

and by using the velocity fluctuations u' and V'_θ from the pseudo-turbulent model analysis, this can be integrated to obtain an estimate of the fluctuating drag force. Base recognizes the early stage of development of this method, including the fact that the effect of the rotor blade on turbulence is not considered, but believes that it is as reliable as the established power spectra methods.

E. Arrays of WECS

As with other power systems, arrays or "parks" or "farms" of wind turbines have been suggested, where suitable terrain allows a large number of units to be clustered to provide larger amounts of power. The problem of optimum spacing arises because of the shadowing effect of one rotor upon another. The simple axial momentum tandem analysis of Section II indicates the momentum deficit in the wind stream, which must be renewed, at least in major part, before siting an adjacent unit. The kinetic energy of the stream is renewed horizontally and vertically by shear and turbulence effects and it is the rate at which this takes place that requires analysis.

Crafoord (1975) uses a model consisting of a logarithmic boundary layer and large-scale roughness elements added to an already rough terrain. He computes the effect of spacing, output size, and total number of units on the relative power output compared to a single unobstructed unit. Figure 46 is given to show the order of his results as the relative power versus number of rows in a quadratic array for different spacings (as separation distance/rotor diameter), with the shaded areas showing the region bounded by 50-kW and 5-MW units. Crafoord emphasizes the crudeness of many of his assumptions, such as the log profiles, the superposition of surface stress and rotor drag, and homogenization of the wind profile 5 rotor diameters downstream, and he also outlines suggestions for further study.

The General Electric Company Wind Energy Mission Analysis (Garate, 1977b) postulates a model in which momentum theory is used to determine the far wake distances and widths as a function of the velocity deficiency. Dissipation of blade tip vortices is introduced to determine a minimum of wake influence, but the effects of vortex flow on mixing rate and velocity recovery are not determined. It is pointed out that the maximum degradational effect occurs at rated wind speed, i.e., the minimum speed for maximum power, as above this value the unused wind

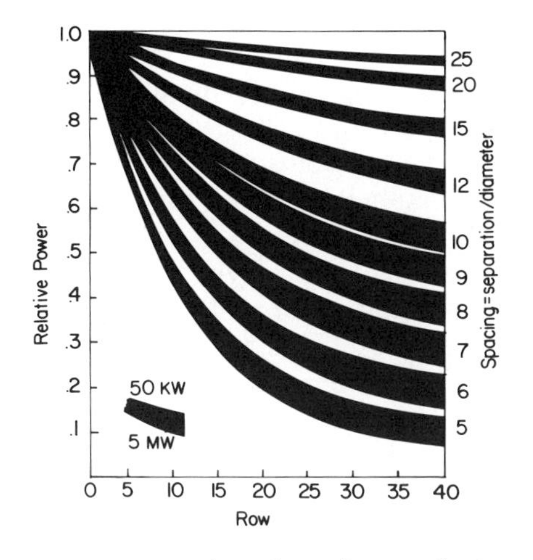

Fig. 46. Relative power versus number of rows in a quadratic array of turbines. (From Crafoord, 1975.)

energy increases the momentum through the rotor and reduces the velocity defect. The results of the analysis suggest that a minimum spacing of 3000 ft for a 200-ft-diam rotor be used, i.e., a "packing density" of 15 rotor diameters, due to the detrimental effects of tip vortices on the energy concentrations in the wake and that momentum theory describes a wake that maintains a significant velocity defect which cannot be effectively reduced with increased spacing above 3000 ft. It is also concluded that the size and pattern of an array is critical to both unit and overall efficiencies and that small numbers of units be used in a pattern to conform to a given wind rose for the best results. Ljungström (1977) reports results of a study of large-scale WECS in relation to wind characteristics and groupings. He concludes that it is economical in land use to use large units of 2–5 MW and close spacing (7–10 diameters), although he states that according to his simple wind tunnel experiments, an array pattern having a "kink angle" of 10° in each row of three units is sufficient to give no interference loss in a steady wind in the near field between two turbines, with a spacing as low as 4–5 diameters. This is the lowest value yet suggested and his actual group studies use the higher values mentioned previously.

It is apparent that results of array studies are tentative, owing to the large number of assumptions and degree of engineering judgment which have to be used in any analysis.

In view of the time, labor, and cost necessary to evaluate wind sites, simple and rapid methods of locating those of good potential are very welcome. One such is contained in a study by Kolm *et al.* (1975) using Landsat pictures to identify geomorphologic features indicating above average wind speeds. Features of sand dunes in Wyoming are being investigated to correlate ground test data with such imagery and only a few preliminary and tentative results are on hand. This study is limited to the particular land features mentioned, but with the ever-increasing technology of remote sensing via satellite and other means, some interesting possibilities are raised for "wind prospecting" in this general manner.

In addition to closely spaced arrays of turbines at a given location there is the possibility of the effect on a given utility system of a number of arrays or even individual units sufficiently far apart so that at any given time the wind characteristics are different. This is important from the point of view of providing some measure of firm power from a number of sites, each subject to the possibility of absence of significant wind. Thus in addition to wind characteristics at a specific location, optimum siting may require a regional analysis, e.g., the pattern of more-or-less regular meteorological conditions moving across country and contributing wind energy not only on a time-limited local basis, but also on a much longer term system basis.

In contrast to horizontal arrays of turbines, Willem (1977) has suggested tall cylindrical arrays of vertical-axis turbines, for example eight columns of 14 units each, with each column spaced uniformly around a circle. Advantages would be utilization of the higher velocity wind at upper elevations, economy of land use, better response to gust conditions, higher rotational speed of a number of small rotors compared to that of one large one, etc. Disadvantages might be cost of a tower structure, maintenance, and possible screening effects.

VII. ENVIRONMENTAL CONSIDERATIONS

Wind power has been hailed as an unlimited source of energy, which is true if not necessarily economically so, and as nonpolluting and environmentally acceptable, which is not true. While the environmental drawbacks are nowhere near so potentially damaging as those of fossil-fuel and nuclear power plants, nevertheless some aspects warrant serious consideration before actual systems arouse resentments and organized opposition. Possible hazards are (1) danger from structural failure, (2) interference with electromagnetic communication systems, (3) noise, (4) aesthetic pollution, and (5) possible biological reactions. There is no pollution of the thermal or chemical nature with which we are familiar from many technologies today.

Danger from structural failure can be negligible if precautions are taken in siting of large units such as those likely to be used by utility companies or operated by governmental bodies. The danger lies in toppling of towers with their accompanying installed turbines and generating system in very high winds, and in failure of rotor blades with consequent damage at a distance from ballistic objects. The former is likely to be very rare, with damage obviated by leaving sufficient clear space around the tower to accommodate the full weight of the structure, but the latter is more difficult to warrant against, because vibration and fluctuating stresses of all sorts still present problems for design engineers, and unpredictable material fatigue is a major potential source of rotor failure. Large units will most likely be sited in open country or in sufficiently large enclosures to avoid liability damage, but this is not true of small units operated by individuals. The enviable reliability of the thousands of farm windmills of former years was not achieved overnight, but here the great majority of sites provided minimal possibility of damage to the public at large. There is a vogue for small units mounted on the top of buildings in towns and cities, and there may well be danger until such small units have achieved statistically acceptable reliability over a period of years. Because there seems to be a viable application of wind power for a number of uses in office buildings, apartment blocks, as well as for private residences, possible damage from mechanical failure will have to receive the proper attention and the likelihood of regulated installation recognized.

Interference with television and FM broadcasts has been recognized as potentially troublesome, although here there may be more of a chance of a technological remedy. The interference effect is the same as that caused by low-flying aircraft, with e.m. signals specularly reflected from the rotor blade surfaces, rather than from aircraft wings. A study has been made by Senior *et al.* (1977) with the object of determining the levels of interference and assessing their impact on siting, using the NASA Plum Brook facility for field testing. The program had three major aspects: (1) laboratory and simulation measurements, (2) field testing, and (3) theoretical investigation including a comprehensive computer study. The simulation study comprised the effect on color television sets of modulation modeled on that which a windmill would impose. It was found that phase (or frequency) modulation had no effect on the television picture, which has an amplitude-moderated signal. There was almost no effect on the sound at even very high levels of modulation, and this was also the case for FM radio reception. However, amplitude modulation had a very considerable effect and it was concluded that at a modulation level of 20%, distortion was severe and unacceptable. This initial finding confirmed that a theoretical investigation and field testing were essential and urgent. A detailed signal analysis together with an analysis of windmill scattering was made, as was a program for computing the primary or direct field of a linearly polarized transmitter, as well as the secondary or windmill scattered field. Distances corresponding to those of a receiving antenna for a modulation index ≥ 0.2 were computed. Field tests were hampered by mechanical difficulties with the Mod-0 turbine and/or lack of wind and only minimal test data were obtained. However, static results were in good agreement with the predicted ones, with the blades acting as specular reflectors, with the predicted magnitude of scattering. Only one opportunity was given for a dynamic test (windmill in operation), and again the test results were in good agreement with the model and theoretical analyses. Thus the conclusion is that interference can be troublesome at locations a few miles distant from a rotor, and further investigations are necessary to obtain a firm picture of the effect of other variables for which there was no time available in this particular program. Smaller units than the Plum Brook rotor of 37.5 m (125 ft) diameter will produce correspondingly less interference, but even here there is the possibility of a considerable effect. There seems to be a viable application of wind power of small-to-medium size for heating (and possibly cooling) of office buildings, apartment blocks and even residences, but any number of such installations on the roof tops in a city in direct competition with antennas for television and radio could result in severe interference. Some alleviation of interference effects appears possible by using a high-performance directional receiving antenna, as all simulations and tests used in the work reported supposed the "worst case," with an omnidirectional antenna and rotor blades at the azimuth for maximum effect. Also, there is the possibility of mitigation by rotor blade materials or coatings, yielding less effect than the metal area of the field test unit.

Noise is a very difficult form of environmental pollution to deal with because of its subjective quality. While the farmer of yesteryear may have listened with satisfaction to his windmill filling the farm pond or lighting his home and cowshed, even small rotors may be very noisy in high winds and large ones can be awesome. The 200-kW test WECS at Clayton, New Mexico has been reported as producing a "fearsome whistling noise" as it approached its maximum speed of 40 rpm (*New York Times*, 1978) at its dedication ceremony. Here again it may well be that small units in built-up areas may be more of a problem than the larger ones in open sites.

The problem of the visual effect of WECS on the countryside is even more difficult to evaluate than noise, because of its qualitative aspect and even of its positive or negative nature. To most people, the Dutch windmill is not considered as a polluting artifact, perhaps in part by its historical background and "romantic" associations, and possibly by the fact that its generally satisfying design provides relief in the flat Netherlands landscape. Likewise to some, (but possibly only engineers), the two- or three-bladed rotor and streamlined transmission and generator casing of the modern windmill is not unpleasing in itself and it is as an unaccustomed intrusion unhallowed by time that makes it unacceptable. Unfortunately, the general necessity of siting WECS in open areas, on hill tops and mountain ridges or on the seacoast, lead to maximum exposure and conditioning effect on a natural landscape. There is thus a very natural opposition to the introduction of an artificial element into "unspoilt" terrain, regardless of any aesthetic merits of the actual design of the intrusion. There is also the accompanying problem of transmission lines, which can be equally or more unacceptable. Actual design may be an important factor, as a few modern artifacts, such as suspension bridges, for example, can be a stimulating aesthetic experience.

Finally there is the possibility of biological effects on the environment by alteration of hitherto natural and accustomed air currents, together with the physical presence of a moving structure. A comprehensive study of such possible environmental effects has been carried by Rogers *et al.* (1976) for the NASA Plum Brook WECS, covering the microclimate, the vegetation, soils, mammals, birds, and insects. A very comprehensive bibliography is given, consisting of over 600 items relevant to the matters covered in the study, which should prove valuable for any similar investigations elsewhere. The general conclusion is that "under many, and, in fact, most environmental configurations the effects will be minimal and often not measurable." Many possible changes are masked by much greater normal variations in the weather and certainly some will require a long period of operation of the WECS in order to establish any such change. It is pointed out that in general, power sites will be "windy" sites and hence subject to above-average variations due to wind conditions. The WECS in question has a rotor diameter of 37.7 m (125 ft) and a rated power output (of the rotor) of 133 kW at 8 m/sec (18 mph) wind speed. Using the axial momentum analysis of Sec II and an assumed actual power coefficient, simple calculations show that the diameter of the wake expands only about 4.6 m (15 ft) beyond the rotor disc diameter as a maximum, which occurs just under the rated wind speed. The hub height is such that alteration to the microclimate due to the wake will be no nearer the ground than about 10 m (34 ft). There are two points which require further longer-term study, (1) the ground-level microclimate alteration induced by turbine and tower, and (2) potential collision of migratory birds and to some extent, insects, with the rotating blades. Other sites may require special attention to such features of the life of fauna. It should also be recognized that although much of the study should be generally valid, it was carried out for a particular site, that the turbine was relatively small (100 kW), and that larger systems and particular sites may introduce special effects due to both height above ground and to altitude.

VIII. STRUCTURAL CONSIDERATIONS

In Section I of this review, it was stated that there might rightly be expectation today that advances in materials, manufacturing methods, and techniques of stress analysis, particularly with respect to vibration and flutter, together with the tremendous capability of detailed design with the computer, would allow considerably more certainty in design and in economy in manufacture than was possible even 25 years ago. At this time, insufficient running time has been achieved on large rotors to be able to draw any conclusions from actual testing, although there is a quantity of analytical work available from recent years.

Dealing first with materials, of the four largest units so far installed or running, the NASA–ERDA 100- and 200-kW propeller units in Ohio and New Mexico each have aluminum blades, as does the 200-kW Darrieus unit in the Magdalen Islands in Canada, while the 2-MW propeller type in Tvind, Denmark, has fiber glass and reinforced polyester, with plywood ribs. Hütter in Germany has long been a proponent of synthetic materials for rotors and his Allgaier 100-kW turbine of 1959 had reinforced fiber glass blades. All designers are interested in fiberglass or plastic blades, as potentially being less costly and lighter than metal. For the 17-m Sandia Darrieus test rotor made by Kaman, there is a mixture of materials: extruded aluminum alloy leading edge spar, noncorrosive polyamide (Nomex) paper honeycomb core after body covered with fiberglass skins, and aluminum alloy trailing edge spline. This construction has resulted from helicopter designs.

Large propeller-type rotors are usually mounted to operate in a downwind position, with a coning angle of a few degrees to ensure that the blades do not foul the tower under the forces generated by rotation and by the most severe wind loading when in a running position. Coning is the tilting outwards of the blades from a plane normal to the shaft. The downwind position ensures that the rotor will not strike the tower if a failure occurs and also minimizes the effect of the moving rotor on the tower. It does, however, allow the possibility of blockage and vibratory effects from the tower on the blade during part of each rotation.

Structural analysis of rotors has been stimulated in very recent years by the NSF–NASA–ERDA–DOE construction plans for 100-, 200-, 1.5-MW, and larger units. The plague of rotor design is that of vibration-induced stresses, either from wind forces directly or induced by the tower structure. In the intervening years from the Smith–Putnam wind turbine of the 1940s, the advances in helicopters have occasioned blade analysis methods which in large part are applicable to wind turbine blades. One such analysis by Friedmann (1976) examines the aeroelastic analysis tools developed for fixed-wing aircraft, which are applicable to wind turbines and identifies the more important differences which are present. He presents a procedure to help in formulating a mathematical model for aeroelastic or structural dynamic analyses. He uses as his main model the configuration of the NASA 100-kW rotor and concludes that a sufficient amount of rotary wing aeroelastic technology exists to solve the problems of the combined rotor–tower aeroelastic system.

The design of the NASA Plum Brook WECS is covered in several publications, together with some follow up of operating experience and comparison of calculated and observed performance. Spera (1975) gives a preliminary analysis for the Mod-0 rotor. He discusses the advantages of articulating (or hinging) as in helicopters, in which a pair of connected blades are fastened to the shaft by means of a pivot pin, allowing a seesaw motion parallel to the shaft, which can reduce internal bonding stresses in the blades and vibratory loads on the tower. He concludes that such a teetering design has substantial advantages over the hingeless rotor with respect to shank stresses, fatigue life, and tower loading, but these are mostly apparent for overload conditions rather than normal running. The analysis showed that it is the weight of such blades (2000 lb each), which is the main source of loads and stresses, much larger than the aerodynamic forces, hence there is every inducement to look for materials and design methods to reduce this weight.

The actual blades were designed and built by the Lockheed-California Co., using a hingeless design based on their analyses and experience with helicopter design (Donham *et al.*, 1975; Donham, 1975). The final 100-kW rotor design was thought to be rugged and capable of much higher power outputs with little or no modification (Donham, 1975). An account of the fabrication of major components, assembly, the final lift operation to the tower, and the installation is given by Puthoff (1976). Following preliminary closed-form vibration analyses, Chamis and Sullivan (1976) performed a complete analysis of the free vibration modes using NASTRAN finite element techniques. This was then compared with experimental results on the actual installation obtained by accelerometers on the components and instrumented impacting devices. Data were taken for the tower alone, one blade alone, the blades mounted on the bed plate, and the complete system. The predicted results for the natural frequencies and modes of vibration were in excellent agreement with those measured.

The Plum Brook WECS was finally assembled in September 1975, became operational in October, and first achieved the design speed of 40 rpm and 100 kW output in December. The major finding was that blade loading was significantly higher than predicted, and this was attributed to the effect of tower shadow (Glasgow and Linscott, 1976). The design loading was based on 25% decrement of wind velocity from tower shadow, but the test data corresponded to that produced by over 90% reduction of wind speed. The loads were sufficient to demand a redesign of the tower structure and wind tunnel tests were made of the shadow (Savino and Wagner, 1976). A model to $\frac{1}{25}$ linear scale closely following the actual tower was tested with and without the control stairway and elevator rails, together with a $\frac{1}{48}$ scale model using tubular members for the main structure and without the stairway and rails. It became clear that it was the latter elements that were causing the severe shadow and the overall results indicated that every effort should be made to reduce the size and number of structural members and the number which met in a point or which line up in the wind direction, and that gusset plates should be avoided if possible. The use of tubular rather than angle and channel sections was also helpful. As a result of these wind tunnel tests and analysis, the stairway and rails were removed, following which running tests showed that blade loadings were reduced to close to their expected values. Modification of the yaw drive–brake system which had been found to induce higher loadings also was successful in reducing measured stresses (Donham *et al.*, 1977). All in all, it was demonstrated that tower construction to minimize wind shadow is a very significant design feature.

Spera and Janetzke (1977) have analyzed three 1500-kW horizontal rotor configurations to determine the effects of upwind or downwind rotor location, coning, and tilted or horizontal rotor axes on the dynamic loading of the blade shank, hub shaft, and yaw drive. Coning and tilt axis were found to have little effect, but an upwind rotor reduced loads significantly.

A novel design called the composite bearingless rotor (CBR) has been developed which is claimed to have significant advantages over more conventional designs, such as lower blade and hub weight, simplicity of manufacturing and maintenance cost, and reduced hub drag due to lower hub profile. An account of the aeroelastic characteristics of CBR blades is given by Bielawa (1977), who presents test results of such a dynamically scaled model rotor and complementary analytic results. The elements of the CBR are a spar constructed of torsionally flexible uniaxial composite material (high-strength fibers in an epoxy matrix), with a sufficiently low transverse shear modulus to produce a longitudinal midsection (the "flexbeam") which is torsionally flexible and replaces the feathering bearings normally used for pitch control. The latter is effected by elastically twisting the flexbeam with the necessary moment transmitted by an aerodynamic shell (torque tube), which is relatively rigid in torsion. It is concluded that, in spite of violating some usual assumptions for conventional aeroelastic analysis, the CBR is a practical rotor system with conventional hingeless rotor aeroelastic characteristics.

The Darrieus-type VAWT presents many new problems because of both the novel design itself and the lack of a comparable technology (e.g., helicopters) on which to base analysis. The continuous nature of the cyclical loading, with blades operating in the wake of each other, and of stalling at lower values of tip-speed ratio λ, makes for a complex stressing analysis. It is the aerodynamic loads which present the problem, as the gravitational and centrifugal loadings are straightforward. The purpose of the curved shape of the Darrieus blades is purely structural, because straight vertical blades disposed circumferentially in a "squirrel cage" pattern around the rotating axis would be much simpler aerodynamically. However, such straight, rigid blades would be subjected to very high bending moments due to centrifugal forces and would require extensive bracing. The skipping rope geometry is close to a catenary, in which the major stress is tensile with negligible bending, and the term troposkien ("turning rope") has been given to the shape assumed by a perfectly flexible cable of uniform density and cross section if its ends are attached to a vertical axis and it is then spun at constant angular velocity about the vertical axis. Because of their small radius of rotation and hence small work contribution, the parts of the blade near the attachment to the central shaft are often straight and hinged to the curved central portion in order to simplify manufacture, assembly, transportation, and installation. The central part can be reasonably approximated by a parabola or even a circular arc if such shapes further simplify manufacture. However, even with the stress-relieving troposkien shape, the loads may be considerable and together with possible vibration modes arising, make cross bracing advisable, at least initially until experience of running larger units has been gained. Such straight blade elements and struts have to be considered carefully in the aerodynamic analysis, as it appears that unless they are airfoil sections, their drag may be sufficient to significantly affect the performance of the unit.

Sandia Laboratories has carried out very extensive investigations of the structural problems of the Darrieus rotor, resulting in a large number of reports in the last few years, and they have also had experience with rotors of 2- and 5-m diameters, as preparation for the 17-m unit. Much of this work is reported by a number of authors in the VAWT Technology Workshop (1976), together with bibliographies giving greater detail. The operation of the Sandia 17-m unit and the Canadian Magdalen Islands 200-kW turbine should afford a considerable amount of information to compare with design data.

Darrieus-type VAWT do not have a tower in the sense of that of the propeller type, although the central support is given that name. The top of the tower, which necessarily must be very slender, is tied to the ground with guy wires, which may or may not require an overhang from the top of the central column, depending on the height/diameter ratio of the rotor. The Canadian 200-kW Magdalen Island turbine has a height/diameter ratio of 1.5, rather than the more usual value of unity in order to obviate such overhang and hence minimize column bending moments (Templin, in VAWT Technology Workshop, 1976). Tower and rotor have strong coupling possibilities, which have been recognized, but again operational experience is required before definitive answers can be given. Sandia Laboratories has investigated tower design, particularly the relative merits of truss or tubular construction (Reuter, in VAWT Technology Workshop, 1976). There is a problem here in that the former yields a significantly lower weight but is likely to be of much greater diameter and thus introduce undesirable aerodynamic effects.

Yee *et al.* (1977) have investigated the effect of nonrigid tower foundations on the vibration characteristics of the NASA Mod-0 turbines. Using the free-vibration analysis of the turbine on a rigid base as a reference point, a simple model was used to examine the effect of soils of varying rigidity. The model used three masses and springs for the tower, machinery, and blades, and a torsional spring for the foundation. It was found that there was a significant effect on natural frequency by soils of less than 5000 psi elastic modulus (cohesive soil or loose sand), but negligible effect from well-graded, dense, granular materials or bedrock. Increase of tower height increased the effect of nonrigidity.

IX. TESTING AND TEST PROCEDURE

A. Background

Experimental work on wind turbines as a whole is a very difficult procedure. The problem with testing using the natural wind itself is the unsteady nature of the flow, and the problem with wind tunnel tests is the low value of Reynolds numbers because of the necessarily large scale ratio required between prototype and model to minimize blockage effects in all but very large tunnels. Large rotors are particularly difficult, not only in the wind tunnel simulation of Reynolds number behavior, but also because of the variation of natural-wind speed over the rotor swept area.

One must note in passing the earliest quantitative tests on windmills, those of John Smeaton (1759–1760) in the middle of the eighteenth century. He had no wind tunnel, of course, but operated his models on a revolving beam, with the rotation effected by an assistant pulling steadily on a rope, and with the load measured by the rate of rise of a weight. He enunciated the laws relating to the square and cube of wind velocity, the effect of number of blades, and generally initiated the practice of using scale models in experimental fluid mechanics. In more modern times, Denmark provides instances of long-term testing and also early use of wind tunnels to investigate blade characteristics (Juul, 1956).

These Danish wind tunnel tests were made over 50 years ago and although useful for comparing the relative merits of airfoil sections and number of blades, they were apparently carried out with blades of 2 m diameter in a wind tunnel 2.6 × 3.5 m, and hence there would have been a significant blockage effect. Golding (1976, with updates by Harris) discusses testing, and there are reports of interest giving experience from four European countries in the Proceedings of the United Nations Conference on New Sources of Energy (1964) held in Rome in 1961. The first of these is by Askegaard (1964), who described the use of a calibrated measuring cylinder placed between tower and machine cabin and equipped with 64 strain gauges on the well-known Gedser Mill in Denmark, which gave information on the effect of wind on the structure itself. Clausnizer (1964) was interested in power output measurements and gave results from an Allgaier–Hütter 10-m rotor in Germany. He showed that the measured power output varies with the length of the time interval over which the mean value is taken, i.e., the measuring time interval δ. For technical evaluation, short-time mean values are suitable, but for economic answers, the output in terms of annual mean values is required. Due to time lags between anemometer and rotor, Clausnizer recommended time intervals shorter than 2.5 min at 50-m spacing of anemometer and rotor, with the expression roughly estimated as $\delta_{min} = d/6\,V_{min}$, where d is the spacing and V_{min} is the lowest wind velocity of interest (values in meters). His results showed that the slope of the power output versus wind speed lines decreases as the time interval Δt increases, and that the cut-in speed decreases with increasing Δt, as a result of the changing frequency distribution within Δt. The power output for the Allgaier constant-speed rotor tied into a network is a linear function of wind speed because $C_L \propto \lambda^2$, i.e., $C_L \propto 1/V^2$ at constant ω, and therefore power is proportional to V. From his results, Clausnizer showed how to obtain, at least approximately, reasonable values of mean power output from annual mean wind speeds in terms of measured short-time values, useful for a first economic evaluation of plants and sites, which can be carried out in a reasonable length of time.

Delafond (1964) discussed test methods applied to the Andreau–Enfield WECS (suction turbine at ground level with air discharged at tips of hollow blades). Initial test results showed a great scatter, but by photographing the instrument panel at one-second intervals, simultaneity was said to be achieved and the large number of measurements allowed average values to be computed. A low-response time anemometer is required and it should be placed at hub height two rotor diameters from the tower, taking care it is not influenced by the wake. Finally, Morrison (1964) recounted experience of many hours of running over ten years, largely to examine rotor stress loads by means of strain gauges. He emphasizes the importance of quick-response anemometers and suggests locking the rotor and use of a fixed direction anemometer to achieve simultaneous data.

Testing of a 100-kW turbine was reported by Hütter (1964b), based on the measuring techniques of Clausnizer, mentioned above, for which power coefficients of 0.45 were obtained at about one fourth of rated output, falling to about 0.20 at full power. This rotor was two bladed, the diameter being 34 m, and of very low weight by existing examples at that time.

There is a paucity of detailed data in the more recent past, because reports of WECS of any size are only just beginning to become available, again due to the difficulty and the long test periods necessary in order to obtain reliable and conclusive data. Reported data from the NASA 100-kW Plum Brook turbine have been confined mainly to structural concerns (see Section VIII), as the early months of testing were hampered by mechanical modifications and poor wind conditions. Sandia Laboratories has carried out considerable work with the Darrieus-type VAWT both in the wind and via wind tunnel tests, and their reports are very useful in pointing out the problems and in including possible routes to solutions.

The behavior of the propeller-type and the Darrieus-type rotors differ considerably with wind speed and rotational speed. The former has in general a relatively flat curve of power coefficient C_p with wind speed (or λ for a fixed blade speed), while that for the latter type is considerably more peaked. Thus the Darrieus rotor has a more peaked torque characteristic for a fixed rpm and is self-regulating, if the load (e.g., synchronous generator) can absorb the peak torque without trouble. On the other hand, the propeller turbine has a torque which increases with wind speed and hence requires a control mechanism such as variable pitch blades to keep the output constant above the rated speed. Figure 47 shows a torque versus rpm curve at a fixed wind speed for a Darrieus-type turbine. Below the peak torque, there are two values of rpm for a given torque, but the left-hand side of the curve (lower rpm, positive slope) is unstable and therefore testing requires a loading device which will allow a constant speed to be maintained, such as a special control of the test generator.

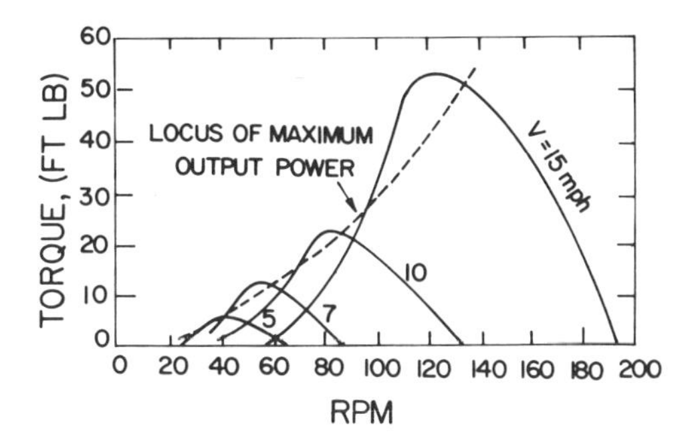

Fig. 47. Torque versus rotational speed for a Darrieus-type turbine. (From Banas and Sullivan, 1977.)

B. Free-Air Testing

The same difficult test conditions in free-air tests are of course present today as they have been in the past, but the development of electronic instrumentation, data collection, and data analysis by machine methods has helped considerably in obtaining more significant results from short-time tests. Sandia Laboratories (Weingarten and Blackwell, 1976) states that the goal of automatic control and data acquisition is to reduce man-power costs by having cost effective automation capable of monitoring the system independently of on-site personnel. They do this via a micro-processor system which interfaces with the turbine's power systems, sensors, anemometers and recording devices, starting up the system when the wind is suitable, monitoring component temperatures, vibration level, speed, and power, and shutting down the system in case of a malfunction. The conventional manual method of instrumentation using strip chart recorders, digitization, and data storage on magnetic tape for input to a computer for processing and plotting is regarded as having severe draw-backs, e.g., excessive chart paper over long runs, accuracy inversely proportional to speed, and time consuming labor in digitizing the information by manual labor. They prefer an automated data acquisition approach, with digital data from the test instruments being processed in the microprocessor, recorded on a paper tape or magnetic cassette, or fed directly to a large computer via a phone link.

Sandia initially found that it was very difficult to correlate wind speed with torque by simple comparison of strip charts because of the unsteadiness of the wind. They developed a computer program called BINS (Sheldahl and Blackwell, 1977), which uses the "method of bins" to average statistically the wind speed–torque data. The latter are recorded at sample rates chosen by the operator, generally from 1–10 samples per second, and the data stored in velocity "bin widths" of 0.5 mph. The data point is counted and the torque value added to the total torque in the appropriate velocity bin. Each bin also contains information constant for each data record, e.g., rpm, temperature, barometer, anemometer identification (see later), etc. If the temperature or barometer changes during a test, the test is ended and the data record stored, and testing then resumed with the new information. Thus tests may be few minutes long or over an hour in duration.

The total torque in each velocity bin i is given by

$$T_i = \sum_{j=1}^{n_i} T_j \tag{9.1}$$

where n_i is the total number of data points in the bin i. Because of varying air density, a normalized summed torque is determined as

$$\overline{T}_i = T_i \, \rho_0/\rho_\infty \tag{9.2}$$

where ρ_0 and ρ_∞ are the reference and the freestream density, respectively, for each data record. Values of $\overline{T}_i$ are summed along with the number of data points n_i for each bin and an average torque is given by

$$Q_i = \sum_{k=1}^{m} (\overline{T}_i)_k \Big/ \sum_{k=1}^{m} (n_i)_k \tag{9.3}$$

where m is the number of data records being summed in the complete data set. Then the power coefficient is given by

$$C_D = \frac{Q_i \omega}{\frac{1}{2}\rho_0 V_{\infty_i}^3 A_s} \tag{9.4}$$

where V_{∞_i} is the average velocity of a bin. The velocity bin width is normally 0.5 mph, but the data may be handled in $\frac{1}{2}$-, 1-, or 2-mph increments.

The Sandia test results (Sheldahl and Blackwell, 1977) indicate that the positioning and response of anemometers used in turbine testing may be critical for obtaining accurate representation of performance. For the tests of their 5-m turbine, six separate anemometers were used and any three could be used together during a single test, thus enabling directly comparative results to be obtained. One anemometer was mounted about 2 m directly above the turbine on the axis, another at the turbine equatorial height about 10 m south of the axis (prevailing winds from east or west, hence the instrument was seldom in the turbine wake), and four other anemometers on a meteorological tower about 20 turbine diameters west of the axis. The first two and one of the last-named group were used extensively, but it was found that the tower-mounted unit gave very poor correlation because of the distance between turbine and anemometer. The top anemometer was affected by the turbine itself, particularly at higher tip-speed ratios, because air was deflected upward by the blades, tending to act as a solid barrier to the air. The side-mounted anemometer was affected by the turbine by an amount less than 1% of the speed. The anemometers responded very quickly by virtue of their small size and could record a different value of wind speed to that experienced by the turbine, which may be only an average value. Also, they responded more quickly to increasing than to decreasing wind speeds, with a net result of a higher than actual indicated speed being recorded. Discrepancy of power coefficient with anemometer position was reduced to a much smaller value if a new coefficient based on the tip speed ωR rather than on V_∞ is used, i.e., $K_p = Q_i\omega/[\frac{1}{2}\rho_0 A_s (R\omega)^3]$. The Sandia tests were all run with a constant turbine speed, because (1) the unit is intended to drive a synchronous generator anyway, and (2) the problem of correlating data because of rotor lag due to inertial effect. Thus K_p may be a better indicator of turbine performance if there is any doubt as to the accuracy of the wind speed or for comparing two or more turbines. The Sandia experience has been summarized in some detail because it reveals the great difficulty in assessing free-air performance on any absolute basis and the need for great care in making comparisons of reported performance under different conditions of test.

Edwards (1977) describes a low-cost recorder for wind energy surveys using two-channel tape decks, one channel recording wind speed via a proportional frequency and the other recording a clock pulse train with time intervals as short as 3–5 sec or as long as a month. Data retrieval is made by a gated counter, with data then transferred to a minicomputer. A simple, cheap, and effective means of acquiring field data with good resolution is claimed.

It should be noted that the DOE (as ERDA) has instituted an experimental test facility for small turbines at Rocky Flats under the management of Rockwell International, where data can be obtained.

C. Wind Tunnel Testing

Wind tunnel testing has been reported on VAWT Darrieus-type turbines by South and Rangi (1972, 1975), Muraca and Guillotte (1976), and by Sandia Laboratories (VAWT, 1976; Blackwell *et al.*, 1976; Weingarten and Blackwell, 1976; Reuter and Sheldahl, 1977). It has its own drawbacks, the major ones being restriction of values of operational Reynolds number attainable for any but small prototype turbines and to the tunnel blockage effect, but there is the great advantage of being able to operate with a steady air speed for any length of time.

The Reynolds number (Re) effect can be very significant because for surfaces with lift (airfoil profiles), the lift coefficient is very dependent on the value of Re. For given values of V_∞ and λ at standard density, Re is proportional to the blade chord and this may be less by an order of magnitude for a model in relation to the prototype. It is perhaps fortunate that lift increases with increase of Reynolds number, hence the prototype performance should be improved over that of the test, and thus the latter values are conservative. Some Sandia results show that C_p may improve by 25–30% for a tripling of Re from about 10^5 to about 3×10^5. There is also the situation that low Reynolds number information on the behavior of airfoils may not be available for design purposes for small rotors or for checking design calculations with wind tunnel model tests.

Because the size of the turbine must be comparatively large to obtain as high a Reynolds number as possible, a large wind tunnel is required to avoid excessive blockage effects. The physical blockage of rotor and associated structure is quite small but the expanded wake area is greater, and the effective velocity becomes greater than the free-air wind velocity. Here again the square law for torque and the cube law for power are effective in magnifying any errors in assessing air velocity. Wilson *et al.* (1974) suggest correction for blockage by means of Glauert's (1935) method for propellers and give curves for correction of the power coefficient C_p against measured torque, with the ratio of tunnel area to rotor swept area A_r as parameter. The Sandia method follows the analysis of Pope and Harper (1966), resulting in simple tunnel blockage factors or velocity increments for the wake e_{wb} and for solid blockage e_{sb}. The Canadian Darrieus model tests were of 12- and 14-ft-diam rotors in a 30 × 30-ft tunnel so the area ratio A_r was about 10. The Sandia model tests were of a 2-m-diam rotor in a 4.6 × 6.1-m tunnel, with A_r again of the order of 10. Thus both blockage correction factors had significance but were not large.

D. Tow Testing

The lack of a sufficiently large wind tunnel and the difficulties of free-air testing have been by-passed in a number of instances by mounting the rotor on an automotive vehicle, such as a truck or jeep, which is driven at constant speed in still air. Extensive use of this method has been made by the Princeton University group in the development of their sailwing rotor and has been reported in some detail (Maughmer, 1976). Irregular results were obtained at first but the test method and instrumentation were eventually refined so that reliable and repeatable data were obtained. Dead calm is highly desirable for the most accurate wind speed measurement, and therefore most testing was carried out in the dawn hours. However, a method of testing was developed which allowed a low cross wind to be taken into account, although this was avoided as much as possible. After trials of other types, a cup anemometer of their own design and calibration mounted ahead of the vehicle was found the most accurate. The sailwing type under testing was one of 3.65 m (12 ft) diameter, and larger rotors of conventional design are known to have been truck mounted in this manner, so that a reasonable size of unit can be tested by use of this method.

X. APPLICATIONS AND SYSTEMS

In order to single out particular performance characteristics, it is convenient for discussion purposes to categorize applications of wind turbines in a number of ways. One is by type of output, i.e., electrical power or mechanical power. A second is by generalized function, i.e., connection to a utility grid system, for heating, for lighting, for pumping, etc. Another is by special siting and duty characteristics, where the economics are of a different order to those normally operative, i.e., for very remote and difficult locations, or for very special duties. These are arbitrary distinctions and the boundaries are mixed, but the discussion here will attempt to follow along the above lines. It is limited to the technical problems rather than the economics of the system, which are discussed later, although it is cost which eventually dictates the technical solution.

The wind turbine is a prime mover having the following major operating characteristics:

(1) It utilizes an uncontrollable natural source of energy which is very variable in level and may be unavailable or available only in unusably small amounts for significant intervals of time; (2) it has a low rotational speed compared to other prime movers (except for the steam engine), possibly requiring a step-up transmission and in order for this speed to remain constant, special control methods may be required or operational penalties may be entailed; (3) for efficient operation, it should be sited in particular locations, with many of the most favorable sites being distant from utilization of the output; and (4) the largest size of an individual unit thought viable in the near future has an output of 2–3 MW.

These characteristics seem quite the antithesis of the requirements of a prime mover for the generation of electrical power, but nevertheless this is the main thrust of wind turbine development.

The major problem is that of variable wind speed and it is the torque characteristics of different turbine types which are the important factor here. At a given wind speed, the propeller-type turbine has a continuously increasing torque with rotational speed, and a spoiling mechanism must come into play at the rated power point in order not to overload the generator. On the other hand, in the Darrieus-type turbine the torque increases up to a maximum and then decreases rapidly, giving a sharp peak before declining at a less rapid rate (see Fig. 47). At fixed rpm, the variation of torque with wind speed is of the same general nature, i.e., monotonic slope for the propeller type and a peaked characteristic for the Darrieus, see Fig. 48. Banas and Sullivan (1977), who give a useful discussion of the system performance of wind turbines, call these characteristics "nested" and "unnested," respectively, with reference to the intersect-

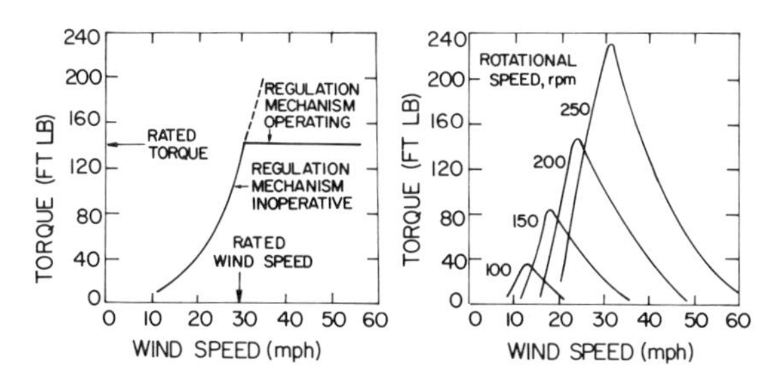

Fig. 48. Torque versus wind speed for a nested and an unnested type. (From Banas and Sullivan, 1977.)

ing or nonintersecting of torque-speed curves. Thus the propeller type must have a power-limiting device while the Darrieus is self-regulating. The power-limiting device is usually the provision of a variable pitch mechanism for the rotor blades, a technology well established for airplane propellers, but requiring further consideration for wind turbines by virtue of the large blade size and the speed of pitch change required to cope with surges due to gusts. On the other hand, propeller types operate at a higher optimum value of the tip-speed/wind-speed ratio λ, possibly 10/1–12/1, while the Darrieus is more usually 5/1–6/1, which reflects on the size and cost of the step-up speed transmission required. (Typically, large propeller turbines operate at 25–40 rpm.)

Wind-driven generators must supply their electrical output at the standard 60 Hz (except for very small local units which may be utilized for dc output or with a special frequency for a particular application). When tied into an electrical network or power grid of large size compared with the contribution of the wind-driven generator, a direct-connected generator–turbine system turbine will be constrained to run at constant speed, within certain limits. With varying wind speed, λ will vary and thus C_p will not be constantly optimized, but the final answer lies in the cost of total energy per annum, not just the running efficiency.

The use of a synchronous generator or an induction generator for constant–speed operation has been a matter of debate for many years and probably will not be settled until much long-term testing has been carried out. The synchronous generator has to run at the utility grid speed, while induction generators will run a little above it, with slip speeds a few percent of the synchronous speed. Smith and Jayadev (1975) and Ramakumar (1977) present informative discussions of various electric generation schemes for WECS, and some of their conclusions are summarized below. A comparison of the synchronous vs induction types includes the following major factors: for the synchronous generator, the main merits are (1) low power excitation requirements, (2) active and reactive power control by field controls and pitch controls, (3) a synchronous tie with the grid, and (4) a reactive value to the system. The main demerits are (1) the strictness of speed control to prevent large fluctuations in power during wind gusts and stability problems unless the WECS power is only very small compared to the grid system, (2) the capability of the machine to draw power from the grid and operate as a motor, and (3) synchronizing equipment to connect the generator to the grid. The advantages of the induction generator are (1) relaxed stability criteria with small speed variations unimportant, (2) synchronizing not necessary, and (3) voltage and frequency completely determined by the grid. The main disadvantages are (1) it needs synchronous machines in parallel to supply reactive kilovolt amperes (20–30% of turbine output), (2) it has a low power factor and may require added capacitance of significant amount, and (3) overspeed control is vital to prevent runaway if maximum torque is exceeded and pull out occurs.

Installed costs of both types are much the same and both require a large step-up gearing from turbine to generator. The induction generator has additional advantages for the Darrieus-type turbine as this is not self-starting (stall at low values of λ), and the induction machine can initially run as a motor and accelerate the turbine until the synchronous speed is reached when the torque is zero, following which it will act as a generator, as the turbine can now supply input power to give the machine a small excess speed over the synchronous speed.

Throughout the years, several other electric generating systems have been proposed beyond the two constant-speed types discussed above, which naturally come to mind initially as the simplest. However, there are possible advantages to generators which can supply constant frequency current at variable rotational speed in order to optimize the turbine performance. A summary overview with bibliography is given by Ramakumar and Hughes (1975), another summary is that of Reitan (1977) on novel schemes of generation, and Smith and Jayadev (1975) comment on other methods.

For small outputs, dc generation is satisfactory and has been widely used in the past for lighting systems with battery storage. However, this is not suited to large outputs and the current would have to be inverted for grid use. There appear to be three methods of variable speed generation which have received significant attention, two of which are under test and development.

The first is called the ac–dc–ac link or *A-D-A* link (Reitan, 1975), with an electronic six-pulse rectifier from the turbine variable frequency alternator feeding a two-pulse inverter to give 60 Hz output. It is said to have

10 years of development behind it up to the 2000-MW, 800-kV level and a power-electronic efficiency of 95% or better. Hence with a 70% or higher alternator efficiency, the overall efficiency should be ≥70%. Its cost, of course, is greater than either of the constant-speed systems, but it is expected by Reitan to compete because of its proved extreme reliability, low maintenance, and minimum of attended operation. However, less optimistic views have been expressed which relegate its economic possibilities to special situations such as long-distance transmission for connection to the grid, or where underground transmission is necessary. A 20-kw experimental program is underway with a variable-speed dynamometer drive to simulate the turbine, together with a 240-ft tower instrumented for comprehensive measurement of wind characteristics.

The second system is that of a field-modulated generator or the "frequency down conversion system" of Ramakumar (1978; see also Ramakumar, 1977; Ramakumar and Hughes, 1975, for summary papers and further references). This consists of a turbine generator giving a modulated frequency much higher than 60 Hz, followed by electronic conditioning to 60 Hz output. This requires a special alternator with three separate phase windings to give single-phase output, with the output phase voltages rectified and combined, yielding a considerable reduction in ripple components. Three such sets must be used to produce three-phase output. The alternator frequency has to be at least 7–8 times the output frequency and a very high step-up gearing ratio is required.

The problem of small and medium power turbines may be different, because the output is intended to be used locally and not fed to the grid. Thus if the application is heating and/or lighting, conversion to electrical power may still be necessary, but together with a storage capacity to provide ability to utilize all the energy produced and to tide over shortage of wind. One such scheme is described Hirst and Rees (1977) which can accommodate a 10/1 range of wind speed for battery charging, with full automatic control to prevent electrical overload via a friction brake with air cooling. The turbine drives a permanent magnet alternator with an epicyclic gear transmission which allows the brake to be operative at a certain voltage. The battery charger is a switching-type power converter to 50 V dc and provision is also made for an inverter to give 240 V, 50 Hz ac (United Kingdom standard). Small commercial wind turbines using proven methods for electric output are divided among alternators with rectifiers for dc and dc generators with inverters for ac.

It is the versatility of electric power, of course, together with its comparatively easy transmission over considerable distances that make it attractive for generation by wind power in spite of the problems it engenders. There are, however, some applications where mechanical power is not only feasible but economically attractive. Before the advent of electrical generation, and in the heyday of the wind turbine, mechanical output was widely used in many ways, as witnessed by the mercantile efflorescence of the Netherlands in the sixteenth through eighteenth centuries. It was indeed the water pumping application which allowed the country to be viable at all, and this application is still a very valid one worldwide. Its attraction is its comparative simplicity and the fact that most often its duty is not one of immediate demand but one which may be performed in random fashion, provided that over a relatively long period there is a high statistical probability of success. The electric motor and the diesel have supplanted the windmill, but this has been in many cases a convenience rather than a significant technological improvement. The windmill pumping water is still a common sight in the less-developed countries of the world, with the rotor generally having sails of cloth or other woven material. For the last fifteen years, the Brace Institute of McGill University, Montreal, has pursued development of small wind turbines for water pumping and other applications, its work including field testing of on-site installations and provision of plans for construction by the individual. A summary of its work, together with a bibliography of their publications is given by LaWand (1973). There is considerable activity in general in the area of low-cost wind turbines of simple construction of locally available materials, by individuals and by groups. One such activity is the cooperative effort by universities and research organizations in the Netherlands (Lysen, 1977); another is an irrigation project at the Omo Mission Station in Ethiopia (Frenkel, 1975).

The low rotational speed of wind turbines is suited to the low-speed reciprocating pump, and the coupling is simple and direct. Windmills for well pumping were in historic use throughout the rural United States before the spread of electric transmission, and such units have continued to be manufactured and are available "off the shelf." The rotors are the multibladed, sheet metal type, the so-called American wind turbine. In recent years the Brace Institute has developed a three-bladed propeller type turbine for irrigation use, with a 32-ft-diam rotor and a gear ratio of 7.2/1 (Chilcott, 1969). Reciprocating pumps have a limited capacity but a step-up gear allows centrifugal pumps to be used, and these have the same general performance parameters as turbines, hence mesh well with the prime mover in this instance.

The pumped storage of water for hydropower installations, now popular in conjunction with off-peak electric power from central stations, is adaptable to wind power and is being studied carefully. One problem here is the compatibility of the reservoir location and a good wind turbine site, while another would seem to be the energy source of the central station. If this is fossil-fuel fired, then there is a possibility of savings, but if nuclear, then the very small fuel cost and the preferred full design output mode of operation militate against a viable system.

Water heating is another suitable application of wind turbines. The most flexible system is by production of electricity and use of resistance heaters, as this can free the tie between energy production and its use. In this case the generator can be as simple and inexpensive as possible, because ac or dc can be used and constant generator speed and voltage are not essential. However, the electrical system can be eliminated by a direct drive to a rotor immersed in a thermally insulated tank of water, the rotor acting simply to dissipate the mechanical energy via turbulence and fluid friction to internal energy. Here again, the attraction is the freedom from on-demand energy by combining the heat generation and storage function in the water itself. The efficiency from turbine shaft to delivered hot water can be very high, limited only by loss of heat from the tank to the surroundings. This is an example of the normally undesirable and irreversible "paddle-wheel" work to provide a useful effect, with the designer being asked to combine a turbine of the highest possible efficiency with a pump of zero efficiency!

A detailed heating study is being carried on at the University of Massachusetts (Cromack *et al.*, 1977) on residential heating, using a specially built full-scale test facility called "Solar Habitat I." The wind turbine is of the propeller type with an ouput of 25 kW at 11.6 m/sec wind speed (26 mph), having a three-bladed rotor of 9.9 m (32.5 ft) diameter. A synchronous generator is used to produce variable voltage and variable frequency, as the turbine is designed to run at a constant value of λ of 7.5. An automatic generator field controller provides the correct field current to produce, but not exceed, the maximum power which the machine is capable of at any given wind speed. The turbine rotor has an electronic pitch controller to give maximum start-up torque for the lowest usable cut-in speed, followed by a one-step change to a different pitch for maximum C_p at $\lambda = 7.5$ ($N = 167$ rpm). If the rpm tends to exceed 167 rpm, then the pitch is again changed to spill the wind. At 180 rpm, an overspeed mechanism is tripped by a flyball mechanism and the blades are set to full feather. Thus for this application it is possible to have a few specified changes of pitch and not a continuously variable control. The experimental program is intended to compare the actual output against computer simulation and some preliminary results are given in the paper.

The original home of the windmill, the farm, can provide a combination of several of the above applications, for instance, irrigation, house and building heating, crop drying, and water heating for dairies and other food-processing uses. The United States Department of Agriculture has a research and development program, first on studies to identify priorities, followed by testing of actual systems, with the accent on using as much existing hardware as possible to develop new applications (Carter, 1977; *Wind Power Dig.,* 1977).

Direct mechanical drive from a wind turbine can be awkward in that for a propeller type the power take-off point is on top of a tower of some height and a long connecting rod to ground level is necessary. Also, there is the inflexibility of load location. This can be circumvented to a considerable extent by using a hydraulic pump and motor system, as instanced by Wesco Ltd. (Sumner, 1975), in what they term a shunt transmission—a mechanical–hydraulic hybrid unit. Pumps are mounted on the tower and driven mechanically by a short transmission shaft from the rotor, the hydraulic fluid then being piped down the tower to either a churning unit for heating or to a hydraulic motor and electrical generator for other uses. By this means of obtaining a near constant output from a variable speed input, they reckon to obtain 60–100% more power. Their immediate application is to greenhouse heating, which ordinarily can require a great deal of fuel. The maximum heating demand in the winter months coincides well with the normally greater wind power available at that time.

There are a number of applications of wind turbines occasioned by geographical circumstances, e.g., remote locations where electric power is unavailable and/or fossil fuels expensive to transport. Some of these are potentially extensive, as, for example, Alaska (Forbes, 1975), Hawaii (Grace, 1975), and in many islands throughout the world. Some involve limited numbers of units but are potentially valuable innovations, e.g., marine navigational aids, distant off-shore oil and gas platforms, electromagnetic communication stations, earthquake data stations, galvanic protectors for pipelines, etc. A detailed study of the navigational-aid application has been made for the U.S. Coast Guard (Herrera and Weiner, 1976), with cost estimates for a variety of power levels and life requirements, comparing wind power, solar photovoltaics, zinc-air batteries, and wave-power generator as sources. A point noted in the report, and which has relevance for other marine applications, is that most wind data measurements in maritime areas are made on ships, on which the anemometer is usually high up in order to obtain unobstructed measurements and where the ship speed itself is a large factor. So the data is for a height usually greater than required for a typical low-power application and has had to be calculated from a vector sum, with possible resultant wider range or error than a corresponding land measurement. Use was made of technical information from some dozen manufacturers of small horizontal-axis turbines to aid in establishing realistic designs. It was concluded that the most economical candidate energy source for a given duty depended on both the power level and the length of the mission, and that wind turbine battery power systems were such a candidate for usage over part of the range particularly for sites with high wind conditions.

An unusual application but one which has been continually discussed is that of the electrolytic conversion of water to make and store hydrogen. Wind power would seem to be highly suited for such a use, first on account of its ability to use all the energy available in the wind capable of being used, because of the undemanding time requirements of electrolysis, and second, the built-in storage capability of compressed or liquid hydrogen. The concept has been bruited for many years but received a boost by the concurrent enthusiasm for a hydrogen economy following the abrupt realization of a real possibility of natural gas and petroleum shortage.

Electrolysis requires dc and this may be provided directly or by conversion from ac generators, depending on relative costs and continuity of the wind turbine and the processing system. In spite of the superficial attractiveness of using wind power, a complete system may be complex and require considerable control outlay. The decomposition of 10 kg of water produces only 1 kg of hydrogen, and 9 kg of oxygen as a by-product. As much use as possible should be made of this oxygen, such as obtaining extra power by expanding it if it is produced at high pressure, and thence using it to cool the hydrogen if the latter is to be liquefied and transported as a cryogenic. The inlet water must be purified and deionized before electrolysis, and some of the hydrogen must be used for providing power for this and for many other auxiliary operations. It would be advantageous if the hydrogen could be pumped as a gas into a pipeline for distribution, either directly for use or mixed with natural gas to conserve the latter. There is optimism with respect to the development of new types of electrolysis cells of high efficiency within the next 5–10 years, these cells operating at high pressure so that compression for storage can be eliminated. Dubey (1977a) discusses these and other aspects together with a cost analysis for the near future. In another paper (Dubey, 1977b) he analyses the use of wind power in the manufacture of ammonium nitrate fertilizer, again by electrolysis of water at high pressure and temperature, combination with nitrogen to form ammonia, and thence to part conversion to nitric acid, with the final nitrate formed by reaction of acid and remaining ammonia.

Another application of the wind-driven electrolysis process is the manufacture of methane (natural gas). The carbon dioxide can be supplied by calcium carbonate, which is released by heating or by a chemical process. A detailed analysis is given by Yang *et al.* 1975 of the Institute of Gas Technology, Chicago.

XI. ECONOMICS OF WIND POWER

The ultimate test of a wind energy conversion system is the cost per unit of output and there is less firm knowledge of this than of many technical aspects. There are to be sure many economic surveys and cost estimates available, by professionals and by the enthusiastic amateur, but it is in the nature of technological innovations of this scale to result in a considerable range of answers. There are many assumptions and choices of inclusions and exclusions to be made and it is, of course, not unusual to find that proponents of wind power tend toward optimism. Perhaps the situation might be summarized by saying that the cost effectiveness at the immediate present appears marginal, or at least uncertain, for the use of wind power as a large-scale substitute for conventional power, but that there are several applications which look more promising and that it is the future that may bring rewards as present energy sources rise in price. But one must guard against the position of estimating on the basis of building now and relying on inflation and rise of fuel cost in the future to produce the savings. On the other hand, how much should be paid for security of energy supply, and environmental values?

In the United States there have been recent governmental study contracts for selection and design of minimum cost wind turbine systems and for mission analyses to identify applications, define goals, evaluate impact on energy users, and identify nontechnical problems arising from WECS. In addition, there have been contracts for construction of complete units or of major components. A great deal of this work has been carried out by companies or divisions of corporations in aerospace engineering, as such organizations have the technical expertise and experience in aerodynamic design. However, the overriding factors in design and manufacture of aerospace vehicles and power systems are to obtain the highest possible degree of technological excellence, as the margin between successful accomplishment and complete catastrophe can be very small, with safety or completion of mission being all important. The aerospace engineer has an inherent tendency toward a high-technology solution because technical performance is dominant and cost often immaterial. This has led to very remarkable progress in materials, design, and techniques, but the thought arises as to whether other groups, possibly less innovative technologically, might by training and experience produce somewhat different answers when the end point is cost effectiveness. One wonders how the automotive industry might respond to some of the tasks in a wind energy program.

The uncertainty in costs is illustrated by the rather startling differences in the original estimates of the Kaman Aerospace Corp, and of the Space

Division of the General Electric Company, the contractors for major investigations into the selection and design of minimum cost wind turbines. Figure 49 is taken from a report by Thomas (1976) of NASA which shows these energy costs in cents per kilowatt hour for turbines from 50 kW to 3 MW. There are not only overall cost differences but also a difference in the effect of output on cost, in that Kaman indicates a specific size for minimum cost, while the General Electric results show a continuous decrease of cost with increase of size. This is attributed to the differing assumptions used for the cost of the rotor, which is a major part of the total system cost. Another factor is the yearly cost of operation, maintenance, taxes, interest on money, etc., which was 16% of capital cost for General Electric, 20% for Kaman. The solid lines are the results of the initial parametric studies, but as preliminary studies of actual designs for 500 and 1500 kW were completed, the energy costs draw closer together, Kaman decreasing and General Electric increasing. Two "spot points" are shown on each diagram, these being for the 500- and 1500-kW designs mentioned, both for the 16% annual cost rate to allow a closer compari-

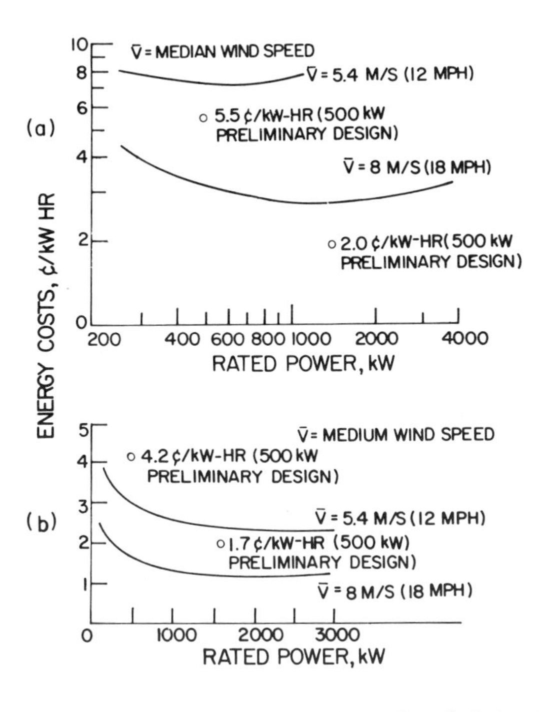

Fig. 49. Comparative energy costs from parametric studies of wind turbines: (a) Kaman and (b) General Electric. (From Thomas, 1976.)

son. For the 500-kW unit, Kaman's figure was 5.5 cents/kW h and General Electric's 4.2 cents/kW h. For the 1500-kW unit, they were 2.0 and 1.7 cents, respectively.

In the same reference, Divone (1976) of ERDA discusses costs and emphasizes the difficulties in making estimates for real utility operation. He does, however, believe that cost figures from General Electric, Lockheed, and Boeing are sufficiently similar to begin to allow confidence in them, and as an example he gives a curve of Lockheed (Fig. 50) of cents per kilowatt hour against power in kilowatts with wind velocity, rotor diameter, and plant factor as parameters. This term "plant factor" (or capacity factor or energy ratio) is an important parameter and is defined as the ratio of annual energy delivered to that which would be delivered if the system operated at its rated power all year. Care is required in the interpretation of "annual energy," as this may be the *actual* energy delivered, or based upon the *estimated* energy corresponding to the *mean* wind speed, or be the energy expected based on a more detailed knowledge of the wind characteristics. This is a function of the rated power, discussed in Section VI, and it is a most important element of first cost and eventually overall cost per kilowatt hour. A higher capacity factor usually entails a larger rotor diameter and thus higher initial cost. Smith (1977), in the paper mentioned in Section VI, relative to siting selection and energy ratio

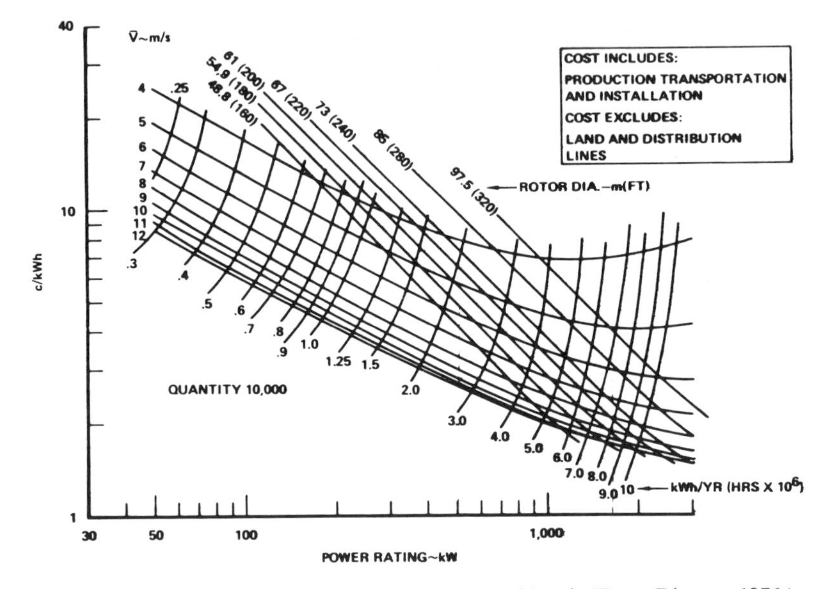

Fig. 50. WECS rating versus energy costs—Lockheed. (From Divone, 1976.)

goes on to discuss cost optimization with some illustrative examples, pointing out the importance of rated output and rotor size.

An important point with respect to total system cost is the contribution of an individual wind turbine to the total utility grid capacity. Because a firm power commitment cannot be made, a single unit cannot be counted as part of system capacity and therefore its contribution can only be that of fuel saving, i.e., when other operating units can be reduced in output and thus the actual fuel cost is saved. In an array of turbines feeding into a utility or if a sufficient number of variously located turbines are interconnected, then some fraction of their rated output can be considered as firm power and hence part of their cost can be regarded as for extension of base-load capacity. Thus the possibility of firm power may be a very important consideration in an economic assessment of wind power for a particular utility grid. The term "incremental heat rate" is an important one in a utility system with respect to continuous monitoring of load change and dispatching decisions. It is defined as the heat saved with the subtraction of each kilowatt hour of generation in the system, expressed as Btu/kW h. Thus the product of incremental heat rate $\times\ 10^{-6}$ and the fuel cost in cents per 10^6 Btu gives the fuel saving value in cents per kilowatt hour. Representative values can be assigned to different units and to modes of operation, and computer control is necessary to operate a system in the most economical manner with a continuously changing incremental heat rate. Different types of plants have different operating characteristics with respect to load changing or cycling ability, with nuclear plants having the least flexibility and fossil-fueled plants having a limited amount, because of boiler stability, thermal stress, etc. It is considerations of this sort that require very careful analysis of WECS loading, in that the value of wind power as a fuel saver can vary a great deal from time to time. Thus simplistic estimated costs of wind-produced energy may contain a number of assumptions which can be misleading. Another factor which varies considerably is transmission costs, as each wind power unit must have its separate lines, overload, safety devices, and so on, and the necessary connections to a grid system may be expensive. This is a factor which comes into decisions on planning of arrays or collections of dispersed units.

Because of the uncertainty of output from an individual wind turbine at any given time, the possibility of storage is obvious. In the past, for small units, batteries were the answer, but this becomes quite uneconomical for large amounts of energy. The general problem of energy storage as a whole has received considerable attention in recent years, because of fuel costs and availability, and possible answers are the use of pumped storage (water or air), electrolysis of water, and flywheels. All these can be used with wind turbines, with the order of listing indicating a seeming order of likelihood of use in the near future.

Water storage in conjunction with the existing pumped storage systems now quite common appears to be a very viable means, as the reservoirs themselves are in existence and their sites are generally in areas and at such elevations as would suit wind power installations. While compressed air storage by itself would be expensive, in conjunction with a conventional power system it would be equivalent to water storage. Such methods are not yet common but they are coming into use or are in a serious design stage. Hydrogen production has been discussed in Section X and storage would entail further expense, but is certainly practicable technically. W. E. Heronemus has been a vigorous proponent of wind power for hydrogen production and a paper on planning methodology for wind power systems by Dambolena *et al.* (1974) uses his concepts, quoting several of his conceptual schemes in the bibliography. Flywheels have interesting possibilities, especially as they also have capability in smoothing out the effect of gusts in normal operation, but need much development and firming of cost before active consideration at this time.

Generalized studies of the economics of wind power are numerous, and particular attention at the present time is being given to the question of arrays, dispersed units, and storage. It is possible only to discuss some of this work, with an attempt to show the trend of analysis and to be representative of the range of investigation.

A wide overview for the United States is provided by the Department of the Interior publication by Todd *et al.* (1977) on cost-effective electric power generation from the wind. Storage, especially hydrostorage, is a major consideration, and use can be made of the combination of favorable wind sites with potential pumped-storage sites. It is concluded that electrical energy can be produced for 10 mills/kW h at the windiest sites with a potential installed capacity of over 100 GW in the 17 western states and much more in the Arctic region of North America. Storage cost is site specific, with an expected average of 5 mills/kW h. Transmission costs of 6 mills/kW h lead to a total delivered cost of about 21 mills/kW h, competitive with new conventional power plants of about 19 mills/kW h.

Justus (1976), in a study of large dispersed arrays of wind turbines for the New England (United States) utility grid, indicates that for 500-kW units, each averaging 190–240 kW output, about 100 kW per unit can be obtained with about 70–85% reliability (seasonal) without storage, by use of a disperse grid of about 500 km and with 45–65% reliability for 200 kW. Larger units of 1500 kW, which need higher winds to operate at full capacity, averaging 240–340 kW output, can have 55–75% reliability for 100 kW and 40–60% for 200 kW. Storage time of 24–48 h would increase

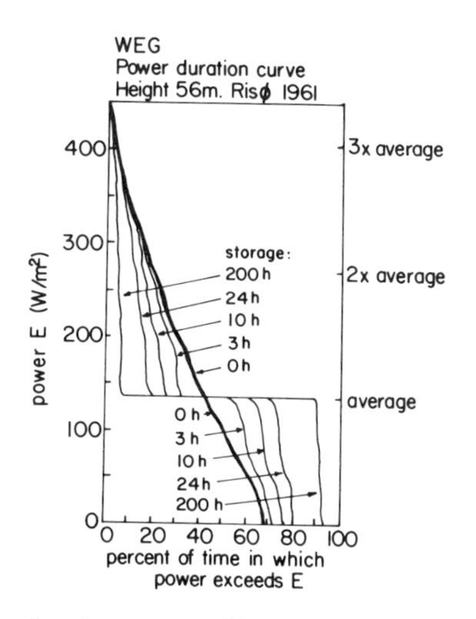

Fig. 51. Power duration curve with storage. (From Sørensen, 1977.)

the 200-kW generator output to about 95% reliability. Another useful result shows that reasonably steady high wind power in winter and high afternoon wind power in summer (peak air conditioning load in summer) would allow significant peak load displacement without storage (see Section VI for discussion of seasonal and diurnal wind variations).

Sørensen (1977), who has made many studies of wind power economics, particularly for Denmark, has considerable belief in the efficacy of wind power. He uses the Gedser Mill (Juul, 1956, 1964) and very complete wind data from Risø, Denmark, to make a power duration curve, power E versus percentage of time in which power exceeds E for the turbine alone, and for different storage capacities expressed as number of hours for which the facility can sustain average power. This is shown in Fig. 51, together with a diagram for a particular nuclear plant (the Vermont Yankee 1974, Fig. 52) of power level as a percentage of the rated power against percentage of time. He concludes that from the nuclear example, 65–70% of time for power above average is acceptable and is reached by the wind unit with a storage capacity of only 10 h. There is a marked gain for storage up to 24 h but a rapidly decreasing advantage above this time. He goes on to calculate the total cost over 25 years, using various assumed figures for inflation increases of cost of operation and maintenance etc., showing that wind power is very competitive. Possibly his nuclear comparison is not what might be expected of such plants in the future and he recognizes the uncertainties of future costs in his assessments, but as he points out, a few years ago estimates of power costs from fossil and

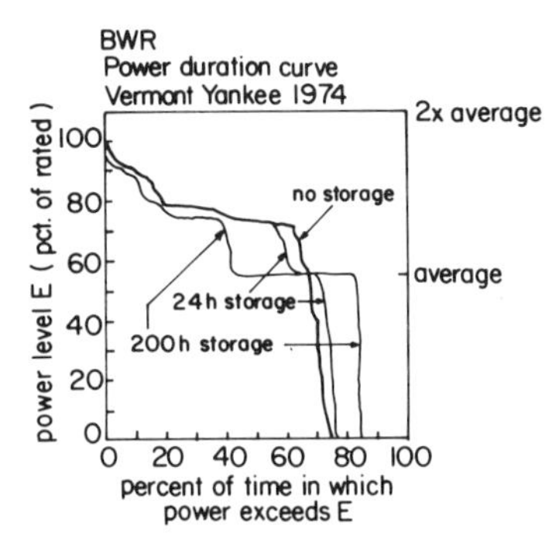

Fig. 52. Power duration curve of a nuclear plant with storage. (From Sørensen, 1977.)

nuclear plants were so much lower that wind power was rejected on that basis.

Molly (1977) makes a study of widely dispersed arrays for providing firm power for West Germany, using the turbine and wind data from the 100-kW Allgaier–Hütter installation at Stötten (Hütter, 1964) as basis for calculations, in terms of a turbine unit of 10,000 m² rotor disk area and 113 m blade diameter, operating at three values of specific output as power/area, 100, 200, and 300 W/m². For an average wind speed of 6.7 m/sec (15 mph), the lowest specific power produces maximum power for 40% of the year, none for 10%. Increase of specific power increases the overall energy output, but for the 500 W/m² unit, the mean power output is increased by only a factor of 2.5, with maximum power for 8% of the time and none for 23%. This means a greater fluctuation of power around the mean value. Hence a compromise value of specific power is necessary and 300 W/m² is suggested for all sites, independent of wind speed. Molly then shows curves of power using different numbers of sites, the effect of site separation distance in terms of mean distance from the "center of gravity" of the sites, and the relationship of power fluctuation with storage capacity.

A paper on design, economic, and system considerations of large WECS by Jorgensen *et al.* (1976) combines the Kaman design experience with that of a public utility (Northeast Utilities Service Company). It summarizes the design work for the NASA–ERDA program and follows with an analysis from the utility viewpoint of WECS as fuel savers and as base load capacity, and the use of storage. The paper is effective in emphasizing the concerns of a utility in detailed economic analysis of incremental heat rate, system capacity, and reliability of power (outages). Another analysis along the same lines by Coste and Lotker (1977) presents the utility viewpoint in discussing fuel saving, availability, capacity factor, storage, and load management, including data for a specific site of known demand, and then estimates generation costs in mills per kilowatt hour against percentage capacity factor for a number of heat engine cycles (capacity factor is defined as the fraction of the year that power is required to be delivered). The conclusion is reached that with storage, a WECS can provide reliable, but costly, power at all capacity factors with a reasonable availability, but that sites with average wind speeds of at least 15 mph may be required for economic reliability. However, contrary to initial expectations, excessive amounts of storage would not be required at such high-wind sites. Another cost analysis emphasizing the interface with utility operation is given by Smith *et al.* (1976).

An analysis by Devine (1977) uses the General Electric 1500-kW turbine design for an assessment of the *net energy* of a WECS to displace fossil fuel in a utility system. This net energy is defined as that remaining for use outside of an energy system after deducting from the gross output of the system all of the energy required for constructing and operating the system, except for the energy content of the primary energy source being processed. His results indicate that the system considered could be a large net producer of energy and should displace a quantity of fossil energy equivalent to that embodied in the machine in considerably less than one year.

A new note is introduced by an economic analysis by Fekete (1977) for a self-contained 5-MW capacity WECS with storage, intended as the sole source of continuous electric power for a remote location. This location is in the Magdalen Islands in the St. Lawrence, and the power would be used for a proposed mining operation, the constraints being independence of any power grid, and that design and manufacture must take place within the next few years by using then known or proven technology. (Note this investigation is independent of the experimental 200-kW Darrieus-type turbine being operated in the Magdalen Islands, as the power supply for inhabitants and tourists is a separate concern.) The storage capacity must be sufficient to allow 5000 kW to be supplied continuously even during the calmest period. The site is an ideal one for wind power, with an annual mean wind speed of 9.3 m/sec (21 mph), with a monthly average during the normally low months of June and July of 7.2 m/sec (16 mph), with the frequency of calm negligibly small, and with an annual probability of a destructive speed of 44.3 m/sec (100 mph) only once every 200 years. The WECS model is a 300-kW NASA–Lewis type and the pumped storage model consists of a brine-holding lagoon near sea level, with cavities deep underground to store the brine after discharging from the hydraulic turbine driving the load generator. The pump head is shown as 442 m and the turbine head as 305 m, and component efficiencies appear conservative. The system was initially proposed as having the load being met only by hydroelectric power, with the WECS used for pumping only, but the low in–out storage efficiency led to a consideration of a "mix" system in which the WECS supplies power directly and only excess energy is stored. The minimum capital cost of the complete system is estimated to be of the order of 65×10^6 dollars and the minimum cost of electricity of the order of 20 cents/kW h (1974 dollars). Another total WECS system study (regarded as interim), this time for a community of 2500 people at present using about 4 MW of power supplied almost wholly from oil and gas fuels, is given by Asmussen *et al.* (1975) for the city of Hart, Michigan. It deals with wind data collection and assessment, power and energy demand and usage, wind turbine performance, hydropower and storage, economic analysis, and environmental concerns. Although for a particular locality with its own specific characteristics, the report indicates the manifold problems and considerations which have to be met in a detailed study of wind power usage for an actual community.

Ramakumar (1977) presents an analysis of wind energy conversion technology relating various types of generator (Section X) to size of turbine and system control. He uses a set of charts showing the interaction of the various costs and operating parameters of mechanical, electrical, and interface components contained in a previous article (Ramakumar, 1976) which present values of break-even points for a range of capital investment parameters. The paper presents no firm conclusions because of the different levels of maturity of the generator schemes, but it does show methods of analysis for cost effectiveness and the nature of trade-offs in mechanical and electrical performance which can be considered in relation to the size of system, type of utilization, and the nature of the wind region.

The Darrieus-type wind turbine is relatively undeveloped, but it has sufficient attractive characteristics to have occasioned interest in many parts of the world. In spite of its design and performance uncertainties, efforts are underway to provide an economic analysis of an "open-ended" nature to allow continual update as new data become available. Banas (1976) presents an outline of economic analysis as part of the Sandia program, which is a comprehensive one in the technical development. He concentrates on aspects related to system design in the area of aerodynamics and structures and identifies turbine cost and transmission cost as the two major contributors to total cost. He uses these as the terms E/A and E/T_R, respectively, where E is the annual energy output, A is the turbine swept area, and T_R is the rated torque, and relates them to performance parameters such as maximum power coefficient and tip-speed ratios corresponding to maximum efficiency, zero efficiency, and maximum power with respect to wind speed. Then using an economic model for the costs of turbine, transmission, and generator together with his performance parameters, features can be identified for cost effectiveness. He gives a reference case for such an analysis, but data are too tentative at the present time for specific use. However, his method shows how on-going analyses can be readily made as firmer information becomes available.

Another analysis of cost of the Darrieus-type rotor comes from Fritzche (1977), as related to German development. His approach and conclusions make an interesting comparison with those of the United States. He considers various materials for the blade in the light of manufacturing methods and costs, and concludes that extrusion, sheet metal forming, bonding, welding, gluing, and riveting are all possible processes. Glass fiber-reinforced plastics (which have been tested for years in Germany) have manufacturing problems characterized by high labor rates, and automation needs large numbers of pieces. He does not consider that the manufacturing of these is highly sophisticated and that developing countries could benefit from their lower labor costs. He compares the amount of material for the rotors of the Darrieus- and propeller-type rotors as the ratio of total blade length and swept area, obtaining figures of 2.25 and 0.69, respectively, with a solidity factor of 0.225 for the Darrieus as an order of magnitude greater than the high-speed horizontal-axis turbine. For the tower, whose size depends on the slenderness of the rotor, he takes the buckling load as proportional to the 1.7th power of the parameter height/radius, hence small height and large radius are preferred (but notice the aerodynamic penalty involved, as mentioned earlier). A cost analysis of a 4-kW turbine with asynchronous generator with production of 50 units showed a breakdown as follows: Blades (three) plus Savonius starting rotors 26%; tower 23%; blade and rotor bearings 11%; transmission and generator 40%. The author's main conclusions are (1) the Darrieus rotor can cover a power range up to 100 kW with today's technology, and (2) low-cost manufacturing methods for small blades are unsuitable for large areas. Compared with the horizontal-axis rotor, (3) the volume of the blades is a magnitude larger, (4) the tower is a rotor radius higher for the same swept areas, (5) the advantage of independence of wind direction decreases for large units because the expense for control does not increase proportionately to power, and (6) the lower tip-speed ratio requires a larger transmission ratio for identical generators. The aerodynamic and structural problems do not limit the economic power range up to about 20 kW but are more important for higher powers, because of the manufacturing costs and low rpm for the generator. The cost of the 4-kW model was estimated at just about the same as that for the propeller type.

XII. STATE OF THE ART

The previous discussion has attempted to summarize the directions and major findings of analysis, research, and development of recent years. It remains to essay briefly some account of existing units and to discuss some experimental designs of potential significance. It is not possible to include all the novel designs which have appeared recently, and the selection may appear arbitrary, but the objective is to mention those which have received a significant analysis and/or test and for which there are published accounts of some detail available.

In spite of the intense activity of the last five years or more, there are very few operating WECS of new design of 100 kW or more output. Pride of place here must be given to the 2-MW unit of Tvind in Denmark, as having the highest design output of any unit to date and of being a remarkable example of initiative and cooperative effort. The rated wind speed is 15 m/sec (31.5 mph). The rotor has three blades of 54 m diameter made of fiberglass and plastic foam, each weighing about 5 tons. The output is unusual in that it is quite specific—80% for heating and 20% for electricity and the rotor runs at variable speed and maximum effectiveness up to the rated power, after which the speed is held constant by pitch control. The generator then delivers current of variable frequency, but only one-quarter of this need be finally supplied at standard frequency via transformation to 440 volts, rectification to dc and inversion to 50-Hz ac current. The remainder is used for heating and any excess ac power is to be sold to the local power company. Reports have been given by Hinrichsen and Cawood (1976), in the *Consulting Engineer* (1976), and most recently by Merriam (1977–1978). More remarkable possibly than the machine itself, is the method of its design, construction and general implementation. It was first put in the planning stage in late 1974 by a school-community college system called the Tvind Schools, in Ulfborg, West Jutland near the North Sea coast. It has been a community effort, going through several design stages, with work being started on the foundations and base while the turbine design was being settled. It is not a state-funded operation, as the genesis, organizational effort, and funds are all from the Tvind community, with consulting help from outside experts. Much voluntary labor has been contributed to the project and all in all, although it will doubtless have its problems in operation, it stands as a remarkable contribution to the wind power field in relation to the seem-

TABLE I

Features of NASA–ERDA WECS

Model Type	No. of Units	Rated Output (kW)	Rated Wind Speed (m/sec)	(mph)	Rotor Diameter (m)	Installed Location	Reported Location
Mod-0	1	100	8	18	38	Plumbrook, Ohio	
Mod-0A	1	200	10	22.5	38	Clayton, New Mexico	
Mod-0A	2	200	10	22.5	38		Block Island, Rhode Island Culebra Island, Puerto Rico
Mod-1	2	1500	8	18	61		?
Mod-2	1	1000–1500	6.3	14	91.5		?

ingly ponderous efforts of national institutions. In the Spring of 1978, it was understood that the unit was undergoing inspection by government officials prior to commissioning.

The NASA–ERDA program is one of progressive construction and operation of a series of WECS of generally increasing power and varying design wind speed. Table I shows some of the salient features but the data may not be precise, as the target seems to be a moving one, and objectives change from time to time. The 100-kW Mod-0 was placed in operation in the fall of 1975 and is reported to have fulfilled its expected performance capability. The interruption of operation soon after the initial running due to the blade stresses induced by tower shadow was very unfortunate, because early testing was expected to help in the design of later models. It seems somewhat surprising that the tower shadow effect was so unexpected, in view of previous history of wind turbine stressing problems caused by cyclical forcing patterns. However, this misadventure has served a purpose in emphasizing such problems for future systems.

The second NASA unit, the 200-kW Mod-0A, was placed in operation in January 1978 on a flat site where the average wind is about 7.6 m/sec (17 mph). No operational reports are yet available. It will be observed from the table that the larger output units have a comparatively low rated wind speed and the Mod-2 is intended to represent a typical WECS that might appear in large quantity in normal sites, if the need for wind power does become firm.

A 200-kW propeller-type WECS has been installed on Cuttyhunk Island, about 14 miles off the coast of Massachusetts. This unit, developed

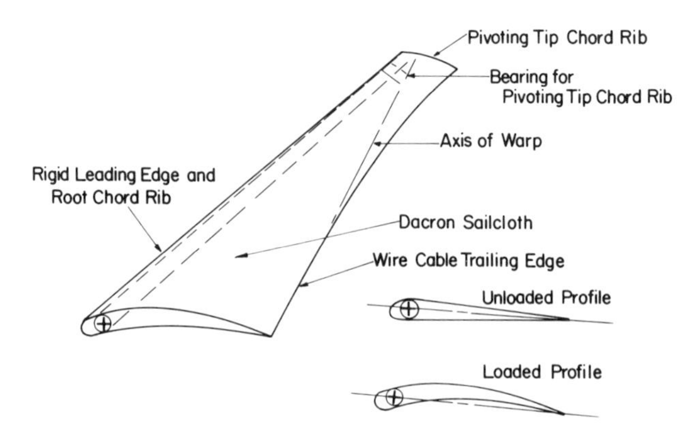

Fig. 53. Sailwing construction. (Courtesy of Grumman Aerospace Corp.)

by WTG Energy Systems (Shepherdson, 1977) is denoted as MP-1-200 and has an output of 200 kW at a rated wind speed of 12.5 m/sec (28 mph). The design is based on the Danish Gedser Mill, which ran for 10 years. It has three 12-m (40-ft) blades turning at 30 rpm, with a 40 : 1 step-up transmission to a synchronous generator. It is intended to supply the island base load, averaging 100 kW in the summer months and 30–40 kW at other times. It is undergoing a year's testing and if found satisfactory, will be purchased by the island governmental body. The projected cost of power is 5.3 cents/kW h, about half the present cost.

A horizontal-axis turbine of different character is the sailwing model initiated by T. E. Sweeney at Princeton University and adopted for a time by the Grumman Aerospace Corp. The sailwing consists of a rigid leading edge and root chord rib, a pivoting tip chord rib, and a wire cable trailing edge, with Dacron sailcloth stretched over the assembly, see Fig. 53. When unloaded (no wind) the fabric surfaces are flat. When loaded, the wing takes on a cambered airfoil shape. A final test report by Maughmer (1976) with the equipment described in Section IX,4 yields the following major conclusions for a two-bladed 12-ft-diam model: (1) The sailwing design is highly competitive in performance with rigid-blade units and has simpler construction and lower costs. (2) Maximum power coefficients up to 0.4 were obtained. (3) A streamlined leading edge pattern is a necessity, giving far superior results to a circular pattern (initially a tubular leading-edge member was used for the sake of simplicity). (4) Three blades were better for performance and for the operating dynamics, but two were used for ease of test changes, etc. [In this connection, it might be observed that the use of three blades appears advantageous in all types, propeller (Juul,

1964), Darrieus (Blackwell *et al.*, 1976), and the sailwing, but two are nearly always used for cost reasons.]. (5) Wing-tip fairings were not beneficial for the particular model tested. (6) Center-body fairings gave a modest improvement in performance, and a significant gain was made by halving the unsupported span of the trailing-edge cable. (7) Finally, it was felt that aerodynamic refinement had reached the point of diminishing returns and that the unit was ready to be put into practical utilization.

Although general activity in the VAWT Darrieus type is considerable, the only WECS of any size is the Canadian Magdalen Island 200-kW unit (Templin and South, 1976). It is intended to be used with the local utility grid which has a diesel-driven electrical capacity of about 26 MW, supplying power at 6 cents/kW h, of which half is the cost of diesel fuel. The rotor has two blades of axial height 36.6 m (120 ft) and equatorial diameter of 24.4 m (80 ft) (see Section VIII on structural considerations), with a blade chord of 0.6 m (2 ft), yielding a solidity Nc/R of 0.1. Each blade is supported by two struts extending horizontally outward from the central column to about 60% of the rotor equatorial radius. The turbine will drive an induction motor, which also serves as the starting motor. A mechanical brake is furnished with centrifugally operated aerodynamic blade spoilers for emergency overspeed protection. The wind conditions at the site are very favorable (Fekete, 1977) and operating experience promises to be very interesting and useful from this type of VAWT.

Sandia Labs have tested 2-, 5- and 17-m Darrieus-type turbines. A summary of the Sandia development as a whole is given by Blackwell *et al.* (1977a), and results of both free-air and wind tunnel tests for the 2- and 5-m units are given by Sheldahl and Blackwell (1977) and Blackwell *et al.* (1976). Results for the 17-m turbine are not yet to hand. The testing of the 2-m turbine in a 4.6 × 6.1 m (15 × 20 ft) wind tunnel for a rotor with NACA 0012 blading at both constant Ω, variable V_∞ and at constant V_∞, variable Ω, yielded the following major findings: (1) A maximum power coefficient $C_{p_{max}}$ of 0.35, increasing with Re. (2) Decrease of solidity σ led to an increased range of tip-speed ratio λ, with 0.2–0.25 as optimum. (3) Three blades were slightly better than two at the same value of solidity (blade area/swept area) and as theory indicates that number of blades enters only as overall solidity, this effect is yet unexplained. (See also Templin and South, 1976, for structural effects.) (4) As indicated by theory, a Darrieus rotor is limited by stalling at tip-speed ratios $\lambda < 2$, but these wind-tunnel tests showed that the mechanical friction was more important than aerodynamic effects, although this is not necessarily true at the higher Reynolds numbers of larger rotors. (5) A thicker airfoil, NACA-0015, together with a modified profile NACA-0012H, gave better results in cascade tests than the NACA-0012 profile. (6) Certain discrepancies with the Canadian NAE wind tunnel rotor tests were revealed, but no explanation was apparent.

Free-air testing of the 5-m VAWT is reported in Sheldahl and Blackwell, 1977). The rotor blades for this model have a circular arc geometry in the equatorial section and two straight sections attaching the circular arc to the axis of rotation. The solidity of the system is 0.26. Some of the major testing aspects, especially the statistical averaging of wind speeds, have been discussed in Section IX. The rotor was run at nearly constant rpm with an induction machine which could either act as starter or generator. The results as judged by power coefficient were disappointing, with a maximum value of C_p being recorded of about 0.27, i.e., lower than the 2-m model wind tunnel tests and those predicted by an aerodynamic computer model. It was concluded that the straight, nonairfoil sections contributed a higher drag than expected and therefore the results were not representative of a possible optimum VAWT. Preliminary test results from the 17-m unit confirm that this conclusion is valid.

Wind tunnel tests were made on four sections, three standard NACA sections, 0009, 0012 and 0015, and one nonstandard, a modified 0012 designated as 0012H. The last-named airfoil and the thicker section 0015 gave improved performance over the 00012.

There are other experimental Darrieus-type VAWT besides the troposkien model, namely, the Loth (1977) cylindrical rotor mentioned in Section V, a circulation-controlled unit of Walters (1975b), the gyromill concept of Brulle and Larsen (1975; Brulle, 1975), and the variable geometry model of Musgrove (1977a, b). The Walters model has circulation-controlled blades obtained by blowing air over the trailing edge from a slit in the upper surface of a hollow blade. This is a manifestation of the Coanda effect and results in an augmentation of the lift coefficient and a delay of the stall condition. The rotor would then consist of straight blades parallel to the axis, which would have the capability of varying performance by controlling the blowing effect over a range of operating conditions (i.e., values of λ).

The gyromill is most readily described here as a straight, vertical-bladed Darrieus type with continuous modulation of the blades to provide optimum output regardless of wind direction. Using a clock dial for direction, with the wind from 12 o'clock, the blade is rotated on its own axis to yield a negative angle of attack from 9 o'clock around to 3 o'clock, where the blade is flipped to give a positive angle of attack around to 9 o'clock again, where it is flipped back to start the negative orientation again. Thus the force vector changes from inward pointing to outward and vice versa for a half cycle each and with continuous modulation yields maximum force from the wind. With each flip of the blades, a vortex is shed and a complete vortex theory has been developed by Larsen along the lines described in Section IV for the conventional propeller type. There are some complex dynamic and vibrational problems in such a design, and the mechanical and control features are liable to be expensive, but overall it is felt that it could compete with conventional turbines. This would seem to lie in the greater degree of power optimization attained with the continuous blade modulation.

The Musgrove variable-geometry vertical-axis turbine consists of two straight blades normally vertical but hinged to a horizontal cross arm, Fig. 54. In high winds and hence higher rotational speed, the blades tend to incline outwards, resisted by a central spring which is attached to the blades by tie wires. This prevents excessive bending moments by centrifugal forces, and as the variable geometry reduces the rate of power increase, the blade stresses are limited automatically.

The Savonius rotor aroused considerable interest at the start of the renascence of wind power a few years ago and it was taken up strongly by the enthusiastic amateur, chiefly because of its very simple construction, i.e., bisecting an oil drum longitudinally, sliding the two pieces transversely to give an S shape with gap and mounting with end plates on a central shaft. However, at the present time, it is not considered generally to be a strong candidate for development, but it should not be completely overlooked because it may have application in particular ways.

It was invented by the Finnish engineer Savonius (1931) some 50 years ago in connection with his studies of the Flettner rotor ship, but did not prosper except for use as a water-current meter. It is a low-speed rotor, $\lambda \approx 1$, and its power coefficient C_p is considered to be low (≤ 0.25), but higher values have been reported. Savonius himself reported a peak value

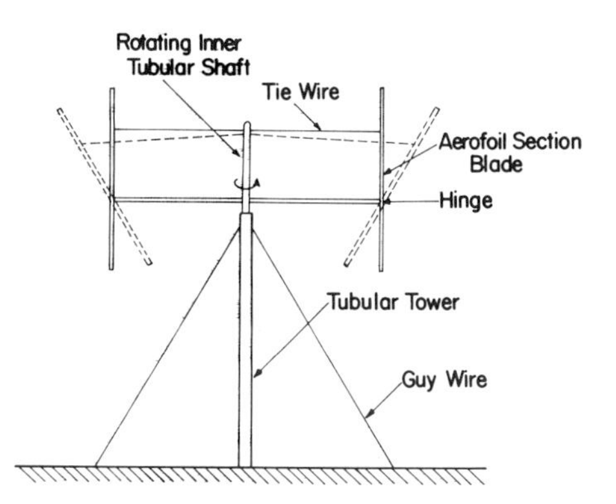

Fig. 54. Variable-geometry vertical-axis wind turbine. (From Musgrove, 1977a.)

of C_p of 0.32 at $\lambda = 0.85$. His tests were performed in a wind tunnel with a model having a cross-sectional area of 0.23 m^2 ($\approx$2.5 ft^2). In recent years, McPherson (1972) has reported wind tunnel tests of a rotor of 0.42 m^2 (4.5 ft^2) projected area (L/D of 2), obtaining a maximum C_p of 0.33, again at $\lambda \approx 0.85$–0.90. (He also reports on other tests of previous years following Savonius' original data.) Turnquist and Appl (1976) performed free-air tests with a rotor 3 m (10 ft) high and 1.8 m (6 ft) diameter, with vanes of curved, but not semicircular, shape following an apparently optimum design of Savonius. They obtained a maximum value of C_p of 0.3 at $\lambda = 1.2$, with the usual broad range of λ for good performance. These results were based on data from apparently steady-flow conditions of reasonable duration, but the authors acknowledged the uncertainties in free-air testing. Manser and Jones (1975) of the University of Queensland, Australia, have reported tests with varying geometry rotors of small size [0.1 m (4 in.) high by 0.1 m (4.3 in.) diameter] in a 0.6 m (24-in.)-diam wind tunnel at a single wind speed of 32 ft/sec. Low values of C_p were obtained ($\approx$0.16–0.17), but a predicted characteristic for the optimum geometry rotor obtained by cross plotting and smoothing indicated a value of 0.235 $\pm$ 10% for $\lambda = 0.9$. They also made water flow tests to obtain flow patterns, one of the major findings being that at the higher tip speeds, no through flow as pictured by Savonius existed, and there was a standing vortex in the intervane passage.

Probably the most detailed experimental study of the Savonius rotor has been that of Sandia Laboratories, reported by Blackwell *et al.* (1977b) which was motivated by the possibility of using its very good starting characteristics for the Darrieus turbine, which is not self-starting. There was concern about reliability of the existing data, much of which was probably due to severe tunnel blockage effects, so a series of tests was undertaken with rotors with two and three buckets, a number of gap spacings (0–0.2, gap/rotor diameter), two rotor heights (1 and 2 m), but with all rotor diameters 0.5 m. The main conclusions were that two-bucket rotors had better aerodynamic performance than three-bucket rotors, except for starting torque, which was much more variable with angular positions, with a very low minimum torque position for the two-bucket arrangement. For the two-bucket rotors, maximum C_p ranged from 0.22 to 0.26, which was about 1.5 times that for the corresponding three-bucket configurations, at a value of λ of about 0.9 for the former and a little higher for the latter. Increasing bucket height gave slightly better performance due to the effect of end plates in reducing end losses and gap width optimized at $s/d = 0.1$–0.15. The Reynolds number was given per unit length of rotor, i.e., $\mathrm{Re} = \rho_\infty V_\infty/\mu_\infty$, and as ρ and μ were essentially constant and the tests were run at only two velocities, 7 m/sec and 14 m/sec, the results are given for values of Re of 4.32×10^5 and 8.64×10^5 m. The effect of Reynolds number was uncertain, and the performance differences at the two values were within the data uncertainty.

The tracked-vehicle designs of Powe (1977; Powe *et al.*, 1974) and Lapin (1976) represent a very different type of wind turbine. Their features were discussed in Section IV and although it does not appear that there are any systems yet constructed, their design performance will be summarized here.

Powe developed a comprehensive mathematical model incorporating a technique to determine optimum carriage speed and airfoil angle of attack for maximum power output, and actual wind spectrum data was used to obtain numerical results via a computer program. His results are given in terms of energy output E per month (kW h)/month, as energy/month per unit airfoil area (kW h)/month m^2, as energy/month per unit swept area (kW h/month m^2), and as energy/month per unit weight of system E/W_t. A range of system parameters was used, ten different systems in all, and although computations were limited because of long executing time, it was possible to determine some major effects. Aspect ratio (increased blade height) gives higher E/W_t as there is less fractional tip loss and the blade is situated at a higher average level of wind speed. However, there is a limiting value of height because of the necessity of increasing carriage weight. Increasing center line height of blading yielded about $\frac{2}{3}$ more energy per unit weight (not including additional weight for elevation of track). Reduction of end turn radius increased E/W_t but increased the velocity loss due to shadowing and these effects are conflicting; hence there is a minimum bend radius for optimum results. Increased length of track increased output per unit weight and change of carriage height had negligible effect. Most of these findings indicate that for this type of system, larger units were more efficient than small units.

Powe made a comparison of the tracked-vehicle airfoil system (TVA) with the conventional HAWT, using an overall coefficient of performance for the latter of about 0.4, with the average velocity V using the Montana wind data and with V defined as equal to $[(1/A)\int_A v^3 dA]^{1/3}$. The HAWT was taken as having rotor diameter equal to TVA blade height, with its hub at the center line of the TVA blades. The ratio of energy output per unit area E/A for the TVA turbine to that for the HAWT is shown in Table II.

On the bases given, the TVA appears competitive. Another illustration of Powe was the overall size of a system for the electrical energy needs of about 15,000 people with a daily average individual requirement of 18 kW h. The track length would be about 8 km, with 6570 airfoil blades 12 m high and 3 m chord mounted on 2190 carriages.

Lapin does not attempt to give absolute values for costs in his work on the wing-driven turbines (WD) similar to those of Powe, and on rotor-

TABLE II
TVA/HAWT Ratio of Energy Output per Unit Area

Center-Line Height (m)	Blade Height or Diameter (m)	$\frac{E/A \text{ TVA}}{E/A \text{ HAWT}}$
2.3	3.0	0.41
4.6	7.6	0.89
7.6	12.2	1.28
9.1	12.2	1.34
18.3	12.2	1.39
18.3	30.5	1.53

driven turbines (RD) based on the Madaras concept, both using the tracked-vehicle model, but makes a comparison of extractive capability between them. This maximum capability for the RD type is about 50% of that for the WD type and occurs at about one-fourth of the characteristic speed V_T/V_∞ where V_T is the linear track speed. However, when the support system losses are taken into account (i.e., rail and carriage friction, aerodynamic drag other than the actual propulsive elements), then the picture changes. The major loss is from rail friction and this varies directly with V_T, and V_T for the WD rotor is about four times that for the RD rotor. The overall effect is that the net power output for WD units is only 20–40% greater than that of the RD units. Lapin presents a considerable number of curves showing performance values for a range of parameters for both types.

The main advantage of the tracked-vehicle systems is that they can be made to give very large outputs, evidenced by Powe's system for a town of 15,000 inhabitants and Lapin's analyses for sizes of several hundred MW. The ground area required is considerable of course and presumably has to be reasonably flat. The components can be readily produced in quantity and the whole design concept allows for modular units of similar components of varying numbers for different outputs. There appears to be no special materials or manufacturing techniques needed, in fact, a railroad or automotive type of production would be appropriate.

A useful, very brief, summary of the main characteristics, advantages and disadvantages of some of the turbine types covered in this review, plus a few others, is given by Weisbrich (1977) in a recent article.

There is considerable activity over many parts of the globe in small wind turbines, i.e., 200 W–25 kW. The majority of this is concerned with the propeller type, some quite conventional, others innovative in electrical generation and controls. Some of it is concerned with marketing to the public, some of it with development particularly directed toward simple types suitable for the less-developed parts of the world. As much of this does not get published in the usual sense or referenced in the scientific and technical field, it is in order to mention a periodical by the name of *Wind Power Digest* (Bristol, Indiana), a popular-type quarterly, intended for the amateur, which does give technical reports on small turbines and special projects, and discusses aeolian affairs in general.

XIII. WIND AND WEATHER PERMITTING

When all is said, how fares wind power? We have to answer, we do not know. We will have to wait and ask the question again when not only all is said, but when more has been done as well. The urgent needs are for operating experience, numbers of wind turbines accumulating hours of running, and data to confirm or confound the computer. If we date the present renascence of wind power from the Stockholm conference of 1972, the quantity of hard information on operational performance is woefully small. The interest is there, the ideas are there, the studies are there, but the hardware is only just appearing.

Running time is the first priority and this involves fully instrumented testing to answer the unresolved questions of reliability of prediction with respect to the effect of wind characteristics and the mechanical reliability of large turbines. Equally important is the running of a wind turbine as part of a utility grid, in order to learn how to manage its loading so as to use it effectively, and to check its behavior in unsteady or unusual conditions.

Of major importance in the near future is the behavior of the vertical-axis Darrieus-type turbines. These are only just beginning to be operated in any but very small sizes and there are many questions to be answered. The horizontal-axis propeller turbine has held pride of place for centuries and continues to do so based almost entirely on the single characteristic of having the highest value of coefficient of performance. But the ultimate answer is the balance sheet, the delivered cost of energy. Design point performance in optimum ambient conditions is one thing, long-term behavior is another, and the motive power of wind is a more uncertain and unpredictable quantity than the familiar sources, fossil fuels and nuclear fuels, and even solar radiation. Hence different types of turbine may extract varying amounts of power due to this factor apart from design-point efficiency. Can the VAWT be built more cheaply? Are its panemone qualities significant? The Canadian 200-kW unit and the Sandia 17-m unit could provide very important information on these questions and perhaps decide the future course of development—if they can be kept running for long periods. In the United States, the propeller type and the Darrieus type are the front runners—what of the various other innovative types?

Of the variants of the basic Darrieus turbine, the giromill and controlled-circulation types appear to be rather too complex for development on any scale at the present time—if the simple Darrieus proves effective then there might be a logical development. The Musgrove variable geometry turbine does appear to have some potential, as it is relatively simple in design although there are certainly some dynamic and structural problems to face.

The Savonius rotor does not seem to have a great deal to offer because of the large quantity of material required for the rotor, the modest power coefficient, and the low rotational speed. It was seen earlier that power coefficients have been quoted in the range of 0.24–0.33, so at the top end it would compete. The comprehensive Sandia tests, however, would indicate that the lower values are perhaps more likely. Its simplicity is attractive and perhaps it might be suitable in some parts of the world if it could be made to work with supported cloth or woven material in place of metal.

The wind concentration types of turbine, that is, the ducted diffusion augmentors of Igra, Oman, and van Holten, the obstruction concentrators of Loth, and the vortex concentrators of Loth, Sforza, and Yen, have not yet had sufficient operational testing as turbine systems to be able to estimate their relative efficacy. The trouble with most of them is that any addition to the structure beyond the basic rotor itself adds to the capital cost and for some types, increased tower loads. The vortex type concentrators are not omnidirectional unless mounted on a turntable, and this again is costly. Thus it reduces to relative cost of rotors and additional structures for a given power level.

The tracked-vehicle wing or rotor concept may seem unattractive at first sight because of the large ground area required. However, the model mentioned in the previous section as suitable for the requirements of 15,000 people would require as replacement about a dozen 2-MW single units of the conventional type, each of which requires a considerable clear area around it. As far as is known at this time, there are no plans for constructing the tracked-vehicle WECS.

The aerodynamics of the HAWT are sufficiently well known to have confidence in design, although there are regions of unsteady and unstable flow which need further investigation. There is still much to do for the Darrieus VAWT, as the cyclical and two-stage aspects of the flow require further analysis and it is a complex flow regime. It is apparent that interaction of tower and rotor is not yet sufficiently predictable for assurance and in spite of modern analysis and computer techniques vibrational problems of many sorts are bound to occur in future prototypes of all sorts. The choice of the straightforward constant-speed generating units is not established, i.e., synchronous or induction machines, and it would be very advantageous if one or the other of the proposed systems which allow variable rotor speeds could be perfected. Again, the techniques are there but the price is not right.

Activity in wind power seems sure to continue to some degree at this juncture, but the rate and extent of its development depends on factors largely outside its technical merits, as no startling breakthrough is likely. These factors are the availability and price of fossil fuels and the environmental concerns with nuclear fuels, of which the most serious are reactor safety and disposal of spent material. It appears at the moment that wind power, like solar power, can be useful and attractive for small outputs and local applications, with large-scale development for use as part of public utility systems proceeding at a modest pace. This development will serve as a longer term insurance in the event of existing sources of energy becoming obviously either critically insufficient or patently uneconomic. Of the several alternative energy sources now being considered, there appears to be a considerable measure of opinion that wind power is probably the most promising type for the longer run, 30–40 years. A report in the journal *Science,* reviewing studies made by two consultant groups of the market penetration by various power technologies within the broad definition of solar energy (wind, direct cooling and heating, thermal power, photovoltaics, biomass, ocean thermal energy conversion, and industry/agriculture), concluded that wind power and biomass can make very substantial energy contributions in the next two decades and that by the year 2020, wind power would become dominant.

REFERENCES

Ashley, H. (1977). Some contributions to aerodynamic theory for vertical axis wind turbines. *Proc. Intersociety Energy Convers. Eng. Conf. (I.E.C.E.C.), 12th, Washington, D.C.* Pap. No. 779272.

Askegaard, V. (1964). Testing of the Gedser wind power plant. *Proc. U.N. Conf. New Sources Energy, Rome, 1961* **7,** 272–277.

Asmussen, J., Fisher, P. D., Park, G. L., and Krauss, O. (1975). "Application Study of Wind Power Technology to the City of Hart, Michigan" COO-2603-1. Michigan State Univ., East Lansing.

Baker, R. W., and Hennessey, J. P., Jr. (1977). *Power Eng.* Mar, pp. 56–57.

Banas, J. F. (1976). Economic considerations. *Proc. Vert-Axis Wind Turbine Technol. Workshop, Sandia Lab.* SAND 76-5586, 23–31.

Banas, J. F., and Sullivan, W. N. (1977). *Wind Technol. J.* **1**(1).

Base, T. E. (1975). *In* "The Effect of Atmospheric Turbulence on Windmill Performances" (T. N. Veziroglu, ed.), Part A, pp. 85–105. Plenum, New York.

BEAMA J. (1955). Nov., 185–188.

Betz, A. (1926). "Wind-Energie und Ihre Ausnutzung Durch Windmühlen." Vanderhoeck and Ruprecht, Göttingen.

Bielawa, R. L. (1977). *J. Am. Helicopter Soc.* **22**(4), 2–9, 19.

Blackwell, B. F., and Feltz, L. V. (1975). "Wind Energy—A Revitalized Pursuit," SAND 75-0166. Sandia Lab., Albuquerque, New Mexico.

Blackwell, B. F., Sheldahl, R. E., and Feltz, L. V. (1976). "Wind Tunnel Performance Data for the Darrieus Wind Turbine with NACA 0012 Blades." SAND 76-0130. Sandia Lab., Albuquerque, New Mexico.

Blackwell, B. F., Sullivan, W. N., Reuter, R. C., and Banas, J. F. (1977a). *J. Energy* **1**(1), 50–64.

Blackwell, B. F., Sheldahl, R. E., and Feltz, L. V. (1976). "Wind Tunnel Performance Data for Two- and Three-Bucket Savonius Rotors," SAND 76-0131. Sandia Lab., Albuquerque, New Mexico.

Bonneville, R. (1974). Wind power projects of the French Electrical Authority. *NASA Tech. Transl.* **TT F-16,057.** (Transl., Kanner, Assoc.)

Brulle, R. V. (1975). "Feasibility Investigation of the Giromill for Generation of Electrical Power," COO-2617-75/1. McDonnell Aircraft Co., St. Louis, Mo.

Brulle, R. V., and Larsen, H. C. (1975). Giromill (cyclogiro wind mill) investigation for generation of electrical power. *Proc. Workshop Wind Energy Convers. Syst., 2nd, Mitre Corp., Washington, D.C.* NSF-RA-N-75-050, 452–460.

Carter, J. (1977). *Wind Power Dig.* No. 8, Spring, pp. 38–40.

Chamis, C. C., and Sullivan, T. L. (1976). Free vibrations of the ERDA-NASA 100 kW wind turbine. *NASA Tech. Memo.* **NASA TM X-71879.**

Changery, M. J. (1975). "Initial Wind Energy Data Assessment Study," NSF/NOAA, NSF-RA-N-75-020. Natl. Clim. Cent., Ashville, North Carolina.

Chilcott, R. E. (1969). *J. R. Aeronaut. Soc.* **73**(700), 333–4.

Clausnizer, A. (1964). Various relationships between wind speed and power output of a wind power plant. *Proc. U.N. Conf. New Sources Energy, Rome, 1961* **7,** 278–284.

Consulting Engineer (1976). **40**², November, 42–43.

Corotis, R. B. (1976). "Stochastic Modelling of Site Wind Characteristics," ERDA/NSF-00357/76/1. Northwestern Univ., Evanston, Illinois.

Coste, W. H., and Lotker, M. (1977). *Power Eng.* May, pp. 48–51.

Crafoord, C. (1975). "An Estimate of the Interaction of a Limited Array of Windmills" Meteorol. Inst., Stockholm.

Cromack, D. E., Heronemus, W. E., and McGowan, J. G. (1977). Design and operational evaluation of 25 kW wind turbine generator for residential heating application. *Proc. I.E.C.E.C., 12th, Washington, D.C.* Pap. No. 779278.

Dambolena, I. G., Kaminsky, F. C., and Rikkers, R. F. (1974). A planning methodology for the analysis and design of wind-power systems. *Proc. I.E.C.E.C., 9th, San Francisco, Calif.* Pap. No. 749009.

Danish Academy of Technical Sciences (1975). "Windpower," ATV. Copenhagen.

Delafond, F. H. (1964). Méthodes d'essais employées sur l'aérogénérateur 100 kW Andreau-Enfield de grand vent. *Proc. U.N. Conf. New Sources Energy, Rome, 1961* **7,** 285–293.

Devine, W. D., Jr. (1977). An energy analysis of a wind energy conversion system for fuel displacement. *Proc. Int. Conf. Alt. Energy Sources, Miami, Fla.*

Divone, L. V. (1976). Federal Wind Energy Program. *Proc. Vert.-Axis Wind Turbine Technol. Workshop, Sandia Lab.* SAND 76-5586, I.1–I.8.

Donham, R. E. (1975). 100-kW hingeless metal wind turbine blade design, analysis and fabrication considerations. *Proc. Workshop Wind Energy Convers. Syst., 2nd, Mitre Corp., Washington, D.C.* NSF-RA-N-75-050, 208–225.

Donham, R. E., Schmidt, J., and Linscott, B. S. (1975). 100-kW hingeless metal wind turbine blade design, analysis, and fabrication. *Annu. Forum, 31st, Am. Helicopter Soc., Washington, D.C.*

Donham, R. E., Anderson, W. D., Linscott, B. S., and Glasgow, J. C. (1977). Experimental data and theoretical analysis of an operating 100 kW wind turbine. *Proc. I.E.C.E.C., 12th, Washington, D.C.* Pap. No. 779273.

Dubey, M. (1977a). Hydrogen from the wind—a clean energy system. *Proc. Int. Conf. Alt. Energy Sources, Miami, Fla.*

Dubey, M. (1977b). Conversion and storage of wind energy as nitrogenous fertilizer. *Proc. I.E.C.E.C., 12th, Washington, D.C.* Pap. No. 779086.

Edwards, P. J. (1977). Low cost recorder for a wind energy survey. *Proc. Int. Symp. Wind Energy Syst., BHRA, Cambridge, Engl., 1976* Pap. S-2.

Eldridge, F. (1976). "Wind Machines," NSF-RA-N-75-051. Mitre Corp., Washington, D.C.

Fekete, G. I. (1977). A self-contained 5,000 kW capacity wind energy conversion system with storage. *Proc. Int. Symp. Wind Energy Syst., BHRA, Cambridge, Engl., 1976* Pap. F2.

Forbes, R. B., ed. (1975). *Proc. Conf. Alaskan Geotherm. Wind Power Resources, Anchorage.*

Frenkel, P. L. (1975). "Food from Windmills." Int. Tech. Publ., London.

Friedmann, P. R. (1976). *J. Am. Helicopter Soc.* **21**(4), 17–27.

Fritzsche, A. A. (1977). Economics of a vertical-axis wind turbine. *Proc. Int. Symp. Wind Energy Syst., BHRA, Cambridge, Engl., 1976* Pap. D1.

Frost, W. (1977a). Analysis of the effect of turbulence on wind turbine generator rotational fluctuations. *Proc. Int. Conf. Alt. Energy Sources, Miami, Fla.*

Frost, W. (1977b). "Analysis of the Effect of Turbulence on Wind Turbine Generator Rotational Fluctuations," Rep. ASD-1. Univ. of Tennessee Space Inst., Knoxville.

Garate, J. A., program manager (1977a). "Wind Energy Mission Analysis," Final Rep., Appendix A, "Meteorological Data and Supporting Analyses." General Electric Co., Schenectady, New York.

Garate, J. A., program manager (1977b). "Wind Energy Mission Analysis," Final Rep., Appendix J. "System Spacing." General Electric Co., Schenectady, New York.

Glasgow, J. C., and Linscott, B. S. (1976). Early operation experience on the ERDA/NASA 100 kW wind turbine. *NASA Tech. Memo.* **NASA TM X-71601.**

Glauert, H. (1935). "Airplane Propellers" (W. F. Durand, ed.), Aerodynamic Theory, Vol. 4. Springer, New York. (Reprinted, Dover, New York, 1963.)

Golding, E. W. (1955). *Proc. Inst. Electr. Eng.* **102,** Part A, No. 6, 677–695.

Golding, E. W. (1976). "The Generation of Electricity by Wind Power" (revised by R. I. Harris). Wiley, New York. (Orig. Ed., Spon, London, 1955.)

Grace, D. J. (1975). The potential for wind energy conversion systems in Hawaii. *Proc. Workshop Wind Energy Convers. Syst., 2nd, Washington, D.C.* NSF-RA-N-75-050, 130–132.

Hennessey, J. P., Jr. (1977). *J. Appl. Meteorol.* **16**(2), 119–128.

Herrera, G., and Weiner, H. (1976). An assessment of wind-powered generation for navigational aids. *Proc. I.E.C.E.C., 12th, Washington, D.C.* Pap. No. 779277.

Hewson, E. W. (1975). *Bull. Am. Meteorol. Soc.* **56**(7), 660–675.

Hinrichsen, D., and Cawood, P. (1976). *New Sci.* June 10, **70,** No. 1004, 567–570.

Hirst, P., and Rees, D. H. (1977). The regulation storage and conversion of wind produced electrical energy at the level of a few hundred watts. *Proc. Int. Symp. Wind Energy Syst., BHRA, Cambridge, Engl., 1976* Pap. F4.

Holme, O. (1977). A contribution to the aerodynamic theory of the vertical-axis wind turbine. *Proc. Int. Symp. Wind Energy Syst., BHRA, Cambridge, Engl., 1976* Pap. C4.

Hütter, U. (1964a). The aerodynamic layout of wing blades of wind-turbines with high tip-speed ratio. *Proc. U.N. Conf. New Sources Energy, Rome, 1961* **7,** 217–228.

Hütter, U. (1964b). *Brennst.-Warme-Kraft* **16,** 333–340. [Engl. transl., *NASA Tech. Transl.* **TT F-15 068** (1973).]

Hütter, U. (1977). *Annu. Rev. Fluid Mech.* **9,** 399–419.

Igra, O. (1974). "Design and Performance of a Turbine Suitable for an Aerogenerator," Rep. No. 1. Dep. Mech. Eng., Univ. of the Negev, Beer Sheva, Israel.

Igra, O. (1976a). *AIAA J.* **14**(10), 1481–1483. [Also as, Shrouds for aerogenerator. *AIAA Aerosp. Sci. Meet., 14th, Washington, D.C.* Pap. No. 76-181 (1976).]

Igra, O. (1976b). *Energy Convers.* **15**(3/4), 143–151. (Short version of Igra, 1974.)

Jorgensen, G. E., Lotker, M., Meier, R. C., and Brierley, D. (1976). *I.E.E.E. Trans. Power Appar. Syst.* **95**(3), 870–878.

Justus, C. G. (1976). Wind energy statistics for large arrays of wind turbines (New England and Central U.S. Regions). *Int. Sol. Energy Soc. Conf., Winnipeg.*

Justus, C. G., and Mikhail, A. (1976). *Geophys. Res. Lett.* **3**(5), 261–264.

Justus, C. G., Hargraves, W. R., and Yalcin, A. (1976). *J. Appl. Meteorol.* **15**(7), 673–678.

Juul, J. (1956). Wind machines. *Proc. New Delhi Symp., Wind Sol. Energy, UNESCO, Paris* pp. 56–73.

Juul, J. (1964). Design of wind power plants in Denmark. *Proc. U.N. Conf. New Sources Energy, Rome, 1961* **7,** 229–240.

Kirschbaum, H. S., Somers, E. V., and Sulzberger, V. T. (1976). *Proc. Am. Power Conf.* **38,** 474–483.

Kolm, K., Marrs, R., Marwitz, J., and Fletcher, J. (1975). "Evaluation of Wind Energy Sites from Aeolian Geomorphologic Features Mapped from Landsat Imagery," ERDA/NSF/00598-75-T1. Univ. of Wyoming, Laramie.

Lacroix, G. (1969). *Tech. Mod.* **40**(5/6), 77–83; **40**(7/8), 105–111.

Lapin, E. E. (1975). Theoretical performance of vertical axis wind turbines. *ASME Pap.* No. 75-WA/Ener-1.

Lapin, E. E. (1976). "Multimegawatt Vertical Axis Windmills—Wing Driven and Rotor Driven," Rep. No. ATR-77(9990)-1. Aerospace Corp., El Segundo, Cal.

LaWand, T. A. (1973). Review of the windpower activities at the Brace Research Institute. *Wind Energy Convers. Syst., Workshop Proc.,* NSF/NASA, NSF-RA-W-73-006, 159–164.

Lissaman, P. B. S. (1977). General performance theory for crosswind axis turbines. *Proc. Int. Symp. Wind Energy Syst., BHRA, Cambridge, Engl., 1976* Pap. C2.

Ljungström, O. (1975). Swedish Wind Energy Program—a three year R & D plan 1975–77. *Proc. Workshop Wind Energy Convers. Syst., 2nd, Washington, D.C.* pp. 140–148.

Ljungström, O. (1977). Large scale wind energy conversion systems (WECS) design and installation as affected by site wind energy characteristics, grouping arrangement and social acceptance. *Proc. Int. Symp. Wind Energy Syst., BHRA, Cambridge, Engl., 1976* Pap. A1.

Loth, J. L. (1975). Wind energy concentrators. *Proc. UMR-MEC Annu. Conf. Energy, 2nd, Univ. Missouri, Rolla* **2,** 92–107.

Loth, J. L. (1977). WVU wind energy concentrators. *Proc. Int. Symp. Wind Energy Syst., BHRA, Cambridge, Engl., 1976* Pap. E2.

Lysen, E. H. (1977). *Ingenieur (The Hague)* **89**(12), 233–239.

MacPherson, R. I. (1972). Design, development and testing of a low head, high efficiency kinetic energy machine. *Annu. Meet. Mar. Technol. Soc., Washington, D.C.*

Manser, B. L., and Jones, C. N. (1975). "Power from Wind and Sea—The Forgotten Panemone," Natl. Conf. Bull. No. 75/9. Univ. of Queensland, Brisbane.

Maughmer, M. D. (1976). "Optimization and Characteristics of a Sailwing Windmill Rotor," Rep. No. AMS 1297 (NSF/RA 760201). Princeton Univ., Princeton, New Jersey.

Merriam, M. (1977–1978). *Wind Power Dig.* No. 11, Winter.

Metz, W. D. (1978). *Science* **200,** May, 636.

Molly, J. P. (1977). Balancing power supply from wind energy converting systems. *Proc. Int. Symp. Wind Energy Syst., BHRA, Cambridge, Engl., 1976* Pap. F1.

Morrison, J. A. (1964). The testing of a wind power plant. *Proc. U.N. Conf. New Sources Energy, Rome, 1961* **7,** 294–304.

Muraca, R. J., and Guillotte, R. J. (1976). Wind tunnel investigation of a 14 foot vertical axis windmill. *NASA Tech. Memo.* **NASA TM X-72663.**

Muraca, R. J., Stephens, M. V., and Dagenhart, J. R. (1975). Theoretical performance of cross-wind axis turbines with results for a catenary vertical axis configuration. *NASA Tech. Memo.* **NASA TM X-72662.**

Musgrove, P. J. (1977a). The variable geometry vertical axis windmill. *Proc. Int. Symp. Wind Energy Syst., BHRA, Cambridge, Engl., 1975* Pap. C7.

Musgrove, P. J. (1977b). *Wind Power Dig.* No. 8, Spring, pp. 27–29.

New York Times (1978). New Mexico town christens its windmill. Jan. 30th, p. A12.

Oman, R. A. (1977). Fluid dynamics of diffuser augmented wind turbines. *Proc. I.E.C.E.C., 12th, Washington, D.C.* Pap. No. 779274.

Oman, R. A., Foreman, K. M., and Gilbert, B. L. (1977). "Investigation of Diffuser-Augmented Wind Turbines," Res. Rep. RE-534. Grumman Aerospace Corp.

Pope, A., and Harper, J. J. (1966). "Low Speed Wind Tunnel Testing." Wiley, New York.

Powe, R. E. (1977). A wind energy conversion system based on the tracked vehicle airfoil concept. *Proc. Int. Symp. Wind Energy Syst., BHRA, Cambridge, Engl., 1976* Pap. B3.

Powe, R. E., Townes, H. W., Bishop, E. H., and Blackketter, D. O. (1974). Technical feasibility study of a wind energy conversion system based on the tracked-vehicle airfoil concept. *Proc. I.E.C.E.C., 9th, San Francisco, Calif.* Pap. No. 749010.

Proc. U.N. Conf. New Sources Energy (1964). *Rome, 1961* Vol. 7, Wind Power.

Puthoff, R. L. (1976). Fabrication and assembly of the ERDA/NASA 100 kilowatt experimental wind turbine. *NASA Tech. Memo.* **NASA TM X-3390.**

Putnam, P. C. (1948). "Power from the Wind." Van Nostrand, New York.

Ramakumar, R. (1976). Wind driven field modulated generator systems. *Proc. I.E.C.E.C., 11th, State Line, Nev.* Pap. No. 76.

Ramakumar, R. (1977). A review of wind-electric conversion technology. *Proc. Int. Conf. Alt. Energy Sources, Miami, Fla.*

Ramakumar, R. (1978). Wind-electric conversion utilizers field modulated generator systems. *Solar Energy* **20,** 109–117.

Ramakumar, R., and Hughes, W. L. (1975). Electrical technology overview and research at Oklahoma State University as applied to wind energy system. *Proc. Workshop Wind Energy Convers. Syst., 2nd, Mitre Corp., Washington, D.C.* ■■.

Reed, J. W. (1976). "Predicting Wind Power at Turbine Level from an Anemometer Record at Arbitrary Height," SAND-76-5397. Sandia Lab., Albuquerque, New Mexico.

Reitan, D. K. (1977). A progress report on employing a non-synchronous AC/DC/AC link in a wind-power application. *Proc. Workshop Wind Energy Convers. Syst., 2nd, Mitre Corp., Washington, D.C., 1975,* 290–297.

Reuter, R., Jr., and Sheldahl, R. E. (1977). "Sandia Vertical-Axis Wind Turbine Project," Tech. Q. Rep., Apr.–June 1976. Sandia Lab., Albuquerque, New Mexico.

Rogers, S. E., Duffy, M. A., Jefferis, J. A., Sticksel, P. R., and Tolle, D. A. (1976). "Evaluation of the Potential Environmental Effects of Wind Energy System Development," ERDA/NSF/07378-75/1. Battelle Columbus Lab., Columbus, Ohio.

Rohrbach, C. (1976). "Experimental and Analytical Research on the Aerodynamics of Wind Turbines," Mid-term Tech. Rep., June 1–Dec. 31, 1975, COO-2615-76-T-1. United Technology Corp., Windsor Locks, Conn.

Rohrbach, C., and Worobel, R. (1975). Performance characteristics of aerodynamically optimum turbines for wind energy generators. *Annu. Forum, 31st, Am. Helicopter Soc., Washington, D.C.* S-996.

Rosenbrock, H. H. (1952). *Aircr. Eng.* **24**(281), 226–227.

Savino, J. M., ed. (1973). *Wind Energy Convers. Syst.; Workshop Proc.* PB 231341, NSF-RA-W-73-006.

Savino, J. M., and Wagner, L. H. (1976). Wind tunnel measurements of the tower shadow on models of the ERDA NASA 100 kW wind turbine tower. *NASA Tech. Memo.* **NASA TM X-73548**.

Savonius, S. J. (1931). *Mech. Eng.* **53**(5), 333–338.

Senior, T. B. A., Sengupta, D. L., and Ferris, J. E. (1977). "TV and FM Interference by Windmills," Final Rep., Jan. 1–Dec. 31, 1976. Univ. of Michigan, Ann Arbor.

Sforza, P. M. (1977). Vortex augmentors for wind energy conversion. *Proc. Int. Symp. Wind Energy Syst., BHRA, Cambridge, Engl., 1976* Pap. E1.

Shankar, P. N. (1976). *Proc. R. Soc. London, Ser. A* **349,** 35–51.

Sheldahl, R. E., and Blackwell, B. F. (1977). "Free-Air Performance Tests of a 5-Meter-Diameter Darrieus Turbine," SAND 77-1063. Sandia Lab., Albuquerque, New Mexico.

Shepherdson, W. (1977). *Wind Power Dig.* No. 10, Fall, pp. 6–11.

Smeaton, J. (1760). On the construction and effects of windmill sails. *In* "An Experimental Enquiry Concerning the Natural Powers of Water and Wind." [*Philos. Trans. R. Soc. London* **51,** Part 1, 138–174 (1759).]

Smith, M. C. (1977). Wind site selection for optimum wind power systems. *Proc. Int. Conf. Alt. Energy Sources, Miami, Fla.*

Smith, R. T., and Jayadev, J. S. (1975). Electrical generation by wind power. *Proc. I.E.C.E.C., 10th, Newark, Del.* Pap. No. 759184.

Smith, R. T., Swanson, R., Johnson, C. C., Ligon, C., Lawrence, J., and Jordan, P. (1976). Operational, cost and technical study of large windpower systems integrated with existing public utilities. *Proc. I.E.C.E.C., 11th, State Line, Nev.* Pap. No. 769302.

Sørensen, B. (1977). Direct and indirect economics of wind energy systems relative to fuel based systems. *Proc. Int. Symp. Wind Energy Syst., BHRA, Cambridge, Engl., 1976* Pap. D2.

South, P., and Rangi, R. S. (1972). "A Wind Tunnel Investigation of a 14 Foot Diameter Vertical-Axis Windmill," NRC, NAE LTR-LA-105. Natl. Res. Counc. Can., Ottawa.

South, P., and Rangi, R. S. (1975). "An Experimental Investigation of a 12 Foot Diameter High Speed Vertical-Axis Wind Turbine," NRC, NAE LTR-LA-166. Natl. Res. Counc. Can., Ottawa.

Spera, D. A. (1975). Structural analysis of wind turbine rotors for NSF-NASA wind power system. *NASA Tech. Memo.* **NASA TM X-3198.**

Spera, D. A., and Janetzke, D. C. (1977). *Wind Tech. J.* **1**(2), 5–10.

Stoddard, F. S. (1977). *Wind Tech.* **1**(1), 3–9.

Strickland, J. H. (1975). "The Darrieus Turbine: A Performance Prediction Model Using Multiple Streamtubes," SAND 75-0431. Sandia Lab., Albuquerque, New Mexico.

Strickland, J. H. (1977). A performance prediction model for the Darrieus turbine. *Proc. Int. Symp. Wind Energy Syst., BHRA, Cambridge, Engl., 1976* Pap. C3.

Sumner, J. (1975). *Engineer* **241**(6248/9), 24–25.

Templin, R. J. (1974). "Aerodynamic Performance Theory for the NRC Vertical-Axis Wind Turbine," Rep. No. LTR-LA-160, N.A.E.

Templin, R. J., and South, P. (1976). Canadian Wind Energy Program. *Proc. Vert.-Axis Wind Turbine Technol. Workshop, Sandia Lab.* SAND 76-5586, I-57–I-85).

Thomas, P. H. (1945). "Electric Power from the Wind." Fed. Power Comm., Washington, D.C.

Thomas, P. H. (1946). "Final Report on the Wind Turbine." PB 25370. War Prod. Rec., Washington, D.C.

Thomas, P. H. (1949). "Aerodynamics of the Wind Turbine." Fed. Power Comm., Washington, D.C.

Thomas, R. L. (1976). Large experimental wind turbines—where we are now. *Proc. Vert.-Axis Wind Turbine Technol. Workshop, Sandia Lab.* SAND 76-5586, I-9–I-32.

Todd, C. J., Eddy, R. L., James, R. C., and Howell, W. E. (1977). "Cost-Effective Electric Power Generation from the Wind." Bur. Reclam., U.S. Dep. Inter., Denver, Colorado.

Turnquist, R. O., and Appl. F. C. (1976). Design and testing of a prototype Savonius wind machine. *Front. Power Technol. Conf., Oklahoma State Univ., Stillwater,* 15.1–15.13.

Vadot, L. (1957). *Houille Blanche,* No. 2, Mar.–Apr., pp. 189–212.

van Bussel, G. J. W. (1977). Windmills with increased power output due to tipvanes. *Proc. Int. Conf. Alt. Energy Sources, Miami, Fla.*

van Holten, T. (1974a). "Performance Analysis of a Windmill with Increased Power Output due to Tipvane Induced Diffusion of the Airstream." Memo M-224. Tech. Hoge., Delft.

van Holten, T. (1974b). "Asymptotic Expressions for the Velocity Field due to a Cylindrical Ring Distribution of Vorticity," Memo M-223. Tech. Hoge., Delft.

van Holten, T. (1977). Windmills with diffuser effect induced by small tipvanes. *Proc. Int. Symp. Wind Energy Syst., BHRA, Cambridge, Engl., 1976* Pap. E3.

VAWT (Vert.-Axis Wind Turbine) Technol. Workshop (1976). ERDA/Sandia Lab., SAND 76-5586.

Vezzani, R. (1950). *Elettrotecnica* **37**(9), 398–419.

Walters, R. E. (1975a). "Innovative Wind Machines," Six Month Rep., Mar. 1–Aug. 31, 1975 (ERDA/NSF/00367-75/T1). West Virginia Univ., Morgantown.

Walters, R. E. (1975b). Innovative vertical axis wind machines. *Proc. Workshop Wind Energy Convers. Syst., 2nd, Mitre Corp. Washington, D.C.* NSF-RA-N-75-050, 443–451.

Weingarten, L. I., and Blackwell, B. F. (1976). "Sandia Vertical-Axis Wind Turbine Program," Tech. Q. Rep., Jan.–Mar. 1976. Sandia Lab., Albuquerque, New Mexico.

Weisbrich, A. L. (1977). Feature review of some advanced and innovative design concepts in wind energy conversion systems. *Proc. Int. Conf. Alt. Energy Sources, Miami, Fla.*

Wentink, T., Jr. (1976). "Study of Alaskan Wind Power and Its Possible Applications," Final Rep., NSF/RANN/SE/AER/-74-00239/FR-76/1. Univ. of Alaska, Fairbanks.

Widger, W. K., Jr. (1976). *Power Eng.* Aug., pp. 58–61.

Willem, R. M. (1977). *Wind Technol. J.* **1**(2), 11–16.

Wilson, R. E., and Lissaman, P. B. S. (1974). "Applied Aerodynamics of Wind Power Machines," PB 238595, Rep. No. NSF-RA-N-74-113. NTIS, Springfield, Virginia.

Wilson, R. E., Lissaman, P. B. S., and Walker, S. N. (1976). "Aerodynamic Performance of Wind Turbines" Oregon State Univ., Corvallis.
Wind Power Dig. (1977). No. 9, Summer, pp. 6–10.
Yang, R. A., Tiedeman, A. F., Jr., Marianowki, A. L., and Camara, E. H. (1975). "Production of Methane Using Offshore Wind Energy," ERDA/NSF/993-75/T1. Inst. Gas Technol., Chicago, Illinois.
Yee, S. T., Chang, T. Y. P., Scavuzzo, R. J., Timmerman, D. H., and Fenton, J. W. (1977). "Vibration Characteristics of a Large Wind Turbine Tower on Non-Rigid Foundations," Rep. ERDA/NASA/1004-77/1. NASA, Washington, D.C.
Yen, J. T. (1977). Tornado-type wind energy system: Basic consideration. *Proc. Int. Symp. Wind Energy Syst., BHRA, Cambridge, Engl., 1976* Pap. E4. [Also as, *ASME Annu. Meet., ASME Pap.* No. 76-WA/Ener-1 (1976).]

Note added in proof: The *Proc. Biennial Conf. Workshop WEES*, 1978, USDOE, Washington, D.C. (Vol. 2, 1978) updates the text but became available after going to press.

Fuels from Biomass

COMPARATIVE STUDY OF THE POTENTIAL IN FIVE COUNTRIES: THE UNITED STATES, BRAZIL, INDIA, SUDAN, AND SWEDEN

Walter Vergara

Department of Agricultural Engineering
Cornell University
Ithaca, New York

David Pimentel

College of Agriculture and Life Sciences
Cornell University
Ithaca, New York

ISBN 0-12-014901-X

I. INTRODUCTION

The rapid depletion of fossil energy supplies, the continued increase in per capita world energy consumption, and rapid growth in the world population emphasize the need to find and develop alternate energy sources. Clearly no one source of energy will supply all world requirements, as natural resources and need vary drastically from one country to another. Several technologies, including biological solar conversion, wind, solar, hydropower, geothermal, and nuclear technologies must be evaluated and the potential of each one developed where technologically appropriate.

In considering all possible energy sources, biomass energy conversion has been suggested as a substantial energy source (Hayes, 1977a; Waterman and Klass, 1977). Today, use of biomass as a source of energy varies and depends in part on the level of development of the countries. In the United States, for example, biomass conversion amounts to about 1% of the United States energy supply (Pimentel *et al.,* 1978), while in Sudan it accounts for as much as 65%. In the United States the energy derived from biomass accounted for about 91% of the energy supply in 1850 (Anonymous, 1977). Of course, in 1850 the United States population was only 23 million and the per capita consumption only one-fourth of the current consumption. Today in many of the underdeveloped countries with low per capita energy consumption, biomass provides most of the basic energy needs, mostly as fuelwood, animal wastes, or crop residues (Makhijani and Poole, 1975).

The main processes by which energy may be obtained from biomass include direct combustion, pyrolysis, hydrogasification, anaerobic digestion, alcoholic fermentation, and biophotolysis. Each technology has advantages and these depend on the biomass source and the type of energy needed.

The major difficulties in utilizing biomass materials for solar energy conversion are: (1) the relatively small percentage (less than 0.1%) of light energy converted into biomass by plants; (2) the relatively sparse and low concentration of biomass per unit area of land and water; (3) the scarcity of additional land suitable for growing plants; and (4) the high moisture content (50–95%) of biomass that makes collection and transport expensive and energy conversion relatively inefficient. All of these factors make biomass energy costly in terms of energy expended in the conversion process and reduce the net energy yield.

Among the advantages of utilizing biomass materials as an energy source are: (1) biomass provides an effective low-sulfur fuel; (2) it provides in some situations an inexpensive source of energy (e.g., fuelwood, dry animal manure, methane gas, alcohol); (3) in some cases processing biomass materials for fuel reduces the environmental hazards from these materials (e.g., biomass from sewage and processing wastes); and (4) the production of biological fuels may be coupled to the synthesis of protein.

This report considers the feasibility of using biomass as an energy source. The analysis assesses the potential for biomass energy conversion for the United States, Brazil, India, Sudan, and Sweden, which represent five different environmental and social conditions. The following aspects of biomass energy conversion are investigated: (1) the potential of biological materials for biomass energy conversion; (2) the role of both agriculture and forestry in solar energy conversion; (3) the potential for biomass energy conversion utilizing combustion, ethanol fermentation, anaerobic digestion, and pyrolysis; and (4) the environmental and social constraints in biological solar energy conversion.

II. COUNTRY STUDIES OF BIOMASS SOURCES AND USES

An analysis is made of the environmental and biomass resources in the United States, Brazil, India, Sudan, and Sweden. In addition, an assessment is made of current biomass energy and fossil energy use in each of these countries.

A. Study Cases

1. Brazil

Brazil is the fourth largest country in the world and is the largest country crossed by the equator. It has a total area of 851 × 10^6 ha. Brazil has a population of 100 million and is increasing at a rate of 3.0% per year (Table I). Still, the population density in Brazil is only 13 persons per square kilometer.

Forest and woods cover over 60.6% of Brazil (Table II). The Amazonas rain forest in Brazil alone covers 500 × 10^6 ha (Volatron, 1976). Arable lands and permanent crops occupy 4% of the total land area but more than twice this amount is estimated to be suitable for cropping (Goldemberg, 1977). Brazil is a net exporter of agricultural products. The main crops include coffee, sugar, tobacco, citrus fruit, bananas, and rice. It is also an important livestock producer. About 46% of the Brazilian population is engaged in agricultural activities.

Brazilian national wealth as measured by the per capita national income

TABLE I

Population, Area, and National Wealth of the United States, Brazil, India, Sudan, and Sweden

Country	Estimated Population in 1976[a] (10^6)	Annual Rate of Increase[a] (%)	Surface Area[a] (10^6 km^2)	Density (habitants/km^2)	Unemployment[a] (10^3 habitants)	National Income[b] (dollars per capita)	Gross National Product[b] (dollars per capita)
United States	213.0	0.8	10.49	20	7830	5923	6200
Brazil	110.0	3.0	8.51	13	968	1190	760
India	613.0	2.1	3.28	182	8918	136	120
Sudan	18.0	2.5	2.50	7	4	143	130
Sweden	8.2	0.4	0.45	18	37	6153	5910

[a] United Nations (1976).
[b] World Bank (1976).

TABLE II

Land Distribution by Uses and Population Engaged in Agriculture in the United States, Brazil, India, Sudan, and Sweden[a]

Country	Total Area (10^6 ha)	Cropland (10^6 ha)	Pasture (10^6 ha)	Forests and Woods (10^6 ha)	Other Land (10^6 ha)	Percentage of Population Engaged in Agriculture
United States	1049	192	300	247	310	4.7
Brazil	851	36	167	511	137	45.6
India	328	167	13	65	83	69.3
Sudan	250	7	24	91	128	82.0
Sweden	45	3	0.5	26	15.5	8.3

[a] United Nations (1976).

(\$1190) and gross national product (\$760) (Table I) is relatively good (World Bank, 1976). The annual compounded growth rate between 1960 and 1970 was 6%. This accelerated growth has resulted in a rapid movement of people from the rural area into urban centers (56% of the population now lives in cities as compared with 36% in 1960 (Serra, 1971). Both urbanization and increase in the gross national product have increased energy consumption in Brazil. Estimated energy consumption is 5.8×10^6 kcal per capita (Tables III and IV.) Energy consumption in Brazil is supplied by coal (3.6%), hydroelectricity (20.8%), petroleum (44.8%), and

TABLE III

Consumption of Commercial Energy in Selected Countries (10^{12} kcal)

Country	Solid Fuels[a]	Liquid Fuels[a]	Natural Gas[a]	Hydroelectric and Nuclear[a]	Total	Per Capita (10^6 kcal)
United States	3588	7720	5643	383	17,330	81.3
Brazil	30	391	5	58	484	4.4
India	609	197	7	27	840	1.4
Sudan	—	15	—	1	16	0.8
Sweden	16	278	—	55	349	42.5

[a] United Nations (1976).

TABLE IV

Actual Estimated Biomass for Energy Conversion (10^6 metric tons)

Country	Fuelwood[a]	Animal Wastes	Crop Wastes	Total Energy Produced (10^{12} kcal)
United States	35.2[f] (131.0)[g]	Negligible	4[b] (16)[g]	147
Brazil	50 (187)[g]	Negligible	5[c] (20)[g]	207
India	60[d] (224)[g]	50[d] (210)[g]	30[d] (126)[g]	560
Sudan	7.5 (28.0)[g]	na[e]	na[g]	28
Sweden	1.0 (3.7)[g]	na[e]	na[g]	3.7

[a] United Nations (1976).
[b] Estimated, mostly sugar cane bagasse.
[c] Derived from data by Azevedo (1976).
[d] Makhijani and Poole (1975).
[e] Not available.
[f] Includes forestry remains used as fuelwood in the lumber industry (Pimentel *et al.*, 1978).
[g] Numbers in parentheses indicate the energy equivalent of the biomass assuming direct incineration and a heat of combustion of 4200 kcal/kg.

biomass (30.5%) (Azevedo, 1976). Although petroleum is the most important energy source in the country, 80% of it has to be imported at an annual cost of $3 billion (Hammond, 1977). Brazilian production of energy amounts to 390 × 10^{12} kcal (3.54 × 10^6 kcal/habitant), biomass being the most important fraction and representing 53% of the total (Table V). Crude petroleum and hydroelectrical energy account for most of the remaining fraction (40%).

The gross national product, national income, and energy resources are not shared by all the population. A disproportionately large amount of commercial energy is used by the relatively well-to-do. For example, only 2% of the electricity generated is consumed in the rural areas where 44% of the population lives (Makhijani and Poole, 1975), and 80% of the population has a lower than average income.

2. *India*

India occupies a total area of 3.28 × 10^6 km². Its current population is about 613 million and is growing at a rate of 2.5% per year (Table I).

TABLE V

Production of Energy (10^{12} kcal)[a]

Country	Coal and Lignite	Crude Petroleum	Natural Gas	Hydro-nuclear	Biomass[b]	Total per Capita (10^6 kcal)
United States	4138.05	5022.60	5094.55	423.57	147	69.1
Brazil	19.59	92.16	7.57	63.82	207	3.5
India	697.80	87.75	8.76	31.80	560	2.3
Sudan	—	—	—	0.096	28	1.6
Sweden	0.08	—	—	61.73	3.7	8.0

[a] United Nations (1976).
[b] Table IV.

India's population density ranks among the highest in the world (182 persons per square kilometer) (Table I).

About 165 million hectares or 50.3% of the country is cultivated land. Forests cover 19.8% of the territory and pastures about 4% (Table II). The main crops include rice, wheat, cotton, tea, and coffee (Food and Agriculture Organization). About 72% of the crop land is irrigated. India's agricultural output during the 1950s and 1960s grew at an annual rate of 4.3%. Production doubled from 51 $\times$ 10^6 tons in 1950 to about 100 $\times$ 10^6 tons in 1970 while population increased by 50% (Makhijani and Poole, 1975). Since then, the trend of rising per capita production has declined (Table VI). From 1970 to 1975 food production increased by 10.5%

TABLE VI

Index of Agricultural Production[a]

Country	Total Production	1961–1965	1970	1975
United States	Food	100	114	136
	All agriculture	100	109	128
Brazil	Food	100	128	151
	All agriculture	100	128	148
India	Food	100	123	136
	All agriculture	100	121	134
Sudan	Food	100	137	175
	All agriculture	100	139	173
Sweden	Food	100	104	113
	All agriculture	100	104	113

[a] United Nations (1976).

(United Nations, 1976) while population has increased by 10.9%. Some of the reasons for the decline are a shortage of fertilizers and insufficient oil and electricity to run the irrigation pumps (Makhijani and Poole, 1975). Consumption of commercial energy amounts to 840×10^{12} kcal or 1.4×10^6 kcal per capita (United Nations, 1976). The most important commercial fuel is coal, which represents 72% of the national consumption (Table III).

Biomass is an important source of fuel in India. About 120 million metric tons of fuelwood, 50 million dry metric tons of dung, and 30 million metric tons of crop wastes are used each year as fuel (Makhijani and Poole, 1975). There is increasing evidence that the demand for the use of India's forest resources for energy is reaching the maximum potential of renovation (Thappar, 1975). From all sources, energy from biomass represents about 68% of the total energy production in India and 40% of the energy consumed.

3. *Sudan*

The Sudan has a total area of 2.5×10^6 km^2 and a population of 17 million. Despite its high rate of population growth (2.5%) it has a low population density (18 persons per square kilometer) (Table I). Two-thirds of the country is either desert, semidesert, or swamp. Nevertheless, its total fertile land is about 84×10^6 ha[1] (Ministry of Agriculture, Food and Natural Resources of the Sudan, 1974), of which only 8.5% is under cultivation.

About 82% of the population of the Sudan is engaged in agricultural activities, either farming or animal husbandry. The per capita national income ($143) and the gross national product ($130) stand among the lowest in the world (Table I). Agriculture accounts for 98% of the Sudanese foreign exchange and provides the raw materials for 90% of the local industry. Main agricultural products include sorghum, millet, cotton, oilseeds, nuts, sugar cane, wheat, cassava, and others (Food and Agriculture Organization, 1976). The Sudan has a climate and soils favorable to the growing of sugar cane (Ministry of Agriculture, Food and Natural Resources of the Sudan, 1974). There are about 24×10^6 ha of pasture in the Sudan. The majority of the cattle are concentrated in the southern and western regions. Per capita consumption of fossil energy is low (0.8×10^6 kcal) and accounts for only 35% of the total energy used. Most of the energy produced in Sudan comes from biomass. Fuelwood is the most important energy resource and this provides 99.7% of all energy

[1] This estimate includes forests, pastures, and so forth.

produced in the country (Table V). Electricity production represents 5% of the total energy use (Royal Swedish Academy of Sciences, 1975) and is heavily concentrated in urban areas.

4. *Sweden*

Total area of Sweden is about 0.45×10^6 km². Solar radiation is of low intensity, about 300 W/m², and the annual distribution over time is irregular (Lönnroth *et al.*, 1977). Forest and woods occupy half of the area of Sweden. Cropland is mainly concentrated in the south and accounts only for 6.6% of the total. Chief agricultural products are crops such as barley, oats, rye, hay potatoes, sugar beets, and wheat. About 8% of the population is engaged in agriculture.

Per capita national income ($6153) and gross national product ($5910) are high, while the rate of growth of the population is only about 0.4% (Table I). Main sources of energy produced in Sweden are hydroelectricity, a variant of solar energy conversion, coal, and biomass, primarily fuelwood. Hydropower supplies about 80% of the electrical energy consumption and 15% of the total energy use.

Sweden imports 81% of its energy, mostly as oil. Energy reserves in the country include uranium (about 25 years supply of energy) and peat (about 100 years of supply).

5. *United States*

The United States occupies an area of 1049×10^6 ha; its population is about 213 million and is growing at an annual rate of 0.8%.

The per capita national income ($5933) and per capita gross national product ($6200) make the United States the wealthiest industrialized country in the world. The United States has 247×10^6 ha of forests and 192×10^6 ha used for crop production. Only a small percentage (5%) of the population is engaged in agricultural activities. Food and agricultural product yields in the United States are among the highest in the world (Food and Agriculture Organization, 1976). On a composite basis the United States is the most productive country (Food and Agriculture Organization, 1976). It is also the largest consumer of energy. The United States alone consumes 34% of the world supply of commercial energy with only 5% of the world population.

Consumption of commercial energy is supplied by liquid fuel, (44.5%), natural gas (32.6%), coal (20.7%), and hydroelectricity and nuclear energy (2.2%) (Table III). Biomass supplies 1.0% of the total consumption of energy (Pimentel *et al.*, 1978).

The United States produces 85% of its total energy demand. Biomass accounts for 1% of the production.

B. Biomass Conversion

The total solar energy fixed by plants has been calculated for the United States, Brazil, India, Sudan, and Sweden (Table VII). The calculated energy fixed was made assuming a productivity independent of the latitude of the country. Solar energy availability, however, does vary with latitude. Plant productivity is increased by an increase in solar energy if all other factors remain constant (Ramirez, 1970). All other factors do not remain constant; therefore, for this analysis the calculated energy fixed per unit plant biomass was assumed to be independent of latitude.

The total amount of biomass produced annually was estimated for each country based on geographic characteristics. The energy equivalent produced per kilogram of biomass was calculated assuming an energy equivalent of 4200 kcal. The sunlight energy fixed by plants in the United States is therefore 13.5×10^{15} kcal; 0.56×10^{15} kcal for Sweden; 5.9×10^{15} kcal for India; 12.7×10^{15} kcal for Brazil; and 2.8×10^{15} kcal for Sudan. The solar energy fixed by plants represents 75% of the commercial energy consumption of the United States, 160% of the Swedish consumption, 2926% of the Brazilian consumption, 702% of the Indian consumption, and 17,467% of the Sudanese consumption. The potential of the biomass produced in each country to supply the country's energy needs varies as a function of the local energy consumption and environmental conditions. Biomass already supplies 65.2% of the energy consumed in Sudan (Tables III and IV) and only 1% of the energy used in the United States. If we take into account the total energy consumed in each country including biomass, the sunlight energy fixed by plants would be equivalent to 75% of the energy consumption for the United States, 160% for Sweden, 848% for Brazil, 421% for India, and 6422% for Sudan.

1. Agricultural and Forestry Production

Biological solar energy conversion is important in each country as measured by the total production of agricultural and forestry products. This energy is included in products of food, fiber, feed, lumber, and pulpwood. An estimate of the energy harvested in agricultural and forestry products in the United States, Brazil, India, Sudan, and Sweden is shown in Table VIII. The ratio of the energy harvested as agricultural products and forestry to the total sunlight fixed gives an idea of the maximum potential for biomass use. The percentages are 50% for the United States, 22% for Sweden, 25% for Brazil, 25% for India, and 17% for Sudan (Table VIII).

TABLE VII

Annual Biomass Production (10^6 metric tons) in the United States, Brazil, India, Sudan, and Sweden[a]

	United States		Sweden		India		Brazil		Sudan	
	Land Area (10^6 ha)	Biomass Production	Land Area (10^6 ha)	Biomass Production	Land Area (10^6 ha)	Biomass Production	Land Area (10^6 ha)	Biomass Production	Land Area (10^6 ha)	Biomass Production
Arable land and production crops	192	1083	3	18	165	990	36	216	7	42
Pasture and grazing land	300	783	0.5	1.5	13	45	167	584	24	84
Forests	247	988	22.5	90	655	262	511	2044	9.5	366
Other	310	163	18.5	9	84	42	137	68	127	63
Total area	1049	—	45	—	328	—	851	—	250	—
Total biomass	—	3017	—	118.5	—	1339	—	2912	—	555
Total energy fixed (10^{15} kcal)	12.7		0.5		5.6		12.2		2.3	
Energy per capita (10^6 kcal)	59.6		59.7		9.1		111		128	

[a] Derived from data presented by the Food and Agriculture Organization (1976).

TABLE VIII Total Annual Amount of Solar Energy Harvested in the Form of Agricultural Crops and Forestry Products (dry)

	United States		Brazil		India		Sudan		Sweden	
	10^6 Metric Tons[a]	10^{12} Kcal[b]	10^6 Metric Tons[a]	10^{12} Kcal[b]	10^6 Metric Tons[a]	10^{12} Kcal[b]	10^6 Metric Tons[a]	10^{12} Kcal[b]	10^6 Metric Tons[a]	10^{12} Kcal[b]
Corn	117	491	14.8	62.1	5.0	1.2	0.02	0.2	—	—
Wheat	45	189	1.3	5.7	21.8	91.5	0.24	0.02	1.3	5.55
Rice	5	21	7.0	29.1	63.5	266.5	0.01	0.04	—	—
Soybeans	35	147	9.2	38.5	0.1	0.4	—	—	—	—
Sorghum	20	84	0.5	1.9	9.5	39.7	2.10	8.6	—	—
Potatoes	3	13	0.3	1.3	1.2	5.2	.01	0.02	0.2	0.8
Cassava	—	—	8.2	34.3	1.9	8.0	0.23	1.40	—	—
Vegetables	2	8	0.1	0.5	0.9	4.0	0.03	0.12	0.02	0.07
Fruits	4	17	0.5	6.4	1.3	5.7	0.04	0.15	0.01	0.01
Nuts	0.4	2	2.1	8.8	7.0	29.4	1.1	4.6	—	—
Oil seeds	5	21	4.1	17.2	6.8	28.6	0.94	3.9	0.33	1.38
Sugar cane	—	—	24.3	101.9	35.6	158.0	0.38	1.6	—	—
Sugar beets	6	26	—	—	—	—	—	—	0.53	2.23
Pulses	—	—	3.0	12.6	12.9	54.2	—	—	—	—
Oats	0.3	1	—	—	—	—	—	—	1.1	4.62
Rye	6.9	29	—	—	—	—	—	—	.35	1.5
Barley	7	29	—	—	—	—	—	—	1.55	6.51
Pasture and others	1072.8	3506	584	2262.6	45	137.5	84	352.8	1.5	6.3
Forested[c]	103.5	565	59.0	322.7	69.0	377.4	8.0	43.7	18.8	102.8
	1559.4	5148	718.4	2905.6	281.5	1207.3	97.1	418.0	25.7	131.8
Total energy per capita (10^6 kcal)	24.1		26.4		2.0		23.2		16.1	

[a] From data presented by the Food and Agriculture Organization (1976).

[b] For food and feed crops food energy values were employed in making the calculations. This makes these estimates lower than if they were calculated as potential heat of combustion.

[c] See Table IX.

III. SOURCES OF BIOMASS AVAILABLE FOR ENERGY CONVERSION

A. Livestock Manures

Livestock manures are a prime source for energy conversion. Manures are produced by cattle, sheep, pigs, horses, camels, goats, poultry, and others. Once dried, manure may be burned and provides thermal energy. When submitted to anaerobic digestion, the methane gas produced is a gas fuel. If the livestock are pastured, as are cattle, sheep, and hogs, the manure is dispersed in fields. To use the manure then, it has to be collected. This is the situation with most livestock in most developing countries.

In India, for example, most cow manure is being utilized. The manure is applied to fields as a source of nutrients for crop production, is burned for cooking, or is used as a substrate for anaerobic digestion (Singh, 1971; Makhijani and Poole, 1975). Small anaerobic digestors may use the manure produced by 3–4 animals to supply the energy needs of a small rural family.

In large livestock feedlots, huge quantities of manure are produced (Loehr, 1975). The use of anaerobic digestion in feedlots may reduce pollution problems and produce an energy source. Livestock manures keep their fertilizer potential when processed by anaerobic digestion (Jewell *et al.*, 1977).

Not all the manure produced is readily available for biomass conversion. If the livestock is dispersed it may be impractical to collect the manure. The estimated amount of manures produced in each of five countries is given in Table XI.

B. Crop Residues

Crop residues have also been considered for biomass energy conversion (Clausen *et al.*, 1977). Crop residues range from none, as in alfalfa, to about 7 tons/ha in corn. Only corn and rice exceed the crop average of about 4 tons of residue per hectare (Environmental Protection Agency, 1973).

A wide range of proposals exist concerning the utilization of crop remains for biomass energy conversion and environmental conservation. Our analysis of the agricultural, environmental, and energetic aspects of the use of crop remains suggests that little or none of these remains should be used for biomass energy conversion. The evidence suggests that crop remains left on the land function to prevent sediment runoff, conserve soil and water, maintain soil carbon ratios and soil structure, and prevent nutrient (N, P, K, etc.) loss.

For example, an estimated 3×10^9 metric tons of soil are washed from United States agricultural land annually into streams, reservoirs, and lakes (Pimentel *et al.*, 1976). In addition, these sediments are considered serious environmental pollutants (Wadleigh, 1968). Unfortunately, the removal of crop remains would increase the soil sediment runoff problem.

Soil erosion rates in corn, cotton, and other row crops average about 45 tons/ha per year (Pimentel *et al.*, 1976), in the United States. Soil erosion is equally serious in other countries (National Academy of Sciences, 1977). With soil being replaced at the rate of only 3.3 tons/ha per year, current erosion rates already are resulting in rapid degradation of the soil (McCracken, 1977). For each 2.5 cm of soil lost, crop yields for such crops as corn, wheat, oats, and soybeans are significantly reduced (Pimentel *et al.*, 1976). Crop remains hold water and soil and therefore are essential to their conservation. [The necessity for crop remains to be left on the land depends upon such factors as slope, crop, soil quality, rainfall, and wind patterns. Crop residue requirements for corn were estimated to range from 4 to 6 tons/ha per year (Lucas, 1977; Cook, 1962).]

Crop residues directly contribute to the maintenance of carbon ratios in the soil, which should be at about a 1.5% level, or equivalent to a soil organic matter content of 2.6% (Lucas, 1977). Another important factor related to retaining crop residues on cropland is soil structure. Crop residues also significantly reduce water runoff rates and increase infiltration rates (Schrader, 1977).

Continual removal of crop remains seriously depletes the nutrients (N, P, and K) that are essential in crop production. For the nine leading crops, Larson *et al.* (1977) calculated that the residues contained about 4×10^6 metric tons of N, 0.5 metric tons of P, and 4×10^6 metric tons of K, or about 40%, 10%, and 80% of N, P, and K, respectively, of current fertilizer applications to *all* crops. Another illustration of the costliness of such a loss is the nitrogen in Iowa corn residues, which amounts to 123 kg/ha (Larson *et al.*, 1977). Just to replace this nitrogen would require a high energy expenditure of nearly 2×10^6 kcal/ha per year, or an equivalent of more than 200 liters of petroleum fuel per hectare.

All available data suggest that, in general, crop remains should not be harvested for biomass energy conversion. Their beneficial roles in agriculture far outweigh their direct energy potential.

C. Fuelwood

Fuelwood represents an important source of energy in many developing countries. In India fuelwood accounts for 16% of all energy consumed

(United Nations, 1976). For Brazil and Sudan the figure is about 26% (Azevedo, 1976) and 65%, respectively. In fact, most of the energy needs of the rural populations in those countries are being met by fuelwood (Openshaw, 1974; Pimentel, 1978). Fuelwood is also an important source of energy for several of the industries in these countries, such as brick and ceramic manufacture, food processing, rubber production, and even steel manufacture (Openshaw, 1974; Azevedo, 1976). Wood can be used for cooking and space heating, and it can also fuel boilers to produce electricity and steam. Cogeneration of electricity and steam may be the best single use for fuelwood (Hayes, 1977b). Even in developed countries like Sweden, fuelwood represents an important fraction (5.6%) of the energy produced.

The increasing demand for fuelwood has led to the deforestation of the land in certain countries. In India, for instance, fuelwood supply is severely limited (Makhijani and Poole, 1975; Thappar, 1975). The shortfall of supply may lead to a reduction in forests. A continual decrease will lead to serious problems, not only of energy supply but also of soil erosion and water supplies (Openshaw, 1974). Proper planning, and replanting of fuelwood plantations have been suggested as solutions to this problem (Thappar, 1975; Azevedo, 1976; Openshaw, 1974).

The potential for fuelwood has been calculated for the selected countries. Brazil has an immense potential mainly due to the Amazonas Basin (Table IX). Sudan also has excellent potential on a per capita basis (Table IX).

TABLE IX

Forest Utilization (10^6 metric tons)

	Total Potential Production[a]	Actual Use		
		Industry	Fuelwood	Total
United States	988	98.7[b]	4.8[c]	103.5
Brazil	2044	9.0[d]	50.0[c]	59.0
India	262	9.0[e]	60.0[e]	69.0
Sudan	366	0.5[b]	7.5[c]	8.0
Sweden	90	17.8[b]	1.0[c]	18.8

[a] Assuming a net productivity of 4 metric tons/ha (Pimentel *et al.*, 1978).
[b] Food and Agriculture Organization (1976).
[c] United Nations (1976).
[d] Volatron (1976).
[e] From data presented by Thappar (1975) and Makhijani and Poole (1975).

Lumber and pulpwood production is important in biomass production. When forests are harvested for timber and pulpwood, from 7 to 20% of the wood remains in the forest as a residue (Franklin, 1973). These forest residues should be available for use in biomass energy conversion.

During the processing of the timber and pulp an estimated 86×10^6 metric tons of residues are produced in the United States (Howlett and Gamache, 1977). At present, however, about half of these residues are used for pulp and the production of synthetic boards. In the future if more residues are used to make boards, etc., the potential from this source would decrease.

D. Sugar Crops

For countries with favorable climatic conditions and large areas of productive but unused land, such as the case of Brazil and Sudan, sugar crops for ethanol production may represent an alternative for biomass conversion. Sugar crops have some potential advantages over trees, grains, or aquatic plants for ethanol fermentation in that much of the biomass is in the form of directly fermentable sugars (Lipinsky and McClure, 1977). Sugar crops are also noticeable for high yields per unit of land area; sugar cane yields 72 metric tons/ha; sweet sorghum yields 45 metric tons/ha; and sugar beets yield 43 metric tons/ha (Gomez da Silva *et al.*, 1976; Lipinsky *et al.*, 1977). The C-4 photosynthetic pathway and the length of the growing season after canopy closure are suggested as the factors of high yields in sugar cane and sweet sorghum (Calvin, 1977; Bull, 1975; Lipinsky and McClure, 1977). Higher sugar cane yields may be achieved by use of close row spacing. Bull (1975) found that canopy closure occurs in 14 weeks when an experimental spacing of 0.45 m was used, as compared with 24 weeks at commercial spacings of 1.4 m.

Brazil has started an ambitious program of ethanol production from sugar cane (Hammond, 1977; Goldemberg, 1977). The goal is to supply enough ethanol to meet Brazil's needs for liquid fuels by the year 2000. If 10% of the unused productive land (Table X) were planted to sugar cane and the yield were 14 metric tons/ha, Brazil could produce 168 million metric tons of sugar crops (Table XI). Similar assumptions for Sudan suggest an annual yield of 108×10^6 metric tons of sugar cane. Using the same assumptions, India could produce 69×10^6 metric tons of sugar cane.

E. Urban Refuse

Urban refuse or municipal solid wastes are another source of biomass. These wastes consist predominantly of paper (40%), yard wastes (20%),

TABLE X

Estimate of Additional Potential Cropland and Forest Land Not Under Use for the United States, Brazil, India, Sudan, and Sweden (10^3 ha)

Land	Land Area	Cropland in Use[a]	Additional Potential Cropland	Forest Land Used for Production of Fuelwood and Lumber[e]	Additional Forest Land
Sudan	237,600	7,495	76,905[b]	2,325	88,675
Brazil	845,651	36,060	120,000[c]	21,500[f]	489,900
India	296,608	167,200	49,458[d]	24,000[g]	41,000
Sweden	41,148	3,051	11,009[d]	21,800	4,674
United States	912,689	192,036	184,053[d]	99,900	148,000

[a] Food and Agriculture Organization (1975, 1976).

[b] Ministry of Agriculture, Food and Natural Resources of the Sudan (1974). This estimate overlaps forests, woods, and pastures.

[c] Goldenberg (1977).

[d] Food and Agriculture Organization (1976). Unused but productive land including wasteland, parks, and gardens.

[e] From data by the Food and Agriculture Organization (1975).

[f] 50×10^6 metric tons of fuelwood (Food and Agriculture Organization, 1975), 9×10^6 metric tons of industrial wood (Volatron, 1976).

[g] 60×10^6 metric tons of fuelwood (Food and Agriculture Organization, 1975; Makhijani and Poole, 1975). 9×10^6 metric tons of industrial wood (Thappar, 1975).

and food wastes (20%) (Clark, 1974; Environmental Protection Agency, 1975). In the United States an estimated 123×10^6 metric tons of solid wastes are produced annually (Environmental Protection Agency, 1975), of which 90×10^6 metric tons are collected. Of these 90 metric tons, 24 metric tons consist of paper, glass, and ferrous metals, which can and should be recycled. The balance remaining available after a portion is recycled is about 66×10^6 metric tons for biomass energy conversion. Urban refuse can be burned or used as a substrate for pyrolysis (Clark, 1974; Freeman, 1973).

The coefficient of urban refuse production per capita of urban population in the United States is 600 kg/yr. To estimate the total municipal solid wastes for the other selected countries it was assumed that per capita wastes averaged one-half those of the United States, or 300 kg per capita per year.

TABLE XI

Sources of Biomass for Energy Conversion (10^6 metric tons)

Source	Sudan	Brazil	India	Sweden	United States[j]
Livestock manure[a]	56.7	391.9	543.8	11.0	255[h]
Crop residues[b]	2.7	39.2	87.1	2.8	430.0
Sugar crops[c]	107.6	168.0	69.0	—	[i]
Sugar crops[d]	54.0	84.0	34.5	—	—
Fuelwood and fuelwood plantations[e]	36.6	204.6	27.0	10.5	60.0
Urban refuse[f]	1.9	36.0	112.9	4.4	123.0
Municipal sewage[g]	1.1	6.7	37.4	0.5	13.0
Others	—	—	—	—	260
Total	206.6	846.5	877.2	29.2	1141

[a] Based on data from the Food and Agriculture Organization (1975) and computing animal waste with estimates by Loehr. 1723 kg dry/year cattle; 199 kg dry/year sheep; 182 kg dry/year pig; and estimates of 1739 kg dry/year horse, and 600 kg dry/year camel.

[b] Based on 52% of crop materials as crop residues.

[c] Calculated as 10% of additional potential cropland and a productivity of 14 metric tons dry/ha, and none for the United States.

[d] Same with 5% and none for United States.

[e] Calculated on 10% of forests and woods and a net productivity of 4 metric tons/ha.

[f] Assuming 61.3 kg dry per capita year (Pimentel *et al.*, 1978; Hecht *et al.*, 1975).

[g] Assuming a production of 300 kg per capita in urban centers of Brazil, India, Sudan, and Sweden and 600 kg for United States.

[h] Anderson (1972).

[i] Lipinsky estimates lead to 232×10^6 metric tons potential sugar cane production.

[j] Data for the United States are from Pimentel *et al.* (1978).

F. Municipal Sewage

Municipal sewage in all countries consists of about 0.02–0.03% solids and hence is more than 99% water. Based on a per capita production of 61.3 kg dry/yr (Pimentel *et al.*, 1978; Hecht *et al.*, 1975), we have estimated the annual production of municipal sewage in the selected countries (Table XI). Sewage sludge is a potential substrate for anaerobic digestion (Bargman, 1966).

G. Fuelwood Plantations

In addition to the previously mentioned supplies of biomass materials, energy crops such as fuelwood can be grown. Such a program, however, does not include the use of forests for the widespread production of fuel wood for two reasons: (1) forest products such as lumber and pulpwood have greater economic value than wood used directly for burning (U.S. Department of Agriculture, 1976); and (2) widespread cutting of forests and intense logging often result in serious environmental problems (Dasmann, 1972).

Certain tree and shrub species, appropriately managed, have several environmental and energetic advantages for energy farming over other plants such as sugar cane, corn, and other crop plants. For example, tree species like poplar and sycamore grow rapidly and produce a large quantity of biomass per hectare if cropped every 3–10 years. The wood is relatively dense and contains only about 50% moisture when freshly cut.

Since these trees sprout from stumps left in place, the land does not have to be cultivated and therefore receives considerable protection from soil erosion. The roots, leaves, and twigs on the soil surface also help protect the land from water erosion and reduce water runoff. In addition, the organic matter from root and leaf production contributes to soil quality and structure. Fuelwood farming, however, would reduce natural plant and animal species diversity because of the planting of a single tree species such as sycamore. This reduced diversity might make the ecosystem less stable than normal (Pimentel, 1961).

If trees like sycamore, poplar, or alder were grown for biomass energy conversion, they could be cut at intervals ranging from 3 to 5 years for sycamore and poplar (Gordon, 1975) and 9–11 years for alder (DeBell, 1975). The yields obtained would depend on the quality of the land and/or the amount of care in the form of cultivation, fertilization, and protection they receive. Average annual production on marginal land is assumed to be about 6 tons of biomass per hectare (Gordon, 1975; DeBell, 1975).

H. Aquatic Plants

Another mentioned possibility for increasing biomass supplies is the production of aquatic plants for energy conversion. Most aquatic plants grow sparsely, contain large quantities of water (90%), and thus are not easily converted to energy. Brown kelp and water hyacinth are two commonly mentioned aquatic plants that might be readily available. North (1977) suggests that brown kelp is too valuable for food and feed to be used as an energy source. Not enough data are available to evaluate the real potential of this source of biomass for energy and therefore it is not included in the calculations.

IV. AVAILABLE TECHNOLOGIES FOR OBTAINING ENERGY FROM BIOMASS

A. Description

1. Ethanol Fermentation

Alcohol fermentation is a widely known microbial technology. Ethyl alcohol can be produced from a variety of sugar-containing materials by fermentation with yeasts or from previously saccharified starch materials. The principles of the fermentation are similar regardless of the substrate employed.

Strains of *Saccharomyces cerevisiae* are usually selected to carry on the fermentation of breaking $C_6H_{12}O_6$ to $2C_2H_5OH + 2CO_2$. The six carbon sugar molecules are converted into two two-carbon alcohol molecules. Sugars fermented by *S. cerevisiae* include glucose, fructose, mannose, galactose, sucrose, maltose, and raffinose.

In the batch process, the substrate, say molasses, is diluted with water to a sugar content of about 20% by weight, acidified to pH 4–5, and mixed in the fermentor with about 5% by volume of yeast culture. Acidity is reduced using ammonia. Increase in temperature due to fermentation is controlled with refrigeration. After alcohol is accumulated to 8–10%, the liquid is distilled, fractionated, and rectified. One gallon of alcohol, 3.79 liters (21,257 kcal) is obtained from 2.5 gallons of cane molasses or the equivalent to 5.85 kg of sugars (21,842 kcal). So there is almost no energy loss in the fermentation process (Calvin, 1977).

When a starchy material, such as corn or barley, is used as substrate, the starch must be converted to fermentable sugars before yeast fermentation because none of the commercial yeasts directly ferment starch. The starchy substrate is mixed with water, cooked to hydrate, and gelatinized to starch and then converted to carbohydrates with amylolytic barley malt.

A major constraint of conventional alcohol production is ethanol or end-product inhibition. Although selected strains of yeast with high alcohol tolerance are customarily employed by the industry, when the ethanol concentration in the fermentor increases about 7–10% the specific production rate is severely suppressed. Therefore, substrates must be diluted to low sugar concentration (10–20%) before fermentation, with consequent increases in costs. The estimated capital costs for alcohol production are about 45% of the total production costs today (Table XIII).

Alcohol fermentations of a large variety of substrates have been documented. They include: cheese whey, sulfite waste liquor, corn and barley mashes (Underkofler, 1954), sugar cane molasses (Lipinsky *et al.*, 1977), sugar beet molasses, several fruits, and starchy products such as potatoes.

Among the sugar crops available as substrates for alcoholic fermentation, sugar cane is the most important. Other crops mentioned as potential substrates for production of alcohol as a fuel or as raw material for chemical feedstocks include sweet sorghum, cassava, and sugar beets (Lipinsky *et al.*, 1977; Hammond, 1977).

The two main by-products of the fermentation are CO_2 and the spent materials, which will contain the nonfermentable fraction of the substrate, the nonfermented sugars, and the yeast cells. This material is often suitable for stock feed (Messell and Butler, 1974).

Production costs of sugar cane in the United States and in Brazil are given in Table XII. The total cost of production per dry metric ton of sugar cane in the United States is $68.03, while in Brazil it is only $35.78. Lower labor costs and lower fertilizer and chemicals cost make up most of the difference. On the other hand, the cost of gasoline in Brazil is $1.80/gal as compared with $0.60/gal in the United States.

A conceptualized ethanol processing facility is calculated to convert 11,067 metric tons of sugar cane into 784,705 liters of (95%) ethanol/day, 139 metric tons of yeast, 1088 metric tons of wet sillage, and 566 metric tons of CO_2 (Lipinsky *et al.*, 1977).

Based on this process, Lipinsky *et al.* (1977) have estimated total annual net costs of $\$88.2 \times 10^6$ or a cost per liter of $0.34 ($1.29/gal). The cost of

TABLE XII

Production Costs of Sugar Cane in the United States and Brazil[a]

	United States[b]	Brazil[c]
Fertilizer and chemicals	13.80	20.81
Labor	26.03	10.46
Land	19.40	36.85
Machinery	8.02	7.32
Financing	2.83	4.21
Others	29.92	20.35
Total cost		
$/metric ton (dry)	68.03	35.78
$/metric ton (wet)	18.71	9.84

[a] Percentage of total cost in dollars.

[b] Lipinksy *et al.* (1977). 1976 prices for Flordia peat soil, 78 metric tons of millable cane yield per hectare.

[c] Calvin (1977). 1974 prices, annual yield 63 metric tons/ha.

TABLE XIII

Processing Costs of Ethanol[a]

Item	Percentage of Production Costs
Sugar cane production	46.4
Juice processing	8.5
Ethanol fermentation	45.1
Total cost of ethanol is $1.29/gal	

[a] From data by Lipinsky *et al.* (1977).

processing the sugar cane juice accounts for 8.5% of the total manufacturing cost, ethanol fermentation for 45.1% (Table XIII). The cost of the sugar cane production contributes 46.4% of the total cost of $1.29/gal.

The two most important reasons for the high capital costs of ethanol production are: (1) the batch nature of the process; and (2) the end-product inhibition of the yeast.

Continuous alcohol fermentations have been tried successfully in the laboratory (Wick *et al.*, 1974; Mor and Fietcher, 1968; Wick and Popper, 1977), and pilot plant operations have been carried out by Rodriguez (1968). This background information provides technological alternatives for optimizing alcohol fermentation. Maldonado *et al.* (1975) have proposed an integrated continuous process for better utilization of sugar crops. It involves the simultaneous production of ethanol, single cell protein, glucose from the enzymatic hydrolysis of bagasse, and pulp for paper manufacture. Moreira *et al.* (1977) have suggested a cogeneration scheme for the production of alcohol and electricity from sugar cane bagasse. Normally, the sugar cane bagasse is used as the boiler's feed for the generation of steam that is later used to supply the heat for the distillation of the water–alcohol mixtures. The cogeneration scheme consists of interposing a high pressure boiler, a steam turbine, and the utilization of process steam at a modest pressure (50 psi) for the distillation. The authors claim that 10^6 Btu/h of process steam at 50 psi can be obtained together with 88 kW of electricity from the utilization of 0.26 metric tons of bagasse per hour. Selling the electricity at market prices in Brazil will result in an extra revenue equivalent to 37.5% of the alcohol cost.

As mentioned before, end product inhibition is one of the major causes of high capital costs. Vacuum fermentation is one way to reduce the problem of ethanol inhibition. Under vacuum operation ethanol is removed as soon as it is formed. Twelvefold increases in productivity have been obtained in the laboratory using this new method (Cyzewsky and

Wilkie, 1977). Another advantage of this process includes the possibility of using concentrated sugar substrates and obtaining concentrated alcohol products (16–20%) which will reduce equipment size and distillation costs.

Ethanol has been considered as a fuel and source of energy by using it as a gasoline extender (Scheller, 1977; Hammond, 1977). Programs for alcohol use for this purpose are under way in Brazil (Goldemberg, 1977) and Nebraska (Scheller, 1976). Scheller (1977) has found that a mixture of 10% liquid volume ethanol in unleaded gasoline undergoes a positive volume change and appears to reduce fuel consumption to about 95% of that for unleaded gasoline alone.

Ethyl alcohol has a caloric content 39% lower than gasoline, but the power of a motor running on alcohol is 18% higher. The overall result is that the consumption of alcohol for the same power is only 1.4% greater than gasoline (Goldemberg, 1977). Ethanol may also be used as a source of light and heat in rural localities in several countries (L'Anson, 1975).

Ethanol can be used as a raw material for the synthesis of many important petrochemicals through the "ethylene route." In fact, most of the ethanol consumed in the United States today is derived from ethylene, but as the price of oil increases ethanol may be in a position of reversing the trend and may become a cheaper way to obtain ethylene and therefore a large array of other valuable chemicals.

Lipinsky *et al.* (1977) have estimated what the synthesis cost of several selected chemicals might be if synthesized from ethanol. However, based on today's chemical market prices none of the chemicals would be cheaper, but as oil prices increase some of the chemicals may offer some economic advantages in the future (Table XIV).

TABLE XIV

Market Price of Selected Chemicals and Estimated Synthesis Cost From Sugar Cane Residues[a]

Chemical	Current Market Price[b]	Estimated Production Cost from Ethanol[c]
Ethanol	378	387
Ethylene	264	—
Ammonia	133	257
Methanol	143	266

[a] Cost calculated in dollars per metric ton.
[b] *Chemical Marketing Reporter* (1976).
[c] Calculated from data reported by Lipinsky *et al.* (1977).

Various authors have suggested that sugar cane could become a major solution to the world shortage of fuel, protein, and fiber (Thomas, 1974). This is doubtful. Annual fossil fuel usage in the United States and Europe now exceeds the total of solar energy fixed by both agriculture and other plant material (Hudson, 1975; Pimentel *et al.*, 1978). Second, a large quantity of energy is required to process the fibrous material into an utilizable fuel; third, the machinery used in sugar cane processing is heavy and expensive. The value of fossil fuel represented by the machinery has been quoted as being as high as 0.3×10^6 kcal per ton of sugar produced (Hudson, 1975).

The magnitude of the energy crisis and the overall energy consumption in countries such as the United States (17.4×10^{15} kcal) make it unlikely that even a highly successful ethanol from biomass program could have more than a minor impact. Nevertheless, the attractiveness of sugar cane or similar crops lies in the fact that it may be used for production of fuel, as a raw material for chemical feedstocks, and also for food and feed.

2. *Pyrolysis*

Pyrolysis or destructive distillation is an irreversible chemical change caused by the action of heat in the absence of oxygen (in the presence of oxygen the name of the process is combustion). Without oxygen the energy input splits the chemical bonds and leaves the energy stored. Some of the misconceptions from the term pyrolysis arise from the fact that many pyrolytic systems use oxygen or air. In those cases, they are more gasification operations than pyrolysis. Nevertheless, the oxygen used is not the total required for combustion. Pure pyrolysis, on the other hand, is frequently sustained by the burning (combustion) of a fraction of the fuel produced or by burning another fuel. Depending on the operational conditions, pyrolysis may yield either solid, liquid, or gaseous fuel. The process variables associated with the pyrolysis reaction include temperature, retention time, and rate of heating (Jewell *et al.*, 1977); one of the most important parameters seems to be the air to feed stock ratio (Tatom *et al.*, 1975).

In a typical process the feed material goes through the following operations (Freeman, 1973): (a) primary shredding, (b) drying the shredded material (heat for drying may come from the combustion gases derived from the pyrolysis), (c) air separation to remove inorganics, (d) shredding to 24 mesh, (e) pyrolysis, carbonaceous material is heated rapidly (1400–3000°F), and (f) storage of energy products.

Typical yields per metric ton of manure (wet) and urban refuse are given in Tables XV and XVI, respectively. The pyrolysis process operates slightly above atmospheric pressure with a maximum temperature of

TABLE XV

Fuel Yields from a Metric Ton of Wet Manure[a] (85% Moisture) Subjected to Pyrolysis[b]

Energy Product	Heat of Combustion	Weight of Energy Product per Wet Metric Ton (metric tons)
Oil	6660 kcal/kg	0.072
Char	6105 kcal/kg	0.027
Gas	10 kcal/liter	0.036

[a] All manure must first have its moisture content reduced to 45% before processing.
[b] From data reported by Freeman (1973).

1800°F. For oil recovery the temperature in the reactor is 1000°F and for gas recovery 1400°F.

The reactions during the pyrolytic processes are complex and the products obtained vary greatly with the nature of the substrate and the conditions employed. Principal compounds identified in the fuel gases from the pyrolysis of cow manure, softwood sawdust, and rice hulls include H_2, N_2, CO, CO_4, CO_2, C_2H_4, C_2H_6 (ethane), C_6H_6 (benzene), and C_7H_8 (toluene) (Crane, 1976); compounds identified in the tar include indene, naphtalene, acenaphtene, 1,2-benzanthreme, and chrysene; compounds in the liquor fraction include methanol, acetaldehyde, acetone, acetic acid, and others.

TABLE XVI

Fuel Yields per Metric Ton of Urban Refuse (75% Moisture)[a] (approximate)

Energy Product	Heat of Combustion[b]	Weight of Energy Product per Metric Ton (metric tons)
Oil	5880 kcal/kg	0.248
Char	5000 kcal/kg	0.15
Gas	11 kcal/liter	0.06

[a] From data reported by Freeman (1973).
[b] The heat of combustion of the final product depends on the operation conditions and the initial substrate.

TABLE XVII

Estimated Economics for 909 Metric Tons per Day Solid Waste Pyrolysis Disposal System[a]

Annual operating costs	
Amortization and interest (25 years at 6%)	$ 953,600
Electric power $0.01/kW h	170,500
Fuel $0.40/10⁶ Btu	155,000
Water $1/1000 gal	200,000
Manpower $4.5/man-hour	749,000
Maintenance	400,000
Miscellaneous	100,000
Total annual operating cost	$ 2,728,100
Cost per metric ton (dollars)	9.7
Capital costs	
Total installed cost	$10,600,000
Contingency at 15%	1,590,000
Total plant cost	12,190,000
Working capital	171,700
Total estimated capital cost (excluding land)	$12,361,700

[a] Adapted from Freeman (1973).

Methanol and charcoal have been obtained on an industrial scale from the pyrolysis of wood. Typical yields are 1–2% of the original dry weight for methanol and 37.0% for charcoal (Bunbury, 1923).

Pyrolysis has been successfully developed for solid waste disposal (Clark, 1974), production of charcoal from wood (Bunbury, 1923), and for coal liquefaction (Beychok, 1975) but only until recently has it been applied to synthetic fuel production from biomass. Laboratory and pilot plant stages have been studied or are under development for the production of fuels from cow manure (Hoffman *et al.*, 1977), municipal refuse (Clark, 1974), rice hulls and cotton gin trash (Crane, 1976), and agricultural and forestry wastes (Tatom *et al.*, 1975; Coyne, 1977).

Alternate pyrolytic-type reactors include vertical shaft reactors, horizontal shaft reactors, rotary kilns, and fluidized beds (Ifeady and Brown, 1975). Among all the reactors, the simplest and least costly is the vertical shaft. In this model, the substrate is fed into the top of the reactor and settles under the influence of gravity. The pyrolysis gases pass upward and are collected at the top. Figure 1 shows a diagram of a vertical shaft reactor.

The cost of pyrolyzing and extracting fuel oil and char from wet fresh cow manure in a 909 MTPD plant is reported at $9.70 per metric ton (Freeman, 1973) (Table XVII). Net costs of pyrolysis of urban refuse range from $4.35 to $8.60 per metric ton (Clark, 1974).

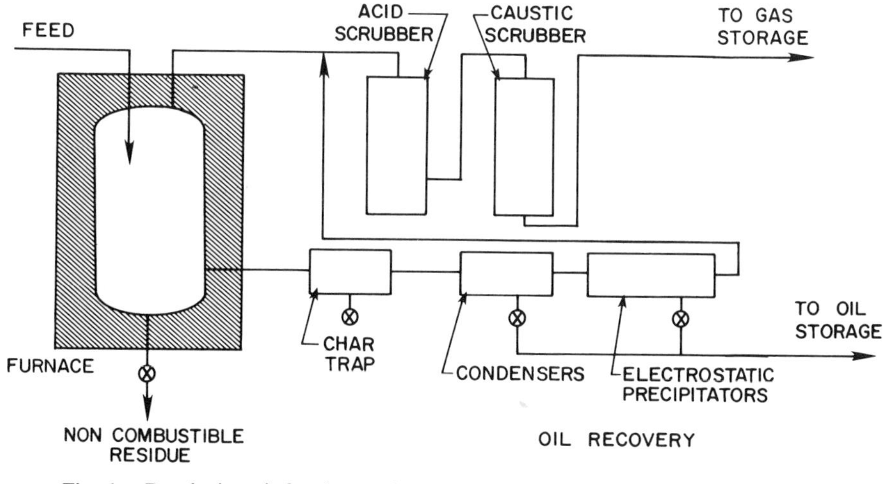

Fig. 1. Pyrolysis unit for the production of fuels. (Adapted from Mantell, 1975.)

As with any other scheme for production of fuels, pyrolysis can be used as a process for the manufacture of chemical feedstocks. As a feedstock the hydrocarbon-rich solid and liquid residues may be the source of ammonia, urea, formaldehyde, and ethanol, to mention just a few. But the economics do not seem to favor such conversions, at least for now (Coyne, 1977).

3. Anaerobic Digestion

Anaerobic digestion or methane digestion of organic substrates is a biochemical process carried out by several microorganisms. Although the complete pathway of reactions has not been completely elucidated, it is known that the process makes use of three major groups of bacteria. Acting sequentially on the original substrate they achieve degradation of the organic molecules into CH_4, CO_2, H_2, and other products in three main steps.

During the first step of the process, insoluble biodegradable molecules are converted into soluble compounds (Fig. 2). Typical reactions are the conversion of polysaccharides and fats into soluble carbohydrates and fatty acids, respectively. The end products of the first step are converted into acetic or propionic acids and CO_2 or acetic acid and H_2 by the acid forming bacteria. In the final step, methanogens utilize the CO_2 and H_2 to produce CH_4 or cleave the acids into CH_4 and CO_2 (Hobson *et al.*, 1974; McCarty, 1964a). There is increasing evidence that H_2 plays a major role in the production of CH_4 (Morris *et al.*, 1975). The end product usually

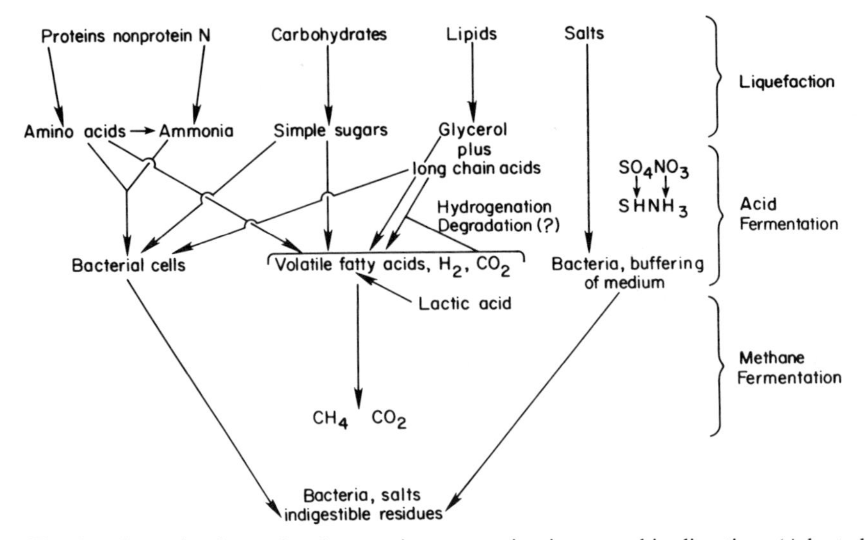

Fig. 2. General scheme for the reactions occurring in anaerobic digestion. (Adapted from Hobson *et al.*, 1974.)

consists of a mixture of CO_2, CH_4, small quantities of H_2, and other gases. The percentage of CH_4 in the final mixture has been reported to vary from 50 to 80% (Bargman, 1966). A typical mixture consists of 65% CH_4 and 35% CO_2 (Lapp, 1974). The residue left after anaerobic digestion consists of (1) nonbiodegradable materials, (2) biodegradable organics not fully digested, and (3) the synthesized microorganisms. Generally the nitrogen is maintained in the residue.

The complex interaction of the bacteria populations involved in the process make it highly sensitive to variable environmental conditions. In particular, the methane bacteria are inhibited by a number of factors; pH and high concentrations of NH_4^+ are among the most common causes of methane bacteria inhibition (McCarty, 1964b; Jewell *et al.*, 1977). A large list of other factors are also known to inhibit or be toxic to the bacterial populations (McCarty, 1964b; Kugleman and Chin, 1971).

Temperature plays a major role in the anaerobic digestion. Methanogens exist and effectively degrade substrates under thermophilic (45°C), mesophilic (20–45°C), and psicrophilic (20°C) conditions (Lundsford and Dunstan, 1958; Scroepfer *et al.*, 1955). The main difference in the activity of methanogens under these temperature conditions is related to the time required to achieve the desired stabilization of organic substrates (Morris *et al.*, 1975). Methane fermentation has been employed to digest a large variety of substrates like sewage sludge (Scroepfer *et al.*, 1955), food processing wastes (Lundsford and Dunstan, 1958), crop residues (Leiu,

1975), animal wastes (Jewell *et al.*, 1977; Fisher *et al.*, 1975), kelp (Klass and Ghosh, 1977), and others. Reported efficiencies of digestion usually are 85–95%.

Anaerobic digestion flow schemes include completely mixed reactors, plug flow reactors, batch reactors, fluidized beds, and anaerobic contact processes. Completely mixed reactors have been extensively used for sewage sludge digestion. Batch experiments and anaerobic contact processes have been used for digestion of crop residues, food processing wastes, and several other substrates. Plug flow reactors are under use or in pilot plant experimentation for the treatment of animal wastes (Abeles, 1977; Jewell *et al.*, 1977). Fluidized beds are under serious consideration for use in anaerobic digestion (Jewell *et al.*, 1978).

Anaerobic digestion of organic wastes may constitute an effective device for pollution control with simultaneous energy generation and nutrient conservation. As a net producer of energy, low mesophilic or psichrophilic operations in warm climates offer the best chances. On the other hand, the increased retention time required for effective digestion represents an increase in volume of the equipment and therefore higher capital costs (Morris *et al.*, 1975). A major advantage of anaerobic digestion is that it utilizes biomass with water content as high as 99%. In a digestion tank, pumping and mixing require a water content of about 84% (Lapp, 1974; Loehr, 1974). Another advantage of anaerobic digesters is the availability of small units that can be operated on individual farms (Costigane *et al.*, 1974; Jewell *et al.*, 1977). Furthermore, the residue has fertilizer value and can be used in crop production (Jewell and Morris, 1974) and pollution impact of the original waste is reduced (Morris *et al.*, 1975).

The primary disadvantages of anaerobic digestion of diluted wastes are the large quantity of waste water that must be disposed of after digestion (Jewell and Morris, 1974) and the cost of gas storage due to the high pressure and low temperatures required for liquefication (Lapp, 1974). In colder climates mesophilic and thermophilic operations may demand more energy to keep the temperature of reaction than the energy represented by the CH_4 obtained (Jewell *et al.*, 1977).

Small anaerobic digestors have been and are under current use in India (Singh, 1971), China (Hayes, 1977b), and several other countries. Large pilot plants and full scale systems have been given more attention in Germany and the United States (Jewell *et al.*, 1977) (see Figs. 3 and 4). The small units have been reported effective as a device for pollution control and energy generation, especially in localities where no other options are available.

Under cost competitive circumstances, with fossil fuels in the United

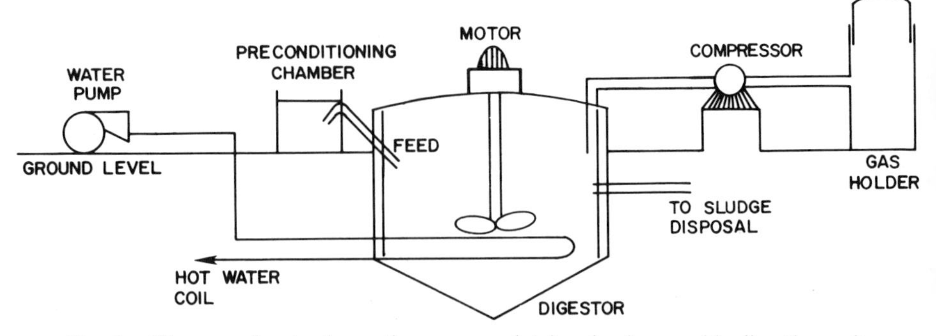

Fig. 3. Diagram of a simple continuous completely mixed anaerobic digestion unit.

States, anaerobic digestion is difficult to justify as an alternative energy source alone (Jewell *et al.,* 1977). Ifeady and Brown (1975) have shown that in the United States anaerobic digestors with capacities ranging from 0.1 to 100 tons/day were not profitable at natural gas market prices of $0.8/10^6$ Btu ($2.5/10^6$ kcal). In the same analysis it is shown that at $2.0/$10^6$ Btu for natural gas ($7.9/10^6$ kcal) plants with capacities of 10–100 tons/day are cost competitive.

Energy production costs in anaerobic digestion operations based on 100-cow free stall dairies range from $10 to $25/10^6$ Btu (Morris *et al.,* 1975), which are not competitive with actual market prices (Table XVIII and Fig. 5). On the other hand, annual cost per cow of alternative liquid

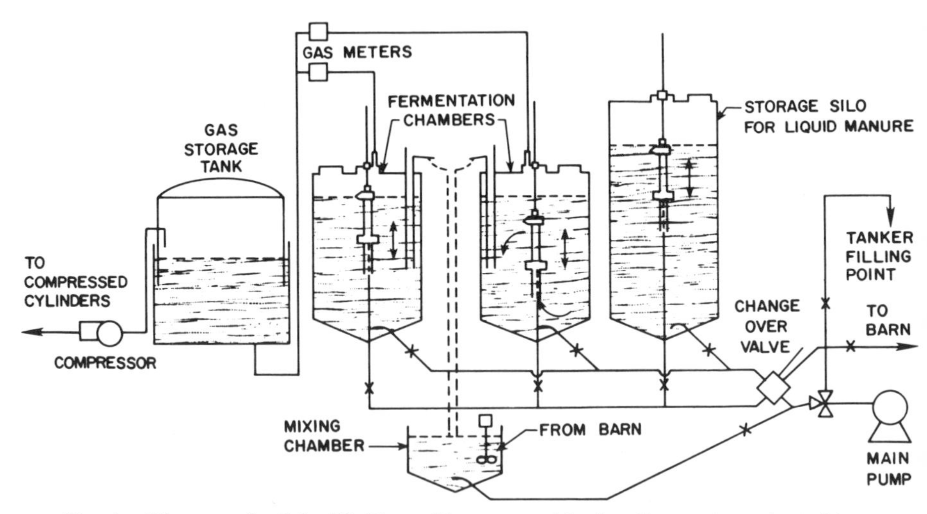

Fig. 4. Diagram of a Schmidt–Eggerglöess anaerobic digestion system adopted in several full scale systems in Europe. (Adapted from Jewell *et al.,* 1976.)

TABLE XVIII

Feasibility Analysis of Alternative Anaerobic Fermentation Systems[a]

	Typical Municipal Digester	Completely Mixed (Mesophilic)	Completely Mixed (Thermo-philic)	Batch Load	Plug Flow
Capital investment total (dollars)	55,000	27,000	21,000	31,000	22,000
Annual costs total (dollars)	15,000	8,000	7,000	9,000	5,000
Net annual gas (10^6 Btu/yr)[b]	600	600	325	755	250
Energy production costs (\$/10^6 Btu)	25	13	22	12	20

[a] Based on a 100-cow free stall dairy operation (Morris *et al.*, 1975).
[b] Natural gas prices are \$1.20/10^6 Btu.

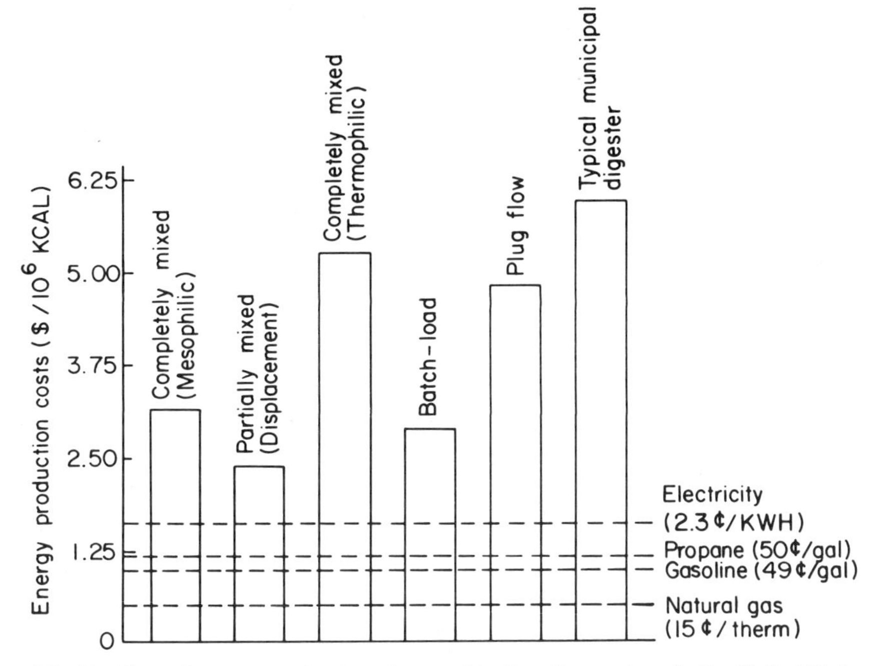

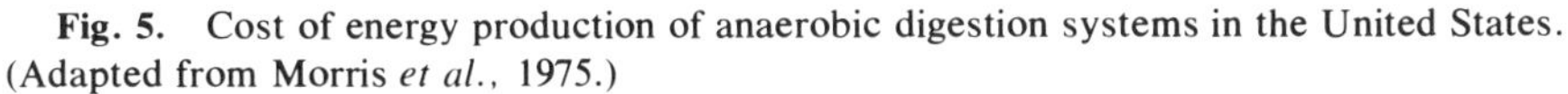
Fig. 5. Cost of energy production of anaerobic digestion systems in the United States. (Adapted from Morris *et al.*, 1975.)

manure handling systems show economical utilization of simple anaerobic digestion systems (Jewell *et al.,* 1977). With increased gas prices and tighter pollution controls anaerobic digestion appears to constitute a successful process for simultaneous treatment of wastes and generation of energy.

B. Inputs and Outputs in Biomass Conversion

To produce, transport, and convert the biomass into usable energy, and then store and distribute the energy all require energy inputs. Thus, it is important to calculate the energy inputs and outputs and determine the net energy returns for biomass energy conversion.

1. Alcohol Production from Sugar Cane

Energy inputs for the production of sugar cane have been calculated by Pimentel (1977), Lipinsky *et al.* (1977), Hudson (1975), and Gomez da Silva *et al.* (1976). Inputs considered include machinery depreciation in terms of energy, fuels, and electricity, herbicides, and pesticides and transportation. Total energy cost estimates range from 78,268 kcal/ton (Gomez da Silva *et al.,* 1976), to 136,509 kcal/ton (Hudson, 1975). Variations are to a degree explained by the fact that the calculations refer to different environments, namely, United States, Brazil, and Barbados. For the United States we chose to work with the data from Pimentel (1977), as it accounts for more inputs. Data from Gomez da Silva *et al.* were used when calculating the energy inputs for sugar cane production in developing countries.

The main differences between the value for sugar cane production in the United States (Pimentel) and Brazil (Gomez da Silva *et al.*) lie in the degree of mechanization used for the production of sugar cane in the United States (Table XIX) and the consequent use of less labor, whereas in the developing countries more labor is employed. Note that fuels for the machinery and fertilizers account for as much as 76% of the total inputs in the United States. Additional consumption of energy is required for the processing of the juice squeezed from the cane. The standard industry practice employs shredders and roller mills in sequence to prepare and crush the sugar cane (Meade, 1976). The fiber left after crushing constitutes the bagasse. Lipinsky *et al.* (1977) have estimated that 126,708 kcal of energy is required to process 1 metric ton of sugar cane (Table XX). An additional 200,750 kcal is required for the fermentation, distillation, and rectification of the ethanol product (Gomez da Silva *et al.,* 1976).

From an energy point of view the two most important outputs from the ethanol facility are the ethanol (heat of combustion, 5260 kcal/liter), which

TABLE XIX

Energy Utilized in Sugar Cane Production

	United States[a] (kcal/metric ton)	Brazil[b] (kcal/metric ton)
Labor	1,192	2,198
Machinery	12,314	9,631
Fuels	71,374	36,598
Fertilizers[c]	28,861	24,743
Irrigation	5,870	
Insecticides	3,615	168
Herbicides	2,866	1,110
Electricity	3,155	
Transportation	1,613	
Others		3,808
Total	130,865	78,268
Yield per hectare in metric tons	86.2	72.0

[a] Pimentel (1977). Other calculations, Hudson (1975): 132,209 kcal/metric ton; Lipinsky *et al.* (1977): 126,708 kcal/metric ton.

[b] Gomez da Silva *et al.* (1976).

[c] Includes nitrogen, phosphorus, and potassium.

can be used as a liquid fuel, and the bagasse (heat of combustion, 1300 kcal/kg), which may be used to generate steam in a boiler by incineration.

Energy obtained from ethanol and bagasse is reported as 672,267 kcal/metric ton of sugar cane (Gomez da Silva *et al.*, 1976) (Table XXI). Net energy obtained from ethanol fermentation is therefore 261,375 kcal/metric ton (982,612 kcal/dry metric ton) or 393,640 kcal/metric ton (1,479,849 kcal/dry metric ton), as reported by Lipinsky *et al.* (1977) and Gomez da Silva *et al.* (1976). Using the data from Pimentel for the agricultural production and the data from Gomez da Silva for the ethanol processing, one arrives at a net energy of 1280 kcal/kg dry, which will be used for the calculations in the United States.

TABLE XX

Energy Utilized in Ethanol Production from Sugar Cane (kcal/metric ton)

Item	Lipinsky *et al.* (1977)	Gomez da Silva *et al.* (1976)
Agricultural production	120,708	78,268
Processing to obtain juice	113,954	—
Fermentation	341,863	200,259
Total inputs	576,525	278,527

TABLE XXI

Energy Available for Utilization from Ethanol Production (kcal/metric ton)

Item	Lipinsky *et al.* (1977)	Gomez da Silva *et al.* (1976)
Ethanol	375,379	347,167
Bagasse	435,708	325,000
Total outputs	811,087	672,267
Net energy available per metric ton	261,375	393,640

Other crops have also been suggested for ethanol fermentation, among them sweet sorghum, cassava, and corn (Table XXII). Table XXIII includes the energy outputs and net energy available for these crops as reported in the literature.

2. *Fuel Production from Pyrolysis of Wastes*

Energy inputs in a pyrolysis process include operational energy, energy inputs in machinery production and maintenance, pretreatment of material, drying of the wet biomass (if it is, say, wet cow manure or wet municipal refuse), pyrolysis, and collection of the organic fuels. The Unihog System (Clark, 1974) for waste production has a capacity of 800 tons of refuse per day and consumes $20,160 \times 10^6$ kcal/yr in fuel for drying and shredding and 7714 MW h for the operation of the shakers and shredders (Clark, 1974). When one accounts for the energy equivalence of machinery, the net energy cost of feed preparation for pyrolysis is 0.211×10^6 kcal/metric ton (0.291×10^6 kcal/dry metric ton) (Table XXIV). Energy is also consumed in the pyrolysis as electricity used to increase the temperature and pressure of the reactor and as energy for the

TABLE XXII

Energy Inputs for Ethanol Production from Agricultural Crops

Crop	Production (kcal/metric ton)	Processing (kcal/metric ton)	Source
Sugar cane	78,268	200,339	Gomez da Silva *et al.* (1976)
Sugar cane	120,708	576,525	Lipinsky *et al.* (1977)
Sweet sorghum	103,711	264,067	Gomez da Silva *et al.* (1976)
Cassava	117,138	306,310	Gomez da Silva *et al.* (1976)
Corn	977,360	2,177,280	Scheller (1977)
Corn	320,440	1,728,720	Lipinsky *et al.* (1977)

TABLE XXIII

Energy Outputs from Ethanol Fermentation and Net Energy Available from Various Crops

	Outputs (kcal/metric ton)	Net Energy Available (kcal/metric ton)	Output-to-Input Ratio
Sugar cane[a]	672,267	393,648	2.41
Sugar cane[b]	811,087	113,854	1.16
Sweet sorghum[a]	704,133[c]	336,355[c]	1.91
Cassava[a]	614,724	191,276	1.45
Corn[b]	1,552,326	−546,834	0.76

[a] Gomez da Silva *et al.* (1976).
[b] Lipinsky *et al.* (1977).
[c] Assuming a yield of 45 metric tons/ha.

production and maintenance of the machinery. All operations accounted for, the energy invested in a 909 metric ton per day Landgard System amounts to 0.578×10^6 kcal/metric ton (0.773×10^6 kcal/dry metric ton) (Table XXV).

The outputs from pyrolysis of refuse include oil (5.884 kcal/kg), char (4996 kcal/kg), and gas (330 kcal/kg) (Freeman, 1973). The oil produced from animal manure or municipal refuse may substitute for No. 6 fuel oil. The charcoal properties vary with the material chosen for pyrolysis and the parameter of operation (Freeman, 1973), but as there are no quality standards for charcoal the possibility of using the char is open. The total energy obtained from the fuels from the pyrolysis amounts to 1.64×10^6 kcal/metric ton (2.19×10^6 kcal/dry metric ton). Therefore, the net energy recovered from the refuse is about 1.417×10^6 kcal/dry metric ton or 1417

TABLE XXIV

Annual Energy Inputs. Preparation of Material for Pyrolysis (776 metric tons refuse/day)[a]

	10^6 kcal
Machinery[b]	1,957
Fuel	20,160
Electricity	20,566
Inputs per Metric Ton[c] (202,000 tons/yr)	0.211

[a] Based on data supplied by Clark (1974).
[b] Based on coefficients for energy value of equipment reported by Vergara (1977).
[c] Assuming 260 days of operation at full capacity.

TABLE XXV

Net Energy Analysis for Pyrolysis of Urban Wastes (plant capacity of 909 metric tons per day)[a]

Inputs	10^6 kcal/year
Substrate production	0[b]
Electricity	45,400
Fuel	97,650
Machinery	5,187[c]
Total	148,237
Inputs per metric ton (10^3 kcal)	578[d]
Outputs[e]	**10^6 kcal/year**
Oil	398,350
Char	21,137
Gas	564
Total	420,051
Outputs per metric ton (10^3 kcal)	1,640
Net energy per metric ton (10^3 kcal)	1,062

[a] From data presented by Freeman (1973).

[b] Based on recovery of energy from waste substrates. If an energy value of production were necessary, the system would not yield a net energy output.

[c] Based on energy coefficients reported by Vergara (1977).

[d] Capacity of 909 metric tons per day, 260 days of operation.

[e] Based on 0.24 metric ton of oil per wet metric ton, 0.15 metric ton of char per wet metric ton, and 0.06 metric ton of gas per wet metric ton (Freeman, 1973).

kcal/kg dry. The heat of combustion of the municipal refuse amounts to 2.939×10^6 kcal/metric ton (Freeman, 1973) (3939 kcal/kg dry), which means an energy efficiency of 36% for the pyrolysis plant with a capacity of 909 metric tons per year.

3. *Anaerobic Digestion of Organic Wastes*

Energy inputs for anaerobic digestion vary with the predominant environmental conditions, technology employed, and character of the digestion (thermophilic, mesophilic, etc.). In a general sense, energy inputs include energy equivalents of machinery depreciation, energy to heat the digestors, and electricity to operate the equipment. It may also include allowances for the collection, transportation, and preparation of the biomass waste and for the storage of the gas.

Estimates of the energy inputs for anaerobic digestion have been reported by Jewell *et al.* (1977) and by Makhijani and Poole (1975). Table

XXVI includes an estimate of the energy inputs and outputs assuming no energy investment in the production of the substrate (animal manure), and using a completely mixed anaerobic digestor and an inverted double drum for gas storage. Heating requirements in tropical countries are nearly

TABLE XXVI

Energy Inputs Using Anaerobic Digestion for Biogas Production from 100 Tons (wet) (13 metric tons dry) of Cattle Manure[a]

Inputs	Quantity	10^3 kcal
Man-hours[b]	20 h	—
Petroleum (heating digester)[c]	61 liters[d]	610[e]
Electricity	2234 kW h[d]	5822[f]
Cement foundation (30-year life)[g]	0.9 kg[d]	2[h]
Steel (gas collector and other equipment with 30-year life)	35 kg[d]	725[i]
Pumps and motors[j]	0.05 kg[d]	1[i]
Steel truck/tractor for transportation (10-year life)	10 kg[d]	200[i]
Petroleum for transport (10-km radius)[b]	34 liters[d]	340[e]
Total Input		7700
Total Output[k]		10.2×10^6 kcal

[a] Retention time in the digester is 20 days. The unit has the capacity to process 1825 tons (wet) per year. Note: the yield in biogas from 100 metric tons is estimated at 10.2 million kcal. Thus, the net yield is 2.5 million kcal.

[b] Estimated.

[c] The heat requirements were calculated including the heat losses to the surroundings, the heat associated with the feed and the effluents, and the heat evolved by the biological reaction.

[d] Vergara *et al.* (1977).

[e] One liter of fuel is assumed to contain 10,000 kcal. Included in this figure are mining, refining, and transport costs.

[f] One kW h = 860 kcal. Based on an energy conversion of fuel to electricity of 33%, thus 1 kW h = 2606 kcal.

[g] The digester was placed underground. Materials used for its construction were concrete and steel. Materials also include a gas storage tank.

[h] One kg of cement = 2000 kcal for production and transport.

[i] One kg of steel = 20,700 kcal for mining, production, and transport.

[j] The design included three electrical devices: a motor to drive the agitator in the digestor; a compressor to store the gas, and a pump to supply hot water.

[k] It is assumed that anaerobic digestion of manure takes place at 35°C with a solids retention time of 20 days. The temperature of the fresh manure is taken at 18°C and the average ambient temperature as 13°C. The manure is assumed to have the following characteristics: production per cow per day, 26.3 kg; total solids, 3.36 kg; biological oxygen demand (BOD), 0.68 kg. The digestion is assumed to transform 83% of the biodegradable material into gas. Gas produced is said to be 65% methane and its heat of combustion is 5720 kcal/m^3 at standard conditions.

TABLE XXVII

Estimated Major Inputs for Electric Power Generation from a 100-MW Power Plant for One Year Requiring Fuel from Biomass of 1.57 × 10^{12} kcal (Federal Power Commission, 1970), Assuming a Net Conversion Efficiency of 25% (Linke, 1977)[a]

Inputs	Quantities	10^6 kcal
Construction quantities		
Steel, Including Equipment (35-year life)	130 tons[b]	2,689
Concrete (35-year life)	76 tons[b]	152
Manhours	21,600 h[c]	—
Operation and maintenance		
Electricity	40 × 10^6 kW h[d]	104,240[f]
Man-hours	100,000 h[e]	—
		2,841

[a] Note that a 25% net efficiency (393 × 10^9 kcal) includes electricity used in operating the plant.

[b] Federal Energy Administration (1974). Figures linearly extrapolated from estimates of concrete and steel needed for plants of 700, 800, and 900 MW, 35-year life expectancy.

[c] Federal Energy Administration (1974). (6.92 man-hours/kw for construction + 0.63 man-hours/kW for design) × (100-MW plant size)/(35-year life).

[d] Electricity used in plant = gross − net generated.

[e] Assuming 0.5 man-hours/kW, 40 h/week, 48 weeks/year.

[f] Excluded from total because this amount of electricity was acccounted for in net efficiency of electrical energy production.

zero. The net energy stored is estimated at 784 kcal/kg of wet manure based on the assumptions employed.

4. *Conversion of Forest Products to Electric Power*

For comparative purposes, estimates on conversion of forest remains to electric power were made by burning the remains in a 100-MW power plant.[2] An advantage of combustion in a power plant is that the forest remains leave only small (about 10%) residue of ash (Skinner, 1974). This ash contains large quantities of potassium, phosphorus, calcium, and other minerals that could be used in fertilizers (assuming no heavy metals are present) and thereby reduce energy used in fertilizer production. The 100-MW power plant produces about 457 × 10^6 kW h (393 × 10^9 kcal) (Federal Power Commission, 1974). Assuming a 25% efficiency of conversion of fuelwood energy into electricity and a total annual operating cost

[2] Forest residues and other combustible materials were converted into electricity for several reasons: (1) electricity is high quality energy; (2) scattered biomass can be collected and converted in an electric power plant located near where the combustible materials are located; and (3) electrical energy can be easily transported to the consumer.

of 2.8 million kcal, the 100-MW plant will need to process about 1.57×10^{12} kcal of fuelwood energy annually (Table XXVII). Or, put another way, with a net yield of 74.2 million kcal per hectare of forest remains, about 21,160 ha would have to be collected to supply one 100-MW power plant. If we assume a forest maturity rate for 30 years, then to supply the annual needs of the power plant, a forested area of 634,800 hectares would be required.

To harvest the trees, machines would "mow" and chop the wood into chips for transport to the power plant for storage and conversion. Drying of the chips was not included because of the relatively high energy expense of drying. Note that the inputs for harvest and transport are relatively substantial and make up almost one-half of the total input (Table XXVIII). However, since much of the machinery contemplated is experimental, increases in efficiency can be expected as technology is improved. Removing about 6 metric tons of biomass from the land will require nutrient replacement. The nutrient requirements were estimated to be nitrogen, 56 kg/ha; phosphorus, 5.5 kg/ha; and potassium, 31 kg/ha (Table XXVIII). In sites where alder can be planted, less nitrogen would be needed because of the nitrogen-fixing property of its roots. Lower

TABLE XXVIII

Average Annual Energy Inputs Needed per Hectare to Produce and Transport Sycamore Fuelwood to an Electrical Plant for Energy Conversion[a]

Inputs	Quantity/Hectare	10^3 kcal
Labor	3 h[b]	—
Machinery (10-year life)	5.5 kg[b]	114
Petroleum	56 liters[b]	560
Phosphorus	5.5 kg[c]	18
Potassium	31 kg[c]	68
Nitrogen	56 kg[c]	850
Electricity	12.5 kW h[c]	38
		1648

[a] Note: harvest is assumed to be 6 metric tons/ha. Assuming this wood contains 50% moisture, then its energy potential is 3740 kcal/kg. Hence, about 22.4×10^6 kcal is produced per acre. Minus the production costs, the net yield is 20.8×10^6 kcal/ha per year.

[b] Rich and Bauer (1975).

[c] Estimated based on the nutrients removed in the wood (Young *et al.*, 1967).

fertilizer application rates also might be possible if the ash residue were returned to the land after combustion.

The 6 metric tons of wood harvested per hectare has 50% moisture and produces about 3740 kcal of energy per kilogram of wood or a total of 22.4×10^6 kcal/ha. The total energy input including the fertilizers for fuelwood farming was estimated to be about 1.6 million kcal (Table XXVIII). Thus, minus the inputs, the net energy yield is 20.7×10^6 kcal/ha.

V. SUMMARY OF ENERGY INPUT–OUTPUT ANALYSIS FOR BIOMASS ENERGY CONVERSION

The potential of biomass energy conversion from various biological materials can be calculated for Brazil, India, Sudan, Sweden, and the United States using the information assembled on input–output energy analysis (Table XXIX). Lifestock manure and municipal sewage are best processed by anaerobic digestion (Table XXVI). The net energy output per kilogram of dry manure is calculated at 784 kcal. This net value is also assumed to be valid for municipal sewage. Pyrolysis of urban refuse

TABLE XXIX

Production of Biofuels from Various Biomass Sources and the Net Energy Output per Kilogram (dry)

Biomass Substrate	Conversion Technology	Biofuels Produced	Net Energy Produced per Kilogram of Dry Biomass (kcal/kg)
Livestock manure	Anaerobic digestion	CH_4	784
Urban refuse	Pyrolysis	Fuel Oil Gas Char	1417
Urban refuse	Incineration		3939
Food processing wastes	Anaerobic digestion	CH_4	730
Sugar crops	Ethanol fermentation	C_2H_5OH	1477
Forest biomass	Incineration	Electricity	935
Forest biomass	Incineration		3740
Municipal sewage	Anaerobic digestion	CH_4	784

TABLE XXX

Estimated Potential for Energy from Biomass (10^{12} kcal)

	Sudan	Brazil	India	Sweden	United States[a]
Livestock manure[b]	22.2	153.0	212.9	4.3	27
Sugar crops[c]	158.9	248.1	101.9	—	100
Fuelwood[d]	68.4	765.2	100.0	39.2	122
Urban refuse[e]	1.2	22.6	77.4	5.5	41
Municipal sewage[f]	0.4	2.6	15.7	0.2	1.3
Others[g]	—	—	—	—	150.7
Total	251.1	1,053.8	508.8	49.2	442.0
Current energy consumption	43	641.	1400.	353	17,330
Percentage biomass energy supply, based on current energy consumption	583.9	164.4	36.3	13.9	2.5

[a] Data for United States are from Pimentel *et al.* (1978).

[b] Assuming that 50% of the residues are easily collected and used in anaerobic digesters.

[c] Assuming 10% of the reported additional fertile land is used for sugarcrops in Sudan, India and Brazil, and 3% for the United States (total area of unused fertile land suitable for growing sugarcane in continental states).

[d] Direct incineration, except for the United States (generation of electricity). Based on 10% use of the additional forest land.

[e] Pyrolysis except for the United States.

[f] Anaerobic digestion.

[g] Includes food processing wastes, forest remains, industrial wastes, and others (Pimentel *et al.*, 1978).

yields 1417 kcal/kg dry as net energy output (Table XXV). This compares with a net energy output of 2939 kcal/kg if urban refuse were incinerated. Ethanol fermentation of sugar cane is calculated to yield 1479 kcal/kg of net energy with high manpower inputs (as in the case of developing countries) or 1327 kcal/kg for mechanized operations. Incineration of forest biomass to produce electricity yields a net energy output of 935 kcal/kg (Table XXVIII). This compares with direct incinerations, which would yield 3740 kcal/kg.

The estimated potential for energy obtained from biomass is about 251.1×10^{12} kcal/yr for Sudan or almost six times its current energy needs (Table XXX). For Brazil the potential is estimated as 1054×10^{12} kcal, which is enough to supply all their energy needs 1.5 times. The potential for the other countries is 36.3% for India, 13.9% for Sweden, and 2.5% for the United States (Table XXX).

VI. ENVIRONMENTAL AND SOCIAL CONSTRAINTS

The use of biomass to reduce dependence on fossil energy is important even if the percentage contribution to the overall energy economy is relatively small, as in the United States (1.9%) and Sweden (13.9%) (Table XXX). For countries like India, Brazil, and Sudan, biomass energy conversion is extremely important.

Biomass conversion, like agricultural production, is relatively labor intensive. The percentage of economic costs for biomass energy conversion attributed to manpower is higher than most industries (Folk and Hanna, 1974). For example, waste management in livestock feedlots used 35–40% of the total labor input in the system (Loehr, 1974). The relatively large input of labor is necessary because most of the resources available for biomass energy conversion are diffuse (Hayes, 1977a). Biomass materials are often widely dispersed, bulky, and contain large quantities of water. Therefore, difficulties exist to harvest the biomass in rural environments and transport the energy to large cities. It results in easier and more efficient utilization of biomass to supply the energy needs of agricultural communities (Hayes, 1977a). Undertaking biomass conversion on a large scale can be expected to result in social changes (labor demand) and have an impact both on the environment and economy.

In general, the use of wastes such as livestock manure, and municipal sewage for biomass generation would help reduce environmental problems associated with the management of these wastes. In addition, the residual materials remaining after processing can be used as fertilizers.

An additional advantage of biogas generation is that numerous, small anaerobic digester units can be located in farming regions where wastes accumulate. For instance, small units might be located on individual farms or shared between two or three farms. A wide distribution of biogas units can substantially reduce transport costs of the wastes for processing and the residues can be effectively recycled to nearby cropland.

There are obvious energy, economic, and agricultural tradeoffs inherent in using livestock manure and food processing wastes for biogas production. All factors should be carefully examined and evaluated before any decisions are made.

Balanced against the benefit of using forest remains is the possible impact this might have on the environment. Of particular concern is the effect that driving heavy equipment over the land and removing residues might have on soil erosion (Dasmann, 1966). The added cost of both machinery and manpower would also be an important consideration.

The use of forestry processing wastes for energy conversion would contribute to the effective disposal of these wastes. The prime constraint to their use is probably the specific cost inputs for processing and distribution.

Considerable environmental benefit would occur in the conversion of urban and industrial wastes for heat and electricity. Because these organic wastes have a low sulfur content generally, their combustion under carefully managed conditions would cause minimal air pollution problems. In addition, combustion would reduce the extensive need for landfills and the residue ash could (dependent on chemical makeup) be used as a fertilizer for crops and lawns. Urban and industrials wastes can also be converted into liquid and solid fuels by using pyrolysis. The pyrolysis of these materials require about 52% (Table XXV) of the combustion energy of the original wastes and therefore is less attractive as an energy source technology. On the other hand, the production of oil, char, and gas gives more freedom of action for the matching of the energy needs with the fuels produced.

The use of sugar crops for the production of ethanol is feasible in areas where availability of fertile land is not a constraint and when plentiful solar radiation and manpower are used as basic energy inputs. This is the case of Brazil and Sudan. Land demand in India for agricultural production and the growing population make less likely the availability of resources for an important ethanol from sugar crops production. The lack of enough land for sugar cane raising and the comparatively lower solar radiation are among the causes for the low potential contribution of sugar crops to provide energy in the United States and Sweden.

Fuelwood is already the most widely used source of energy in Sudan (65%) and contributes substantially to the energy needs of Brazil (29%), India (16%), and Sweden (1%). Indiscriminate use of the forest resources can lead to a reduction in the forest lands with consequent increases in soil erosion and decreased water availability. Unfortunately, an important fraction of the forest lands in India (Thappar, 1975) and Brazil (Volatron, 1976; Azevedo, 1976) have been already destroyed. Any future use of fuelwood must be accompanied with sound reforestation programs and/or fuelwood farming.

Fuelwood farming is advantageous because these crops and their management result in minimal soil erosion and water runoff. To be economically practical, however, the generating plants used for conversion should be located close to the fuelwood source to minimize transport costs and other economic constraints. The monocultures of sycamore or other energy crops would, of course, tend to reduce the diversity of the plant and animal species in the natural ecosystem and, as mentioned, this may have an undesirable environmental impact.

The use of crop remains for biomass conversion was analyzed but is generally not recommended because of the undesirable environmental impact that their removal has on soil erosion, water runoff, soil structure, and the soil carbon ratio. Nevertheless, crop residues must be weighed on a site-specific basis before any definitive decision is taken.

Biological solar energy conversion offers the developing countries an opportunity to meet some of their energy needs (Hayes, 1977a). The uneven distribution of wealth in developing countries constitutes a serious obstacle for the effective utilization of renewable energy sources. Social changes are essential if biomass energy conversion is to benefit the immense number of poor in the developing countries.

VII. CONCLUSIONS

The potential of biomass to supply the national energy needs was investigated for the United States, Brazil, India, Sudan, and Sweden. The analysis suggested that biomass could easily meet all energy requirements in Brazil and Sudan if adequately managed and developed. Among the factors that contribute to this potential are: (1) the large areas of fertile but unused land in both countries, (2) the high solar radiation, (3) the large fraction of the population living in rural areas and engaged in agriculture, and (4) the comparatively low per capita energy consumption. Both the use of energy crops and utilization of organic residues are feasible options for biomass energy conversion in both countries.

The already large demand on the natural resources of India for the production of food and fiber and the high population density make it impossible to allocate large areas of land for the production of energy crops. Nevertheless, the organic wastes produced in India should be utilized to maximize their value as energy sources and fertilizers. The estimated potential is equal to nearly 40% of the energy consumed in the country.

The potential of biomass energy conversion is not enough to supply the energy requirements in Sweden and the United States. In Sweden, the land area available, the low solar radiation, and the high per capita energy consumption preclude biomass from becoming an important supplier of energy. The potential for the United States is 329×10^{12} kcal. This is comparable to the potential in Sudan and India but short of a significant contribution to the energy requirements of the country.

In general, it is concluded that biomass energy conversion has the best chances of becoming, and should be developed as, an important energy source among tropical countries with large quantities of natural resources (land) not yet fully utilized. Effective management of the biomass resources in Brazil and Sudan should enable both countries to become more independent in energy terms and be a simultaneous factor of development.

REFERENCES

Abeles, T. P. (1977). Design and engineering considerations in plug flow farm digesters. A preliminary analysis. *Symp. Pap., Clean Fuels Biomass Wastes, Orlando, Fla.* pp. 417–423. Inst. Gas Technol., Chicago, Illinois.

Anderson, L. L. (1972). Energy potential from organic wastes: A review of the quantities and sources. U.S. Bur. Mines. Inform. Circ. 8549, 16 pp., Washington, D.C.

Anonymous (1977). "The National Energy Plan," Executive Office of the President, Energy Policy and Planning. U.S. Gov. Print. Off., Washington, D.C.

Azevedo, P. (1976). *Bras. Florestal* **7**(27), 3–8.

Bargman, R. D., chm. (1966). "Anaerobic Sludge Digestion," Manual of Practice, No. 16. Water Pollut. Control Fed., Washington, D.C.

Beychok, M. R. (1975). "Process and Environmental Technology for Producing SNG and Liquid Fuels," Environ. Prot. Agency Rep. 660/2-75-011. Natl. Environ. Res. Cent., Corvallis, Oregon.

Bull, T. A. (1975). *Agron. J.* **67**, 421–426.

Bunbury, H. M. (1923). "The Destructive Distillation of Wood." Van Nostrand, New York.

Calvin, M. (1977). Energy and materials via photosynthesis. *In* "Living Systems as Energy Converters" (R. Buvet, M. J. Allen, and J. P. Massue, eds.), pp. 231–259. North-Holland Publ., Amsterdam.

Chemical Marketing Reporter (1976). Dec. 20, p. 209.

Clark, R. H. (1974). Solid wastes and resource recovery in Canada. *Waste Recycl. Environ., 7th Symp.* pp. 53–88, R. Soc. Can., Ottawa.

Clausen, E. C., Sitton, O. C., and Gaddy, J. L. (1977). *Process Biochem.* **12**(7), 5–7, 30.

Cook, R. L. (1962). "Soil Management for Conservation and Production." Wiley, New York.

Costigane, W. D., Sharer, J. M., and Silveston, P. L. (1974). *In* "Uses of Agricultural Wastes; Food, Fuel, Fertilizers" (R. J. Catania, ed.), pp. 101–131. Univ. of Regina, Saskatchewan, Canada.

Coyne, J. K. (1977). Pyrolysis of solid wastes for production of gaseous fuels and chemical feedstocks. *Symp. Pap., Clean Fuels Biomass Wastes, Orlando, Fla.* pp. 237–248. Inst. Gas Technol., Chicago, Illinois.

Crane, T. H. (1976). Pyrolysis of agricultural wastes for resource recovery. *In* "Agricultural Residues as a Source of Energy," pp. 22–33. Univ. of California, Davis.

Cyzewsky, G. R., and Wilkie, C. R. (1977). *Biotechnol. Bioeng.* **19**(8), 1125–1144.

Dasmann, R. F. (1972). "Environmental Conservation." Wiley, New York.

DeBell, D. S. (1975). *Iowa State J. Res.* **49**(3), 345.

Environmental Protection Agency (1973). "International Research and Technology Corporation, Problems and Opportunities in Management of Combustible Wastes," EPA Rep. 620/2-73-056. U.S. Gov. Print. Off., Washington, D.C.

Environmental Protection Agency (1975). "Resource Recovery and Waste Reduction. Third Report to Congress," Off. Solid Waste Manage. Programs. U.S. Gov. Print. Off., Washington, D.C.

Federal Energy Administration (1974). "Independence Blueprint Final Task Report." U.S. Gov. Print. Off., Washington, D.C.

Federal Power Commission (1970). "The 1970 National Power Survey," Part IV. U.S. Gov. Print. Off., Washington, D.C.

Fisher, J. R., Sievers, D. M., and Fulhage, C. D. (1975). Anaerobic digestion of swine wastes. *In* "Energy Agriculture and Waste Management" (W. J. Jewell, ed.), pp. 307–316. Ann Arbor Sci. Publ., Ann Arbor, Michigan.

Folk, H., and Hanna, B. (1974). An energy, pollution and employment policy model. *In* "Energy: Demand Conservation and Institutional Problems" (M. S. Macrakis, ed.), pp. 159–173. M.I.T. Press, Cambridge, Mass.

Food and Agriculture Organization (1975). "Production Yearbook," Vol. 29. Food Agric. Organ. U. N., Rome.

Food and Agriculture Organization (1976). "Production Yearbook," Vol. 30. Food Agric. Organ. U. N., Rome.

Franklin, W. E. (1973). "Paper Recycling; the Art of the Possible, 1970–1985. Report of the Midwest Research Institute for Solid Waste Council." Midwest Res. Inst., Kansas City, Missouri.

Freeman, H. M. (1973). "Problems and Opportunities in Management of Combustible Solid Wastes." Int. Res. Technol. Corp., Washington, D.C.

Goldemberg, J. (1977). "Brazil. Energy Options and Current Outlook." Inst. Phys., Univ. of Sao Paulo, Sao Paulo.

Gomez da Silva, J., Serra, G. E., Moreira, J. R., and Goncalves, J. C. (1976). *Braz. Acucareiro* No. 6, pp. 8–21.

Gordon, J. C. (1975). *Iowa State J. Res.* **49**(3), 267.

Hammond, A. L. (1977). *Science* **195,** 564–566.

Hayes, D. (1977a). "Energy for Development: Third World Options," Worldwatch Pap. No. 15. Worldwatch Inst., Washington, D.C.

Hayes, D. (1977b). *Bio Science* **27**(7), 540–546.

Hecht, N. L., Duvall, D. S., and Rahidi, A. S. (1975). "Characterization and Utilization of Utility Sludges and Ashes," Vol. 2, EPA Rep. 670/2-75-0336. U.S. Gov. Print. Off., Washington, D.C.

Hobson, P. N., Bousfield, S., and Summers, R. (1974). *Crit. Rev. Environ. Control* **4**(2), 131–191.

Hoffman, W. J., Halligan, J. E., Peterson, R. L., and de La Garja, E. (1977). Ammonia synthesis gas and petrochemicals from cattle feedlot manure. *Symp. Pap., Clean Fuels Biomass Wastes, Orlando, Fla.* pp. 249–253. Inst. Gas Technol., Chicago, Illinois.

Howlett, K., and Gamache, A. (1977). Silvicultural biomass farms. *Technol. Rep.* **6**.

Hudson, J. C. (1975). *Span* **18**(1), 12–14.

Ifeady, C. N., and Brown, J. B., Jr. (1975). Technologies suitable for the recovery of energy from livestock manure. *In* "Energy Agriculture and Waste Management" (W. J. Jewell, ed.), pp. 373–395. Ann Arbor Sci. Publ., Ann Arbor, Michigan.

Jewell, W. J., and Morris, G. R. (1974). The economic and technical feasibility of methane generation from agricultural waste. *In* "Uses of Agricultural Wastes: Food Fuel, Fertilizer," (R. J. Catania, ed.), pp. 132–164. Univ. of Regina, Saskatchewan, Canada.

Jewell, W. J., Davis, H. R., Gunkel, W. W., Lathwell, D. J., Martin, J. H., Jr., McCarty, T. R., Morris, G. R., Price, D. R., and Williams, D. W. (1977). "Bioconversion of Agricultural Wastes for Pollution Control and Energy Conservation," Div. Sol. Energy, ERDA. Natl. Tech. Inf. Serv., Springfield, Virginia.

Jewell, W. J., Capeher, H. R., Dell'Orto, S., Fanfoni, K. J., Hayes, T. D., Leuschner, A. P., Miller, T. L., Sherman, D. F., Van Soest, P. J., Wolin, M. J., and Wujcik, W. J. (1978). "Anaerobic Fermentation of Agricultural Residue: Potential for Improvement and Implementation." Cornell Univ., Ithaca, New York.

Klass, D. L., and Ghosh, S. (1977). The anaerobic digestion of macrocystis pyrifera under mesophilic conditions. *Symp. Pap., Clean Fuels Biomass Wastes, Orlando, Fla.* pp. 323–351. Inst. Gas Technol., Chicago, Illinois.

Kugleman, I. J., and Chin, K. K. (1971). Toxicity, synergism and antagonism in anaerobic waste treatment processes. *In* "Anaerobic Biological Treatment Processes" (F. G. Pohland, ed.), Advances in Chemistry Series, No. 105, pp. 55–90. Am. Chem. Soc., Washington, D.C.

L'Anson, J. A. P. (1975). *Ann. Technol. Agric.* **24**(3/4), 317–324.

Lapp, H. M. (1974). Methane production from agricultural by-products: An overview. *In* "Uses of Agricultural Wastes; Food, Fuel and Fertilizer" (P. J. Catania, ed.), pp. 5–11. Univ. of Regina, Saskatchewan, Canada.

Larson, W. E., Holt, R. F., and Carlson,.C. W. (1977). Residues for resource conservation soil productivity and environmental protection. Unpublished manuscript.

Leiu, S. S. (1975). Production of methane from farm wastes. *In* "Workshop on Biogas Technology for Asia and the Pacific." Oct. 13–18, Manila, Philippines, pp. 18–28. Econ. Social Comm. Asia Pacific.

Linke, S. (1977). Personal communication.

Lipinsky, E. S., and McClure, T. A. (1977). Using sugar crops to capture solar energy in biological solar energy conversion. *In* "Biological Solar Energy Conversion" (A. Mitsui, S. Miyachi, A. S. Pietro, and S. Tamura, eds.), pp. 397–410. Academic Press, New York.

Lipinsky, E. S., McClure, T. A., Nathan, R. A., Anderson, T. L., Sheppard, W. J., and Lawhon, W. T. (1977). "Systems Study of Fuels from Sugarcane, Sweet Sorghum, and Sugar Beets," Vol. 1. Batelle Columbus Lab., Columbus, Ohio.

Loehr, R. C. (1974). "Agricultural Waste Management." Academic Press, New York.

Lönnroth, M., Steen, P., and Johansson, T. B. (1977). "Energy in Transition. A Report on Energy Policy and Future Options." Bohlislaningens AB, Udevalla, Sweden.

Lucas, R. C. (1977). Personal communication.

Lundsford, J. V., and Dunstan, G. H. (1958). Thermophilic anaerobic stabilization of pea blancer wastes. *In* "Biological Treatment of Sewage and Industrial Wastes, Vol. 2, Anaerobic Digestion and Solids—Liquid Separation" (J. McCabe and W. W. Eckenfelder, Jr., eds.), pp. 107–114. Reinhold, New York.

McCarty, P. L. (1964a). *Public Works* **95**(9), 107–112.

McCarty, P. L. (1964b). *Public Works* **95**(11), 91–99.

McCracken, R. J. (1977). Personal communication.

Makhijani, A., and Poole, A. (1975). "Energy and Agriculture in the Third World." Ballinger, Cambridge, Massachusetts.

Maldonado, O., Espinosa, R., Rolz, C., and Humphrey, A. E. (1975). *Ann. Technol. Agric.* **24**(3/4), 335–342.

Mantell, C. L. (1975). Solid Wastes: Origin, Collection, Processing and Disposal." Wiley, New York.

Meade, G. P. (1976). "Cane Sugar Handbook," 10th Ed. Wiley, New York.

Messell, H., and Butler, S. T. (1974). "Solar Energy." Shakespeare Head Press, Sydney.

Ministry of Agriculture, Food and Natural Resources of the Sudan (1974). "Food and the Sudan." Khartoum Univ. Press, Khartoum.

Mor, J. R., and Fietcher, A. (1968). *Biotechnol. Bioeng.* **10**(6), 787–803.

Moreira, J. R., Goldemberģ, J., and Serra, G. E. (1977). "A Cogeneration Scheme for the Production of Alcohol and Electricity from Sugar Cane." Inst. Fis., Univ. de Sao Paulo, Sao Paulo.

Morris, G. R., Jewell, W. J., and Cassler, G. L. (1975). Alternative animal waste anaerobic fermentation designs and their costs. *In* "Energy Agriculture and Waste Management" (W. J. Jewell, ed.), pp. 317–335. Ann Arbor Sci. Publ., Ann Arbor, Michigan.

National Academy of Sciences (1977). "Supporting Papers: World Food and Nutrition Study," Vol. 2. Natl. Acad. Sci., Washington, D.C.

North, W. J. (1977). Personal communication.

Openshaw, K. (1974). *New Sci.* **61,** 271–272.

Pimentel, D. (1961). *Ann. Entomol. Soc. Am.* **54,** 76–86.

Pimentel, D. (1977). Personal communication.

Pimentel, D., chm. (1978). "Energy Needs, Uses and Resources in the Food Systems of Developing Countries. Workshop Report." New York State Coll. Agric. Life Sci., Cornell Univ., Ithaca, New York.

Pimentel, D., Terhune, E. C., Dyson-Hudson, R., Rochereau, S., Samis, R., Smith, E. A., Denman, D., Rerfschneider, D., and Shepard, M. (1976). *Science* **194,** 149.

Pimentel, D., Nafus, D., Vergara, W., Papaj, D., Jaconetta, L., Wolfe, M., Olswig, L., Frech, K., Loye, M., and Mendoza, M. (1978). "Solar Energy Conversion by Living Systems and the Energy Economy." New York State Coll. Agric. Life Sci., Cornell Univ., Ithaca, New York.

Ramirez, J. M. (1970). "Estimates of Potential Productivity from the Climatology of Solar Energy." Dep. Atmos. Sci., Univ. of Missouri, Columbia.

Rich, J. R., and Bauer, R. H. (1975). "The Feasibility of Generating Electricity in the State of Vermont Using Wood as Fuel: A Study of the State of Vermont." Agency Environ. Conserv., Dep. For. Parks, Concord, Vermont.

Rodriguez, B. V. (1968). "Continuous fermentation rum pilot plant. Agricultural Experimental station. U.P.R. Rio Piedras, Puerto Rico, *Jt. Meet. Am. Inst. Chem. Eng. Inst. Ing. Quim. Puerto Rico, 2nd, Tampa, Fla.* Preprint No. 21.

Royal Swedish Academy of Sciences (1975). "Energy Uses. Report of Group 8." Aspakagardan, Sweden.

Scheller, W. A. (1976). Net energy analysis of ethanol production. *Am. Chem. Soc. Meet., 169th, Philadelphia, Pa.* Preprint No. 21.

Scheller, W. A. (1977). The use of ethanol-gasoline mixtures for automotive fuel. *Symp. Pap., Clean Fuels Biomass Wastes, Orlando, Fla.* pp. 185–200. Inst. Gas Technol., Chicago, Illinois.

Schrader, W. D. (1977). "Biomass. A Cash Crop for the Future?" Midwest Res. Inst. Battelle Columbus Lab., Columbus, Ohio.

Scroepfer, G. J., Fuller, W. J., Johnson, A. S., Ziemke, N. R., and Anderson, J. J. (1955). *Sewage Ind. Wastes* **27,** 460–486.

Serra, J. (1971). "El Milagro Brasilero: Realidad a Mito." Quimantu, Santiago de Chile.

Singh, R. B. (1971). "Biogas Plants." Gobar Gas Inst., Uttar Pradesh.

Skinner, J. H. (1974). Demonstration of systems for the recovery of material and energies from waste. *Nat. Mater. Conservation Symp.* pp. 53–63. ASCM, Philadelphia, Pa.

Tatom, J. W., Colcord, A. R., Knight, J. A., Elstom, L. W., and Har-oz, P. H. (1975). A mobile pyrolytic system—agricultural and forestry waste into clean fuels. *In* "Energy Agriculture and Waste Management" (W. J. Jewell, ed.), pp. 271–288. Ann Arbor Sci. Publ. Ann Arbor, Michigan.

Thappar, S. D. (1975). "India's Forest Resources," Birla Inst. Sci. Res. McMillan Co. India, Ltd., Delhi.

Thomas, G. (1974). *Proc. Queensl. Soc. Sugar Cane Technol.* 41.

Underkofler, L. A. (1954). "Industrial Fermentations." Chem. Publ. Co., New York.

United Nations (1976). "Statistical Yearbook," Vol. 30. United Nations, New York.

U.S. Department of Agriculture (1976). *Agricultural Statistics 1976*. U.S. Gov. Print. Off., Washington, D.C.

Veraga, W. (1977). Energy analysis of food processing: The case of vegetables processed in New York State. Unpublished master's thesis, Cornell Univ., Ithaca, N.Y.

Veraga, W., Welsch, F. and Serviansky, D. (1977). Utilization of food processing waste, Dept. Food Science, Cornell Univ., Ithaca, N.Y.

Volatron, B. (1976). *Bois For. Trop*. 166, 55–70.

Wadleigh, C. H. (1968). Waste in relation to agriculture and forestry. *U.S. Dep. Agric., Misc. Publ*. No. 1065.

Waterman, W. W., and Klass, D. L. (1977). Biomass as a long range source of hydrocarbons. *Symp. Pap., Clean Fuels Biomass Wastes, Orlando, Fla*. pp. 2–12. Inst. Gas Technol., Chicago, Illinois.

Wick, E., and Popper, K. (1977). *Biotechnol. Bioeng*. **19**(2), 235–246.

Wick, E., Popper, K., and Graham, R. P. (1974). *Biotechnol. Bioeng*. **16**(12), 1611–1631.

World Bank (1976). "World Tables 1976." John Hopkins Press, Baltimore, Maryland.

Young, H. E., Carpenter, P. N., and Altenberger, R. A. (1967). *Maine Agric. Exp. Stn., Tech. Bull*. No. 20.

ADVANCES IN ENERGY SYSTEMS AND TECHNOLOGY, VOL. 1

Geothermal Energy

Vasel Roberts

Electric Power Research Institute
Palo Alto, California

ISBN 0-12-014901-X

I. INTRODUCTION

A. Early Developments

Geothermal energy is the natural heat of the earth and is a truly international source of energy, available as an indigenous source of energy in all countries on earth. At present, geothermal energy is being sought primarily for the production of electric power, but its utilization in other forms, such as space and industrial heating, may be of great value in the economy of many nations seeking development of alternative energy sources and conservation of fossil and nuclear fuels.

Geothermal energy has been used for centuries. Early uses included space heating and therapeutic baths (balneology). Current applications include agriculture, municipal and industrial heating, and the generation of electric energy.

Attempts to harness geothermal energy to produce electric power date back to the beginning of this century. The first experimental electric power generator was operated at Larderello, Italy, in 1904, using natural geothermal steam that seeped through rock fractures to the surface. The first power plant was completed in 1913 in Italy, the pioneer in developing geothermal energy for power generation. In a number of other countries, including Japan, the United States, and Indonesia, attempts to develop geothermal energy for power production were made in the early 1920s. Geothermal energy developments have since occurred in a number of countries, reaching a total world electric capacity of about 1.3 GW (e)[1] in 1976. However, the incentive for developing geothermal energy was small when fossil fuel was inexpensive and presumably in abundant supply. Furthermore, energy planners have traditionally been reluctant to depend on an energy resource when its quantity could not be precisely predicted or its cost was not known well in advance. After the energy crisis of 1973, when fossil fuel costs rose markedly compared with the period immediately preceding the crisis, a reexamination of all existing energy resources brought about a renewed awareness and interest in geothermal energy for both electric and nonelectric uses.

[1] (e) represents electric.

B. Current Trends

This very large energy resource, available in one form or another in every country on earth, could play an increasingly important role in the overall energy supply. Some countries fortunate enough to be in the major geothermal belts of the world may be able to supply a significant portion of their energy requirements from geothermal resources. However, even countries in areas of normal temperature gradients may find that certain geologic, climatic, and economic circumstances may combine to create a favorable economic setting for utilizing hot subsurface water.

The degree of penetration of any natural resource into the marketplace is always a function of dynamic economic relationships that are sometimes difficult to determine in advance. The concept of reserves is intimately linked to these relationships, where reserve is that portion of the resource base known to exist and to be producible and marketable in competition with other resources. In the case of geothermal energy, neither the reserves nor the resource base is well known at this time. As a first approximation, the potential of geothermal energy is examined on the basis of those geothermal, geologic and geophysical parameters that can be quantified in a general sense, and engineering estimates of energy recovery and conversion from past experience and practice. In general the estimates derived by this method tend to be somewhat higher than those projected by various individuals and organizations around the world.

The geothermal resource base underlying the continental land masses of the world to a depth of 3 km and at temperatures higher than 15°C was calculated to be 1.2×10^{13} GW h (th).[2] The sum of the estimates reported by individual countries was 5.8×10^{11} GW h (th). The difference lies in the gross assumptions of the general method used in the first approach, the general incompleteness of detailed assessments of geothermal potential, and conservative estimates of specific resources in individual nations. Only about 2% of this resource base is high enough in temperature to be considered for electric power generation, and only about 1 part in 10,000 can be converted to electric energy with present technology. Although the overall recovery and conversion efficiencies are very small. The electric energy potential with current technology is substantial, roughly 1.04×10^{9} GWh (e). Potential geothermal energy utilization throughout the world is summarized in Tables I and II. The basis for the summary is developed in the sections on resources, energy conversion, and nonelectric utilization.

[2] th represents thermal.

TABLE I

World Geothermal Resource Base—Calculated

Temperature Class (°C)	Resource Base[a] [GW h (th)]	Potential Utilization[b] Electric [GW h (e)]	Thermal [GW h (th)]
Under 100	1.05×10^{13}	0	7.3×10^{11}
100–150	1.07×10^{12}	0	7.5×10^{10}
150–250	2.7×10^{11}	4.8×10^{9}	1.9×10^{10}
Over 250	1.4×10^{10}	2.5×10^{8}	9.7×10^{8}
	1.2×10^{13}[c]	5.1×10^{9}	8.2×10^{11}

[a] Resource base is the energy stored at temperatures above 15°C to a depth of 3 km.
[b] It is assumed that 20% of these quantities are producible with current technology.
[c] For comparison the sum of the national estimates of resources base was 5.8×10^{11} GW h (th).

The potential for nonelectric uses is much greater than the potential for electric power generation in terms of total heat; but the growth of nonelectric applications may be slower than electric applications because of the necessity of compatible industries and of populations to be near the resource. However, geothermal energy is a truly international energy resource because it is present in all countries of earth. Thus, it should be carefully considered in evaluation of alternative energy sources throughout the world.

TABLE II

Estimated World Consumption of Geothermal Energy

	Electric Capacity [GW (e)]		Thermal Capacity [GW (th)]	
Year	Calculated	National Estimates[a]	Calculated	National Estimates
1977	—	1.3	—	7
1985	170	12	25,000	33
2000	500	90	80,000	76
2020	1000	Unavailable	120,000	210

[a] National estimates were obtained from the literature and contact with many of the individual countries, then summed to obtain the totals shown in the table.

II. GEOTHERMAL RESOURCES

Temperatures of the earth's crust increase with depth at varying rates, depending on location. The normal temperature gradient is about 25°C/km depth. Thus, in a region where the ambient surface temperature (mean annual temperature) is 15°C, it is expected that a random hole drilled to a depth of 1 km will encounter a temperature of 40°C (the ambient temperature plus the temperature increase due to the normal temperature gradient in the region). However, in some regions of the world the temperature gradient is much greater than normal, increasing in places to a gradient as high as 1°C/m. In those areas associated with recent volcanism, relatively high temperatures may be encountered at shallow depth.

A. Resource Types

Geothermal resources are divided into three basic types: liquid-dominated and vapor-dominated hydrothermal systems, hot dry rock systems, and geopressured systems. These types are characterized by their thermodynamic and hydrologic properties, as discussed below.

1. Liquid-Dominated Systems

Liquid-dominated geothermal resources are those controlled by the presence of circulating liquids (water or brine) that can transport the thermal energy of the rock from deep regions to near-surface regions by natural circulation. The temperature of known liquid-dominated geothermal systems varies from ambient or slightly above to as high as 360°C. Most investigators agree that liquid-dominated resource types are far more abundant than the vapor-dominated ones. Certainly far more liquid-dominated systems have been discovered than vapor-dominated systems.

Usually the temperature in the best liquid-dominated geothermal regions increases rapidly with depth until the top of the convection cell is reached. Further increases in temperature with depth will be slight until a depth is reached where liquid domination ceases. Temperatures seldom exceed the boiling point of water at prevailing hydrostatic pressures.

2. Vapor-Dominated Systems

Vapor-dominated systems, sometimes referred to as dry steam fields, are relatively rare. However, the most important and successful geothermal power developments in the world today are associated with the development of vapor-dominated systems (Larderello, Italy; The Geysers, California; and Matsukawa, Japan). In such systems the continuous phase within the pore space in the near-surface region is that of steam, while in the deeper regions water is presumed to be present. Temperatures are typically in the range of 150–220°C. Production of steam from this type of reservoir is relatively simple, and quite often a slight superheating of the steam occurs during production.

3. *Hot Dry Rock Systems*

Since temperatures increase with depth, independent of hydrothermal convection, it is reasonable to assume that much more heat is stored in the rock matrix than in the circulating water. Since porosity generally decreases with depth, it may be further assumed that the rock-to-water ratio increases and that vast volumes of hot dry rock exist at greater depths within the earth's crust. Research efforts are under way to develop methods of introducing cold surface water into such hot dry rock systems with artificially induced fractures and extracting heated water through a pattern of holes drilled in the vicinity of the injection holes. Because of the high rock-to-water ratio, calculations suggest that the heat reserve in known geothermal systems is much larger than the heat contained in the fluids only. It is conceivable that many of the hydrothermal systems known today could become depleted of geofluids long before the heat reserve itself has been exhausted. Thus, it is possible that today's liquid-dominated geothermal systems may be further exploited as hot dry rock systems at some time in the future.

4. *Geopressure Systems*

Geopressured reservoirs are generally located in deep sedimentary strata in geologic regions where sediment compaction has taken place over geologic periods of time and where an effective shale cap has formed. Under conditions of shale compression, in which water is squeezed out of the shale matrix into adjacent sand bodies, an internal pressure greater than the ordinary hydrostatic pressure at that depth is imparted to the water. In extreme cases of geopressure, water pressures approach those of the overall weight of the overlying rocks (close to lithostatic pressure). This overpressured water system, known as a geopressured geothermal resource, is often characterized by higher-than-normal temperature gradients because of the increased specific heat capacity of the overpressured rock-water system. Temperatures as high as 237°C have been encountered in some geopressured zones in the Gulf Coast of the United States, with wellhead pressures in excess of 7.6×10^7 Pa (11,000 psi). In addition, geopressured fluids typically contain anomalously high concentrations of dissolved methane gas. Practically all large synclinal basins of the world contain some geopressured zones. In the

United States alone, geopressured geothermal resources cover an area of more than 200,000 km^2 in the states of Texas and Louisiana.

B. Geographic Distribution of Resources

Geothermal energy resources are concentrated to a large extent along certain well-defined belts, shown in Fig. 1. These belts of higher geothermal potential are associated with earthquake activity and recent volcanism. These belts of high volcanic activity, seismicity, and hot spring activity are associated with geologic plate boundaries and cover approximately 10% of the earth's surface. The most outstanding geothermal belt of the world is the Circum-Pacific Belt (the so-called Belt of Fire). One branch of the belt extends through Central America to the western portion of South America and then to Antarctica. Another branch of the belt runs through the East Pacific region through New Zealand, New Guinea, and Indonesia and then branches out north, running through the Philippines, Japan, and Eastern Siberia; it then turns eastward through the Aleutians back to northern Canada. Another outstanding geothermal belt runs through East Africa, Ethiopia, Kenya, Tanzania, and gradually weakens southward. The western branch of the Circum-Pacific Belt extends

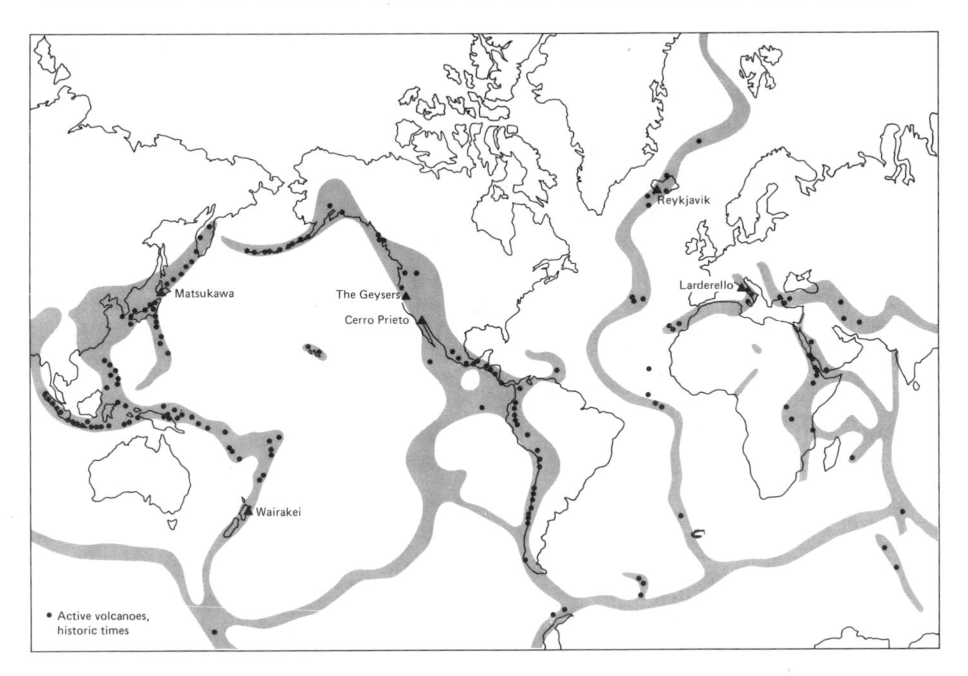

Fig. 1. Map of the geothermal regions of the world.

through northern India, Iran, Turkey, and Greece. A continuation of that belt runs through Italy and probably through North Africa. The Mid-Atlantic Ridge Belt usually runs underwater. However, where islands have formed after volcanic eruptions associated with the separation of the major earth plates, considerable geothermal activity is to be found, as in Iceland or in the Azores chain. Geopressured resources are also abundant in many major sedimentary regions of the world, including the Gulf Coast of the United States, the North Sea, western India, and the Persian Gulf.

C. Geothermal Resource Base of the World

1. Analytic Estimates

Active geothermal exploration is proceeding in practically all countries that are crossed by the major geothermal belts. In many countries where fossil fuel resources are not known to exist or are in short supply, exploration of hot water resources for direct heat purposes is currently under way. Estimates of the geothermal resource base for nations of the world were derived by a simple analysis involving a number of assumptions with respect to heat content and volumes of thermal resources.

a. Assumptions. The following assumptions were used in calculating the resource base:

1. All "normal" areas of the world have a vertical temperature gradient of 25°C/km.
2. There are certain geothermal belts in the world where the vertical temperature gradient is higher than normal. Either part or all of any country may lie in a geothermal belt. Let A denote the total area of a country, of which a fraction X lies in a geothermal belt. X may range from 0 to 1.
3. Of the geothermal area (XA) in a country, 90% ($0.9XA$) is assumed to have a temperature gradient of 40°C/km depth and 10% ($0.1XA$) is assumed to have a temperature gradient of 80°C/km.
4. The average ambient atmospheric temperature of all countries is assumed to be 15°C.
5. For the purpose of this analysis, the geothermal resource base of a country is defined as the total heat contained in subsurface rocks and fluids to a depth of 3 km and at temperatures above 15°C.
6. Four temperature resource classes are considered:
Class 1: <100′C; Class 2: 100–150°C; Class 3: 150–250°C; Class 4: >250°C.
7. Although the heat capacity of rock depends on lithology, porosity, and water content, an average volumetric specific heat (C_v) of 0.6 cal/cm³ °C is an acceptable value for most areas.

8. The resource base Q associated with a particular temperature class is given by

$$Q = AHC_v(T - 15) \quad (2.1)$$

where A denotes the area underlain by the resource, H the thickness (depth range) of the resource, C_v the volumetric specific heat of the subsurface material, and T the average temperature of the resource. Taking A in cm^2, H in cm, $C_v = 0.6$ cal/cm³ °C, and T in °C,

$$Q = 0.6AH(T - 15) \quad (2.2)$$

where Q is in calories.

9. In any temperature class, 20% of the total resource base is stored in the water and steam and the remainder is stored in the rock matrix. The rationale is as follows: Assuming the porosity range of subsurface material to be from 5 to 20%, the amount of water in 1 cm^3 of porous rock varies from 0.05 to 0.2 cm^3. Assuming an approximate volumetric specific heat capacity of from 0.05 to 0.2 cal/°C, compared with 0.6 cal/°C for the 1-cm^3 rock-fluid system (assumption 7), the fraction of the resource base stored in water is from 0.05/0.6 to 0.2/0.6, or from 8.3 to 33.3%, with the average assumed to be 20%. The remaining 80% of the resource base is assumed to be stored in the rock.

b. Calculation Procedure.

1. Data on the surface area of each country were obtained from *The Times Atlas of the World* (1974) and converted to square centimeters.

2. The value X for each country was determined by estimating areal fractions in Fig. 1 by expert, but subjective, judgment.

3. Resource base in normal areas with 25°C/km temperature gradients: Area of normal gradient = $(A - XA)$ cm^2. Temperature gradient = 25°C/km. Temperature (T) at any depth (h) is given by $T = 15 + 25h$ (e.g., 3 km, $T = 90$°C).

Thus, the entire resource in a normal area belongs to Class 1 (i.e., <100°C). The average temperature to a 3-km depth in the normal areas is $(15 + 90)/2$, or 52.5°C. Using Q^0 to denote normal areas and Q_1, Q_2, Q_3, and Q_4 to denote the four resource temperature classes, the resource base of the normal areas (in calories) can be calculated as follows:

$$Q_1^0 = 0.6(A - XA)3(52.5 - 15)10^5$$

or

$$Q_1^0 = 6.75 \times 10^6(A - XA) \quad (2.3)$$

$$Q_2^0 = 0 \quad (2.4)$$

$$Q_3^0 = 0 \tag{2.5}$$

$$Q_4^0 = 0 \tag{2.6}$$

The total resource base in normal areas = $Q_1^0 + Q_2^0 + Q_3^0 + Q_4^0$. Of this resource base, 20% is classified as hot water or steam and 80% is in the dry rock category.

4. Resource base in geothermal areas with 40°C/km temperature gradients: Area under consideration = $0.9XA$ cm². Temperature gradient = 40°C/km. Temperature (T) at any depth (h) is given by $T = 15 + 40h$, or $h = (T - 15)/40$.

The depth to the bottom of the Class 1 resource is (100 − 15)/40, or 2.125 km. The average temperature between the surface and this depth is (15 + 100)/2, or 57.5°C. Denoting these areas by Q^1 and using Q_1, Q_2, Q_3, and Q_4 to denote resource classes, and using Eq. (2.2),

$$Q_1^1 = 0.6(0.9XA)(2.125)(57.5 - 15)10^5 \tag{2.7}$$

or

$$Q_1^1 = (4.88 \times 10^6)XA$$

The depth to the bottom of the Class 2 resouce is (150 − 15)/40 or 3.38 km. Therefore, only part of the Class 2 resource falls within the 3-km depth range. The temperature at 3-km depth is 15 + (40)(3), or 135°C. The average temperature of the resource between depths 2.125 and 3 km is (100 + 135)/2 or 117.5°C and the corresponding thickness $h = (3 - 2.125)$ or 0.875 km. Therefore, using Eq. (2.2),

$$Q_2^1 = 0.6(0.9XA)0.876(117.5 - 15)10^5$$

or

$$Q_2^1 = (4.84 \times 10^6)XA \tag{2.8}$$

$$Q_3^1 = 0 \tag{2.9}$$

$$Q_4^1 = 0 \tag{2.10}$$

The total resource base in 90% of the geothermal area in a country is $Q_1^1 + Q_2^1 + Q_3^1 + Q_4^1$. Of this resource, 20% belongs to the hot water or steam classification and 80% is in the dry rock classification.

5. Resource base in geothermal areas with 80°C/km temperature gradients: Area under consideration = $0.1XA$ cm². Temperature gradient = 80°vC/km. Temperature (T) at any depth (h), in km, is given by $T = 15 + 80h$, or $h = (T - 15)/80$.

The depth to the bottom of the Class 1 resource is (100 − 15)/80, or 1.063 km. The average temperature between the surface and this depth is

(15 + 100)/2, or 57.5°C. Denoting these areas by Q^2 and using Q_1, Q_2, Q_3, and Q_4 to denote resource classes:

$$Q_1^2 = 0.6(0.1XA)1.063(57.5 - 15)10^5 \tag{2.11}$$

or

$$Q_1^2 = (2.7 \times 10^5)XA$$

The depth to the bottom of the Class 2 resource is (150 − 15)/80, or 1.688 km. The average temperature between 1.063 km and this depth is (100 + 150)/2, or 125°C, and the corresponding thickness $h = 1.688 - 1.063 = 0.625$ km. Using Eq. (2.2),

$$Q_2^2 = 0.6(0.1XA)0.625(125 - 15)10^5 \tag{2.12}$$

or

$$Q_2^2 = (4.125 \times 10^5)XA$$

The depth to the bottom of the Class 3 resource is (250 − 15)/80, or 2.938 km. The average temperature between 1.688 km and 2.938 km is (150 + 250)/2, or 200°C, and the corresponding thickness $h = 2.938 - 1.688$, or 1.250 km. Using Eq. (2.2),

$$Q_3^2 = 0.6(0.1XA)1.250(200 - 15)10^5 \tag{2.13}$$

or

$$Q_3^2 = (1.388 \times 10^6)XA$$

At 3 km, the temperature is 15 + 80(3), or 255°C. Therefore, the average temperature between 2.938 km and 3 km is (250 + 255)/2, or 252.5°C, the corresponding thickness h being (3 − 2.938), or 0.062 km. Using Eq. (2.2),

$$Q_4^2 = 0.6(0.1XA)0.062(252.5 - 15)10^5 \tag{2.14}$$

or

$$Q_4^2 = (8.55 \times 10^4)XA$$

The total resource base in 10% of the geothermal area of a country = $Q_1^2 + Q_2^2 + Q_3^2 + Q_4^2$. Of this resource 20% belongs to the hot water and steam classification and 80% belongs to the dry rock classification.

6. The resource base of a country, in different classes, is given as follows:

$$\text{Class 1:}\quad Q_1 = Q_1^0 + Q_1^1 + Q_1^2 \tag{2.15}$$

$$\text{Class 2:}\quad Q_2 = Q_2^0 + Q_2^1 + Q_2^2 \tag{2.16}$$

$$\text{Class 3:} \quad Q_3 = Q_3^0 + Q_3^1 + Q_3^2 \tag{2.17}$$

$$\text{Class 4:} \quad Q_4 = Q_4^0 + Q_4^1 + Q_4^2 \tag{2.18}$$

7. The total resource base of a country is given by

$$Q = Q_1 + Q_2 + Q_3 + Q_4 \tag{2.19}$$

The total hydrothermal (hot water and steam) resource of a country is 0.2 Q cal. The total hot dry rock resource of a country is 0.8 Q cal. The resource base was calculated in this manner for various countries of the world. The result is shown in Table III.

c. Calculated Values of Resource Base. In Table III it is noted that the resource base in the less than 100°C class is by far the largest quantity. Although this class is too low in temperature for electric power generation, it may be useful for space heating and other direct heat uses, provided the economics of drilling and producing this portion of the resource is favorable in some particular regions of the world. The resource in the 100–150°C class is not likely to be economical for electric power generation in many nations. However, this part of the resource can be used for direct heat purposes. The total worldwide resource base (in the continental masses to 3 km depth) for these two temperature classes is 9.0×10^{24} and 9.2×10^{23} cal, respectively. The resources in the 150–250°C and 250°C classes are currently competitive with fossil fuels for electric power generation; the higher the temperature, the better are their thermodynamic and economic efficiencies. The resource bases in these two classes are 2.3×10^{23} and 1.2×10^{22} cal, respectively.

d. National Estimates of Resource Base. Only a few countries have estimated their geothermal resource base. The estimates that are available are shown in Table IV. In many cases the resource base reported in Table IV is much lower than that estimated in Table III by the volumetric heat content method. Such discrepancies are inherent and are expected in estimates of this nature. The difference lies in the gross assumption of the volumetric heat method, the general incompleteness of geothermal assessments, and perhaps some conservatism in official estimates.

III. ELECTRICAL POWER POTENTIAL

Although the geothermal resource base computed in Section II is very large, only a small fraction is suitable for conversion to electric energy. The comparatively low temperature and the energy extraction difficulties result in an overall conversion efficiency that is small relative to that attainable with fossil fuels. The overall conversion efficiency is limited by the first and second laws of thermodynamics and by the inefficiencies associated with energy extraction and conversion to electrical power.

A. Analytical Approach

To facilitate analytical estimates, the inefficiencies are grouped as follows:

F_a is the available energy fraction. This factor includes an energy recovery component expressed as the ratio of recoverable energy to resource base, corrected for thermal efficiency.

F_w is the wellhead fraction. This factor accounts for the friction, work, and heat flow losses associated with energy extraction.

F_p is the power plant fraction. This factor accounts for losses between the wellhead and the power plant.

F_b is the busbar fraction. This factor accounts for losses associated with the conversion from heat to electricity. It does not include the thermal conversion efficiency associated with the second law of thermodynamics, which was included in F_a.

The electrical energy delivered to the busbar, E_b, is calculated by the equation

$$E_b = F_a F_w F_p F_b Q \tag{3.1}$$

where Q is the resource base.

For a given geothermal field of volume V, the geothermal resource base can be estimated in terms of the average specific heat and the difference between the initial reservoir temperature and the reference datum temperature:

$$Q/V = C_v(T_1 - T_0) \tag{3.2}$$

where C_v is the volumetric specific heat (0.6 cal/cm^3°C, from Section II), T_1 is the initial reservoir temperature, and T_0 is the reference datum temperature (15°C, from Section II).

The thermodynamic model of the typical geothermal system, assuming no heat recharge, is illustrated by the absolute temperature–entropy diagram of Fig. 2. The nomenclature of Fig. 2 is used in the derivation of equations in this section.

1. Energy Recovery

Assuming that T_2 is the temperature at which the geothermal reservoir production ceases to be economical, the ideal fraction of the energy per

TABLE III

Geothermal Resource Base of Countries of the World—Calculated

	Area (km^2)	X^a	Resource Base (cal)				
			Class 1 (<100°C)	Class 2 (100–150°C)	Class 3 (150–250°C)	Class 4 (>250°C)	Total
Algeria[b]	2.4×10^6	0.007	1.6×10^{23}	8.5×10^{20}	2.3×10^{20}	1.4×10^{19}	1.6×10^{23}
Angola[b]	1.3×10^6	0	8.4×10^{22}	0	0	0	8.4×10^{22}
Argentina[b]	2.8×10^6	0.03	1.9×10^{23}	4.4×10^{21}	1.2×10^{21}	7.0×10^{19}	2.0×10^{23}
Australia	7.7×10^6	0	5.2×10^{23}	0	0	0	5.2×10^{23}
Austria	8.4×10^4	0	5.7×10^{21}	0	0	0	5.7×10^{21}
Bangladesh	1.4×10^5	0	9.6×10^{21}	0	0	0	9.6×10^{21}
Barbados[b]	4.3×10^2	1.00	2.2×10^{19}	2.3×10^{19}	6.0×10^{18}	3.7×10^{17}	5.1×10^{19}
Belgium	3.1×10^4	0	2.1×10^{21}	0	0	0	2.1×10^{21}
Bolivia[b]	1.1×10^6	0.30	6.9×10^{22}	1.7×10^{22}	4.6×10^{21}	2.8×10^{20}	9.1×10^{22}
Brazil[b]	8.5×10^6	0.05	5.7×10^{23}	2.2×10^{22}	5.9×10^{21}	3.6×10^{20}	6.0×10^{23}
Bulgaria	1.1×10^5	0	7.5×10^{21}	5.8×10^{19}	1.5×10^{19}	9.5×10^{17}	7.6×10^{21}
Burundi[b]	2.8×10^4	1.00	1.4×10^{21}	1.5×10^{21}	3.8×10^{20}	2.4×10^{19}	3.3×10^{21}
Cameroun[b]	4.8×10^5	0.10	3.1×10^{22}	2.5×10^{21}	6.6×10^{20}	4.1×10^{19}	3.5×10^{22}
Canada[b]	1.0×10^6	0.20	6.4×10^{23}	1.1×10^{23}	2.8×10^{22}	1.7×10^{21}	7.8×10^{23}
Chad[b]	1.3×10^6	0.10	8.4×10^{22}	6.7×10^{21}	1.8×10^{21}	1.1×10^{20}	9.2×10^{22}
Chile[b]	7.6×10^5	1.00	3.9×10^{22}	4.0×10^{22}	1.1×10^{22}	6.5×10^{20}	9.0×10^{22}
China[b]	9.6×10^6	0.25	6.1×10^{23}	1.3×10^{23}	3.3×10^{22}	2.0×10^{21}	7.7×10^{23}
Colombia[b]	1.1×10^6	0.30	7.1×10^{22}	1.8×10^{22}	4.8×10^{21}	2.9×10^{20}	9.4×10^{22}
Costa Rica	5.1×10^4	1.00	2.6×10^{21}	2.7×10^{21}	7.1×10^{20}	4.4×10^{19}	6.0×10^{21}
Cuba	1.1×10^5	0	7.7×10^{21}	0	0	0	7.7×10^{21}
Cyprus	9.3×10^3	0	6.2×10^{20}	0	0	0	6.2×10^{20}
Czechoslovakia[b]	1.3×10^5	0.10	8.4×10^{21}	6.7×10^{20}	1.8×10^{20}	1.1×10^{19}	9.2×10^{21}
Dahomey	1.2×10^5	0	7.8×10^{21}	0	0	0	7.8×10^{21}
Denmark	4.3×10^4	0	2.9×10^{21}	0	0	0	2.9×10^{21}

Dominican Republic[b]	4.8×10^{4}	0.25	3.1×10^{21}	6.4×10^{20}	1.7×10^{20}	1.0×10^{19}	3.9×10^{21}
Ecuador	4.6×10^{5}	1.00	2.4×10^{22}	2.4×10^{22}	6.2×10^{21}	3.9×10^{20}	5.4×10^{22}
Egypt	9.5×10^{5}	0	6.4×10^{22}	0	0	0	6.4×10^{22}
El Salvador	2.1×10^{4}	1.00	1.1×10^{21}	1.1×10^{21}	2.9×10^{20}	1.8×10^{19}	2.5×10^{21}
England	1.3×10^{5}	0	8.8×10^{21}	0	0	0	8.8×10^{21}
Ethiopia	1.2×10^{6}	0.60	7.1×10^{22}	3.9×10^{22}	1.0×10^{22}	6.3×10^{20}	1.2×10^{23}
Fiji[b]	1.8×10^{4}	0.20	1.2×10^{21}	1.9×10^{20}	5.1×10^{19}	3.1×10^{18}	1.4×10^{21}
Finland	3.6×10^{5}	0	2.4×10^{22}	0	0	0	2.4×10^{22}
France	5.5×10^{5}	0	3.7×10^{22}	0	0	0	3.7×10^{22}
East Germany	1.1×10^{5}	0	7.3×10^{21}	0	0	0	7.3×10^{21}
West Germany	2.5×10^{5}	0	1.7×10^{22}	0	0	0	1.6×10^{22}
Ghana	2.4×10^{5}	0	1.6×10^{22}	0	0	0	1.6×10^{22}
Greece	1.3×10^{5}	0.30	8.3×10^{21}	2.1×10^{21}	5.5×10^{20}	3.4×10^{19}	1.1×10^{22}
Greenland	2.2×10^{6}	0	1.5×10^{23}	0	0	0	1.5×10^{23}
Guadeloupe	1.8×10^{3}	1.00	9.1×10^{19}	9.3×10^{19}	2.5×10^{19}	1.5×10^{18}	2.1×10^{20}
Guatemala	1.1×10^{5}	1.00	3.4×10^{21}	5.7×10^{21}	1.5×10^{21}	9.3×10^{19}	1.1×10^{22}
Haiti[b]	2.8×10^{4}	0.025	1.9×10^{21}	3.7×10^{19}	9.5×10^{18}	6.0×10^{17}	1.9×10^{21}
Honduras[b]	1.1×10^{5}	0.50	6.7×10^{21}	2.9×10^{21}	7.8×10^{20}	4.8×10^{19}	1.0×10^{22}
Hungary	9.3×10^{4}	0	6.3×10^{21}	0	0	0	6.3×10^{21}
Iceland	1.0×10^{5}	1.00	5.3×10^{21}	5.4×10^{21}	1.4×10^{21}	8.8×10^{19}	1.2×10^{22}
India[b]	3.3×10^{6}	0.02	2.2×10^{23}	3.4×10^{21}	9.0×10^{20}	5.6×10^{19}	2.2×10^{23}
Indonesia	1.9×10^{6}	1.00	9.8×10^{22}	1.0×10^{22}	2.6×10^{22}	1.6×10^{21}	2.3×10^{23}
Iran[b]	1.7×10^{6}	0.20	1.1×10^{23}	1.7×10^{22}	4.6×10^{21}	2.8×10^{20}	1.3×10^{23}
Iraq	4.3×10^{5}	0	2.9×10^{22}	0	0	0	2.9×10^{22}
Ireland	7.0×10^{4}	0	4.7×10^{21}	0	0	0	4.7×10^{21}
Israel[b]	2.1×10^{4}	0.025	1.4×10^{21}	2.8×10^{19}	7.3×10^{18}	4.5×10^{17}	1.4×10^{21}
Italy	3.0×10^{5}	0.50	1.8×10^{22}	7.9×10^{21}	2.1×10^{21}	1.3×10^{20}	2.8×10^{22}
Ivory Coast	3.2×10^{5}	0	2.2×10^{22}	0	0	0	2.2×10^{22}
Jamaica	1.1×10^{4}	0	7.7×10^{20}	0	0	0	7.7×10^{20}
Japan	3.7×10^{5}	1.00	1.9×10^{22}	1.9×10^{22}	5.1×10^{21}	3.2×10^{20}	4.4×10^{22}
Jordan[b]	9.8×10^{4}	0.025	6.6×10^{21}	1.3×10^{20}	3.5×10^{19}	2.1×10^{18}	6.8×10^{21}
Kenya	5.8×10^{5}	0.60	3.4×10^{22}	1.8×10^{22}	4.9×10^{21}	3.0×10^{20}	5.7×10^{22}
Korea[b] (North and South)	2.2×10^{5}	0.50	1.3×10^{22}	5.8×10^{21}	1.5×10^{21}	9.4×10^{19}	2.1×10^{22}

(continued)

TABLE III (*continued*)

	Area (km^2)	X[a]	Resource Base (cal) Class 1 (<100°C)	Class 2 (100–150°C)	Class 3 (150–250°C)	Class 4 (>250°C)	Total
Kuwait	2.0×10^4	0	1.4×10^{21}	0	0	0	1.4×10^{21}
Liberia	1.1×10^5	0	7.5×10^{21}	0	0	0	7.5×10^{21}
Libya	1.8×10^6	0	1.2×10^{21}	0	0	0	1.2×10^{21}
Luxemburg	2.6×10^3	0	1.8×10^{20}	0	0	0	1.8×10^{20}
Malawi[b]	9.5×10^4	0.01	6.4×10^{21}	5.0×10^{19}	1.3×10^{19}	8.1×10^{17}	6.5×10^{21}
Malagasy Republic[b]	5.9×10^5	0.005	3.9×10^{22}	1.5×10^{20}	4.1×10^{19}	2.5×10^{18}	3.9×10^{22}
Malaysia	3.3×10^5	0.02	2.2×10^{22}	3.4×10^{20}	9.2×10^{19}	5.6×10^{18}	2.2×10^{22}
Mali	1.2×10^6	0	8.1×10^{22}	0	0	0	8.1×10^{22}
Martinique	1.1×10^3	1.00	5.7×10^{19}	5.8×10^{19}	1.5×10^{19}	9.4×10^{17}	1.3×10^{20}
Mexico	2.0×10^6	0.60	1.1×10^{23}	6.2×10^{22}	1.6×10^{22}	1.0×10^{21}	1.9×10^{23}
Morocco	4.4×10^5	0.04	3.0×10^{22}	9.3×10^{20}	2.5×10^{20}	1.0×10^{21}	1.9×10^{23}
Nepal	1.4×10^5	0.10	9.3×10^{21}	7.4×10^{20}	2.0×10^{20}	1.2×10^{19}	1.0×10^{22}
Netherlands	3.6×10^4	0	2.4×10^{21}	0	0	0	2.4×10^{21}
New Guinea	2.4×10^5	0.60	1.4×10^{22}	7.5×10^{21}	2.0×10^{21}	1.2×10^{20}	2.3×10^{22}
New Hebrides	1.5×10^4	1.00	7.6×10^{20}	7.8×10^{20}	2.1×10^{20}	1.3×10^{19}	1.8×10^{21}
New Zealand	2.7×10^5	0.50	1.6×10^{22}	7.0×10^{21}	1.9×10^{21}	1.1×10^{20}	2.5×10^{22}
Nicaragua	1.5×10^5	1.00	7.6×10^{21}	7.8×10^{21}	2.1×10^{21}	1.3×10^{20}	1.8×10^{22}
Nigeria	9.2×10^5	0	6.2×10^{22}	0	0	0	6.2×10^{22}
Norway	3.2×10^5	0	2.2×10^{22}	0	0	0	2.2×10^{22}
Pakistan	8.4×10^6	0.01	5.6×10^{23}	4.4×10^{21}	1.2×10^{21}	7.2×10^{19}	5.6×10^{23}
Panama[b]	7.6×10^4	1.00	3.9×10^{21}	4.0×10^{21}	1.1×10^{21}	6.5×10^{19}	9.0×10^{21}
Paraguay[b]	4.1×10^5	0.10	2.7×10^{22}	2.1×10^{21}	5.7×10^{20}	3.5×10^{19}	3.0×10^{22}
Peru[b]	1.3×10^6	1.00	6.6×10^{22}	6.8×10^{22}	1.8×10^{22}	1.1×10^{21}	1.5×10^{23}
Philippines	3.0×10^5	1.00	1.5×10^{22}	1.6×10^{22}	4.2×10^{21}	2.6×10^{20}	3.6×10^{22}
Poland	3.1×10^5	0	2.1×10^{22}	0	0	0	2.1×10^{22}
Portugal	9.2×10^4	0.05	6.1×10^{21}	2.4×10^{20}	6.4×10^{19}	3.9×10^{18}	6.4×10^{21}

Rumania	2.4×10^{5}	0.01	1.6×10^{22}	1.2×10^{20}	3.3×10^{19}	2.0×10^{18}	1.6×10^{22}
Saudi Arabia[b]	2.3×10^{6}	0.02	1.5×10^{23}	2.4×10^{21}	6.3×10^{20}	3.9×10^{19}	1.5×10^{23}
Senegal	2.0×10^{5}	0	1.3×10^{22}	0	0	0	1.3×10^{22}
Sierra Leon	7.2×10^{4}	0	4.9×10^{21}	0	0	0	4.9×10^{21}
South Africa	1.2×10^{6}	0	8.2×10^{22}	0	0	0	8.2×10^{22}
Spain	5.1×10^{5}	0.05	3.4×10^{22}	1.3×10^{21}	3.5×10^{20}	2.2×10^{19}	3.5×10^{22}
Sri Lanka	6.6×10^{4}	0	4.4×10^{21}	0	0	0	4.4×10^{21}
Sudan	2.5×10^{6}	0	1.7×10^{23}	0	0	0	1.7×10^{23}
Sweden	4.5×10^{5}	0	3.0×10^{22}	0	0	0	3.0×10^{22}
Switzerland	4.1×10^{4}	0	2.8×10^{21}	0	0	0	2.8×10^{21}
Taiwan	3.6×10^{4}	1.00	1.8×10^{21}	1.9×10^{21}	5.0×10^{20}	3.1×10^{19}	4.2×10^{21}
Tanzania	9.4×10^{5}	0.03	6.3×10^{22}	1.5×10^{21}	3.9×10^{20}	2.4×10^{19}	6.3×10^{22}
Thailand	5.1×10^{5}	0	3.5×10^{22}	0	0	0	3.5×10^{22}
Trinidad and Tobago	5.1×10^{3}	1.00	2.6×10^{20}	2.7×10^{20}	7.1×10^{19}	4.4×10^{18}	6.1×10^{20}
Tunisia	1.6×10^{5}	0.01	1.1×10^{22}	8.6×10^{19}	2.3×10^{19}	1.4×10^{18}	1.1×10^{22}
Turkey	7.8×10^{5}	0.50	4.6×10^{22}	2.0×10^{22}	5.4×10^{21}	3.3×10^{20}	7.3×10^{22}
USSR	2.2×10^{7}	0.05	1.5×10^{24}	5.8×10^{22}	1.5×10^{22}	9.5×10^{20}	1.6×10^{24}
United States	9.4×10^{6}	0.25	9.9×10^{23}	1.2×10^{23}	3.2×10^{22}	2.0×10^{21}	1.1×10^{24}
Uruguay	1.9×10^{5}	0.01	1.2×10^{22}	9.8×10^{19}	2.6×10^{19}	1.6×10^{18}	1.2×10^{22}
Vietnam (North and South)	3.3×10^{5}	0.50	2.0×10^{22}	8.8×10^{21}	2.3×10^{21}	1.4×10^{20}	3.1×10^{22}
Venezuela	9.1×10^{5}	0.20	5.9×10^{22}	0	2.5×10^{21}	1.6×10^{20}	7.1×10^{22}
Yugoslavia	2.6×10^{5}	0	1.7×10^{22}	0	0	0	1.7×10^{22}
Zaire	2.3×10^{6}	0.01	1.6×10^{23}	1.2×10^{21}	3.2×10^{20}	2.0×10^{19}	1.6×10^{23}
Zambia	7.5×10^{5}	0.10	5.0×10^{22}	4.0×10^{21}	1.6×10^{21}	6.4×10^{19}	5.5×10^{22}

[a] X represents the fraction of area underlain by geothermal deposits.
[b] The country occurs in conjectured geothermal belt not yet proven by exploration.

TABLE IV

Geothermal Resource Base—National Estimates

Country	Resource Base (cal)
Canada	4×10^{23}
England	2×10^{22}
El Salvador	4.5×10^{18}
Fiji	8×10^{17}
West Germany	3.4×10^{22}
Hungary	1.3×10^{22}
India	1.9×10^{19}
Mexico	4.1×10^{19}
New Zealand	3×10^{19}
Nicaragua	1.3×10^{20}
Norway	3.1×10^{22}
Spain	1.2×10^{17}
Sweden	3.0×10^{17}
Taiwan	3.0×10^{16}
	5.0×10^{23}

unit volume that can be produced is given by

$$E_{1-2}/V = C_v(T_1 - T_2)$$

Thus the fraction of the heat that can ideally be recovered is

$$E_{1-2}/Q = (T_1 - T_2)/(T_1 - T_0) \tag{3.3}$$

2. *Thermal Efficiency*

The thermal efficiency represents the fraction of the thermal energy E_{1-2} that can be converted to mechanical or electrical energy under the second law of thermodynamics. In Fig. 2, the shaded area represents the differential of ideal work, δWk_i. The energy bounded by the same interval of the curve down to 0°K is δE. The Carnot efficiency is

$$\eta_{\text{Carnot}} = \delta Wk_i/\delta E = 1 - T_0/T_H \tag{3.4}$$

where $\delta E = C_v \delta T_H$. Integration of δWk_i in Eq. (3.4) between the limits $T_1 \geq T_H \geq T_2$ results in

$$Wk_i/E_{1-2} = 1 - T_0/T_{1-2,\ln} \tag{3.5}$$

where $T_{1-2,\ln}$ is defined by

$$T_{1-2,\ln} = (T_1 - T_2)/\ln(T_1/T_2) \tag{3.6}$$

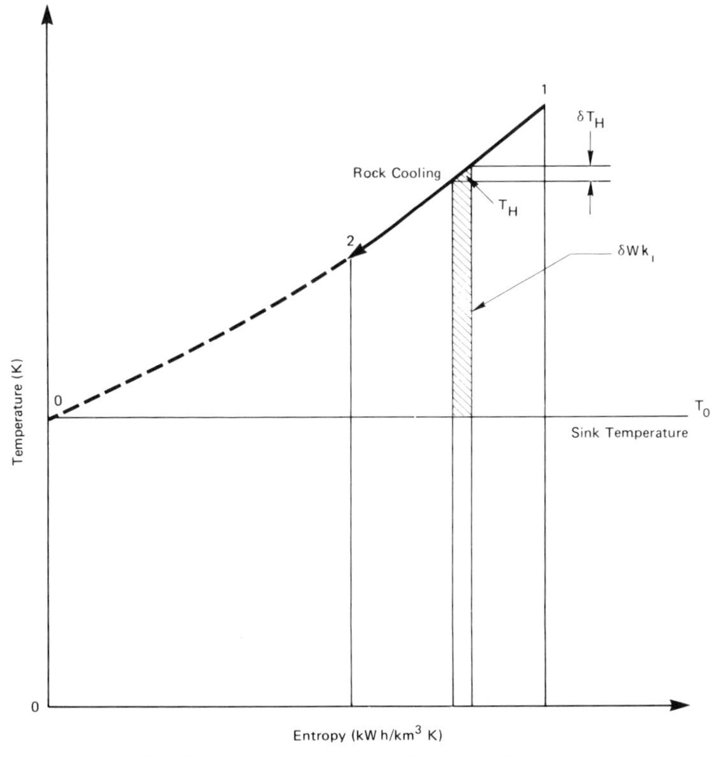

Fig. 2. Geothermal reservoir thermodynamics.

3. *Available Energy Fraction*

An expression for the fraction of the rock thermal energy that is available for conversion to electric energy, F_a is obtained by combining Eqs. (3.3) and (3.5),

$$F_a = \frac{Wk_i}{Q} = \frac{T_1 - T_2}{T_1 - T_0}\left(1 - \frac{T_0}{T_{1-2,\ln}}\right) \tag{3.7}$$

The available energy fraction may also be expressed as functions of the temperature ratios, $a = T_1/T_2$ and $b = T_1/T_0$, as

$$F_a = \left(\frac{a}{b}\right)\left(\frac{a-1}{b-1}\right) - \frac{\ln a}{b-1} \tag{3.8}$$

Figures 3 and 4 represent Eqs. (3.7) and (3.8), respectively, for $F_a = f(a)$ for $b = T_1/T_0 = 600/288$. F_a is the product of the components E_{1-2}/Q from Eq. (3.3) and Wk_i/E_{1-2} from Eq. (3.5). These components are also shown graphically in Fig. 3. Figure 4 shows F_a as a function of the initial

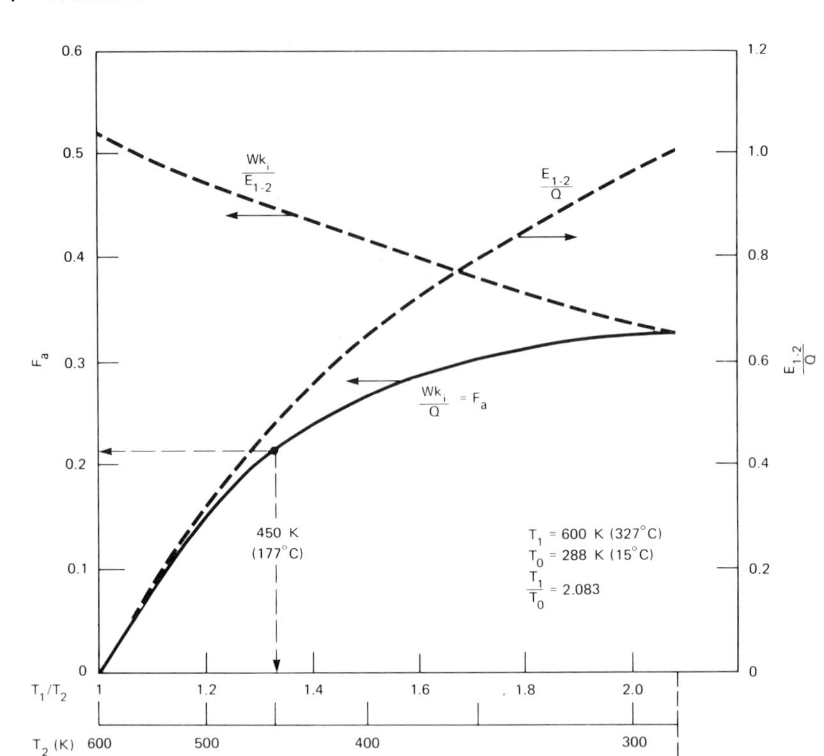

Fig. 3. Available energy fraction.

resource temperature T_1 and the lower production limit temperature T_2. Both figures show the strong influence of a large T_1 on F_a and a small T_2 on the available energy fraction. For example, for $T_1 = 600°K$ and $T_2 = 450°K$ (177°C), the available energy fraction is $F_a = 0.215$. However for $T_1 = 500°K$, F_a (from Fig. 4) is reduced by a factor of 2.3 to only 0.093. Furthermore, if the minimum usable resource temperature T_2 can be reduced from 177 to 150°C, Fig. 4 shows for $T_1 = 500°K$ that the fraction F_a increases from 0.093 to 0.136, improvement by a factor of 1.5.

One further factor affecting the "ideal" available energy fraction is the sink temperature T_0, shown in Fig. 2. If, because of local climatic conditions, the local sink or available coolant temperature T differs from T_0, Eq. (3.7) is modified to

$$F_a = \frac{T_1 - T_2}{T_1 - T_0}\left(1 - \frac{T_0(T/T_0)}{T_{1-2,\ln}}\right) \tag{3.9}$$

The sensitivity of F_a to the ratio T/T_0 is noted in Fig. 5. A reference point was selected corresponding to an initial reservoir temperature

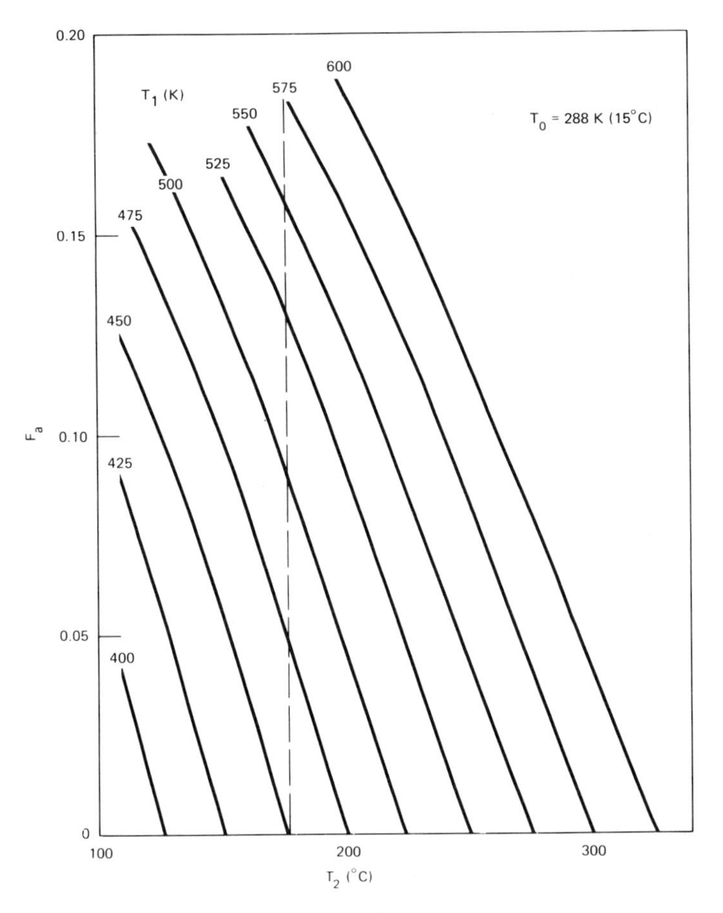

Fig. 4. Available energy as a function of initial and final reservoir temperatures.

$T_1 = 227°C$ (550°K) and a lower production limit temperature $T_2 = 127°C$. For $T = T_0 = 15°C$ (288°K), the available energy fraction F_a for the reference case is 0.168, as noted in Fig. 4. The data in Fig. 5 show the strong sensitivity of F_a to both T_1 and T_2 compared with the reference case, but a relatively low sensitivity of F_a to T.

4. *Wellhead Fraction*

The wellhead fraction F_w accounts for nonuniformity of rock temperature T_2 in a large-scale reservoir, resulting from long conduction paths in impermeable rock masses that inhibit heat transfer to circulating fluids. It also accounts for fluid friction heat losses in the well bore during production. The wellhead fraction is thus related to the efficiency of mining the available energy F_aQ. It can vary with the well and reservoir production

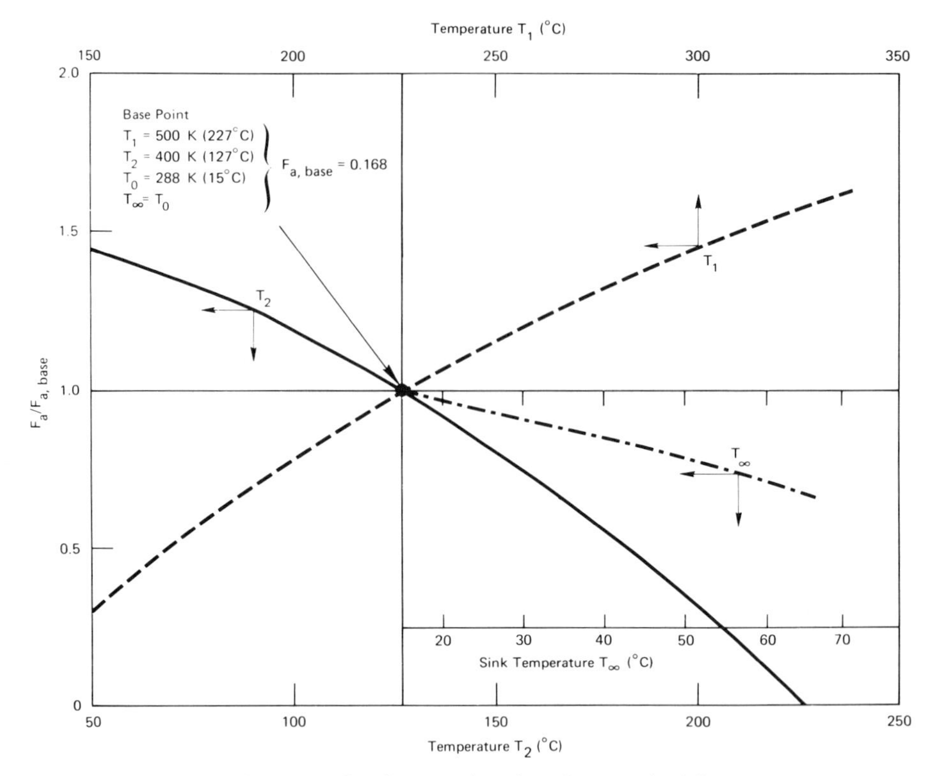

Fig. 5. Available energy fraction as a function of reservoir sink temperatures.

techniques employed. Enhanced extraction may be achieved by explosive or hydraulic fracturing or by injection of colder fluids to "sweep" the thermal energy to the well bore. Thus the wellhead fraction F_w, in contrast to the available energy fraction F_a, which is determined by thermodynamic principles, is very much dependent on judgment and experience. It is also dependent on the type of geothermal reservoir and its porosity, permeability, and rock-size distribution. Based on expected values for these characteristics, values for F_w are given below in Table V.

5. *Power Plant Fraction*

The power plant input fraction F_p considers the piping from wells to power plant, the mode of generation (such as single or multiple flash), heat exchanger losses for binary cycle turbines, and other temperature or pressure losses. Based on experience with vapor-dominated systems at The Geysers (United States) and with liquid-dominated systems at Wairakei (New Zealand), the values for the power plant fraction given in Table VI are used.

TABLE V

Values of F_w for Different Reservoir Types

Reservoir Type	Wellhead Fraction F_w	
	Range	Average
Vapor dominated	0.2–0.4	0.3
Liquid dominated	0.2–0.6	0.4
Hot dry rock	0.01–0.15	0.08
Geopressured	0.2–0.8	0.5

TABLE VI

Values of F_p for Different Reservoir Types

Reservoir Type	Power Plant Input Fraction F_p
Vapor dominated	0.95
Liquid dominated[a]	0.6
Hot dry rock[a]	0.6
Geopressured	0.6

[a] For either flash steam or binary cycle operation.

6. *Busbar Fraction*

The busbar fraction F_b reflects the efficiency of conversion of the net available energy delivered to a turbine (or other expander) to electric power at the busbar. At The Geysers and at Wairakei, the busbar fraction F_b is about 0.72. This is probably typical of most hydrothermal systems. Additional pumping power requirements will lower the expected value of F_b, whereas the additional hydraulic energy component of geopressured resources may increase the conversion efficiency. Expected values of F_b are given in Table VII.

TABLE VII

Values of F_b for Different Reservoir Types

Reservoir Type	Busbar Fraction F_b
Vapor dominated	0.72
Liquid dominated	0.72
Hot dry rock	0.65
Geopressured	0.75

TABLE VIII

Overall Efficiencies for Different Reservoir Types

Reservoir Type	F_b	F_p	F_w	F_a	η_0
Vapor dominated	0.72	0.95	0.3	0.09	0.018
Liquid dominated	0.72	0.6	0.4	0.09	0.016
Hot dry rock	0.65	0.6	0.08	0.185	0.0058
Geopressured	0.75	0.6	0.5	0.009	0.020

7. *Overall Efficiency*

In summary, the overall efficiency for producing electric power from a given geothermal reservoir is given by

$$\eta_0 = E_b/Q = F_a F_w F_p F_b \tag{3.10}$$

Typical values are given in the tabulation for hydrothermal (and geopressured) reservoirs with $T_1 = 227°C$ and $T_2 = 177°C$, and for hot dry rock reservoirs with $T_1 = 302°C$ and $T_2 = 177°C$.

These results show that the overall efficiency for producing electric power from hydrothermal systems is limited to about 2% of the resource base, primarily because of the small magnitudes of the factors F_a and F_w. Hot rock systems will have a smaller overall efficiency because of the smaller F_w. Although nature has placed a limit on the value of F_a, it is possible to increase the value of F_w through stimulation. Geopressured resources should have higher conversion factors for comparable operating temperatures T_1 and T_2 because of the larger wellhead fraction.

B. Calculated Estimate of World Potential for Geothermal Electricity

For the purpose of estimation, only that portion of the geothermal resource base above 150°C is considered suitable for electric power production, and it is assumed that the overall conversion efficiency as derived herein applies uniformly. Since current technology is not adequate for the development of petrothermal (hot dry rock) geothermal resources, the estimates herein apply to hydrothermal resources only. For these calculations it is assumed that about 20% of the resource base is contained in hydrothermal systems.

1. *Calculation Method*

The following equation was used to calculate hydrothermal electric power potential for hydrothermal systems:

$$E_b = 0.2Q\eta_0 \tag{3.11}$$

for which the terms have already been defined. The coefficient 0.2 is the 20% hydrothermal fraction assumed previously.

2. *Sample Calculation*

Calculation of the national geothermal power potential for Iceland illustrates the method used. The resource base used for conversion to electric power is obtained from Table III as

$$Q = 1.4 \times 10^{21} \text{ (Class 3)} + 8.8 \times 10^{19} \text{ (Class 4)} = 1.5 \times 10^{21} \text{ cal (th)}$$

$\eta_0 = 0.017$ (overall efficiency, taken as the average of the estimates for liquid- and vapor-dominated systems).

The electrical potential is given by Eq. (3.11) as

$$E_b = 0.2(1.5 \times 10^{21})0.017 = 5.1 \times 10^{18} \text{ cal} = 5.9 \times 10^6 \text{ GW h (e)}$$

Further assumptions were made to estimate the growth pattern of installed electric power capacity for the years 1985, 2000, and 2020: (1) By the year 2020, about 90% of E_b is assumed committed for use by the year 2100; (2) by the year 2000, one-half is assumed committed; and (3) by the year 1985, one-sixth is committed. For the total $E_b = 5.9 \times 10^6$ GW h (e) in the Iceland example, the installed-capacity timetable becomes:

Year	Installed Capacity [GW (e)]
1985	1.0
2000	3.0
2020	6.0

The use rate would essentially exhaust the hydrothermal-based $E_b = 5.9 \times 10^6$ GW h (e) by the year 2100, requiring plant replacement after 2020 to maintain power capability, assuming there is no heat recharge to the reservoir. However, it is expected that technology to exploit the remaining 80% of the energy in the formation rock (the hot dry rock resource base), as well as to drill to deeper depths, will have become available by then.

3. *Calculated Values of Electric Power Potential*

Similar calculations of the potential for electric power potential were made for the 28 countries, as shown in Table IX. Collectively, these countries contain more than 95% of the worldwide total of 1.0×10^9 GW h (e). The worldwide potential for installed power capacity for the year

TABLE IX

Geothermal Electrical Potential—Calculated

Country	Electric Power Potential [GW h (e)]	Electrical Capacity Potential [GW (e)]		
		1985	2000	2020
China	1.4×10^8	23	70	140
United States	1.3×10^8	23	70	140
Indonesia	1.1×10^8	20	60	120
Peru	7.5×10^7	13	38	77
Mexico	6.7×10^7	12	36	72
USSR	6.3×10^7	11	33	65
Chile	4.7×10^7	8	25	49
Ethiopia	4.4×10^7	8	23	45
Ecuador	2.6×10^7	5	14	27
Brazil	2.5×10^7	4	13	26
Turkey	2.3×10^7	4	12	24
Japan	2.1×10^7	4	11	22
Kenya	2.1×10^7	4	11	22
Columbia	2.0×10^7	4	11	21
Bolivia	1.9×10^7	4	10	20
Iran	1.9×10^7	4	10	20
Philippines	1.7×10^7	3	9	18
Venezuela	1.0×10^7	2	6	11
Vietnam	9.6×10^6	2	5	10
Nicaragua	8.7×10^6	1.5	5	9
Italy	8.6×10^6	1.5	5	9
New Guinea	8.3×10^6	1.5	5	9
New Zealand	7.9×10^6	1.4	4	8
Chad	7.5×10^6	1.3	4	8
Zambia	6.5×10^6	1.1	3.4	7
Guatemala	6.3×10^6	1.1	3.2	6.5
Korea	6.3×10^6	1.1	3.2	6.5
Iceland	5.9×10^6	1	3	6
	9.5×10^8	169.5	502.8	998.0

2020 is 1.0×10^9 GW h (e). If only about 25% of this potential is realized, perhaps limited by such constraints as local markets, competing energy sources, and site-specific difficulties, the capacity of 2.2×10^8 GW h (e) would be very impressive, equivalent to a supply of 9×10^6 barrels of oil per day at a conversion efficiency of 42%. These estimates were derived for a conservative abandonment temperature, $T_2 = 177°C$. The dependence of available energy on T_2 was noted in Fig. 5. If conversion technology permits a lower T_2 (e.g., 150°C), the calculated capacity of 2.2×10^8 GW h (e) would be increased by a factor of 1.45.

C. National Estimates of Electric Power Potential

A further observation is of interest. Comparison of these estimates with the projections from various countries shows a general expectation of utilizing only a small fraction of the calculated electric power potential. For the Iceland example, the comparison shows the following:

Year	Iceland's Estimate [GW (e)]	This Estimate [GW (e)]
1985	0.150	1.0
2000	0.500	3.0
2020	0.800	6.0

The sevenfold difference may reflect a conservative attitude toward the constraints that may limit the anticipated domestic market for electric energy due to the availability of other energy resources, such as hydropower, or site-specific considerations, such as inaccessibility. Japan, which has about four times the energy potential of Iceland ($E_b = 2.1 \times 10^7$ GW h (e), projects a much greater utilization factor. Estimates by various countries are shown in Table X for comparison with Table V.

Although the contrast between the volumetrically calculated and national estimates shown in Table V is evident, the growth shown in Table VI is of the order of 25% per annum through 1985 and from 11% to 20% per annum through 2000. These growth rates are encouraging for an emerging alternative energy source.

D. Economics of Geothermal Power

Since many factors affect the cost of geothermal power, and experience is still very limited, it is not possible to generalize about these costs. The major cost factors are:

- ☐ Exploration
- ☐ Reservoir development
- ☐ Geothermal fluid purity and temperature quality
- ☐ Site-specific considerations
- ☐ Power plant
- ☐ Waste geothermal fluid disposal
- ☐ Environmental protection
- ☐ Operation and maintenance
- ☐ Taxes
- ☐ Royalties
- ☐ Lead time from reservoir discovery to power generation

TABLE X

Electrical Generating Capacity—National Estimates

	Capacity [GW (e)]		
	Installed	Estimated	
Country	1976	1985	2000
United States	0.502	3	>20
Italy	0.421	0.8	—
New Zealand	0.190	0.4	1.4
Japan	0.068	2.0	50
Mexico	0.075	0.4–1.4	1.5–20
USSR	0.06	—	—
Iceland	0.0025	0.15	0.5
Turkey	0.0005	0.4	1.0
Canada	—	0.01	—
Costa Rica	—	0.1	—
El Salvador	0.06	0.18	—
Guatemala	—	0.1	—
Honduras	—	0.1	—
Nicaragua	—	0.15–0.22	0.3–0.4
Panama	—	0.06	—
Argentina	—	0.02	—
Portugal	—	0.03	0.1
Spain	—	0.025	0.2
Kenya	—	0.03	0.06–0.09
Indonesia	—	0.03–0.1	0.5–6
Philippines	—	0.3	—
Taiwan	—	0.05	0.2
	1.325	8.3–9.3	14.8–100

Many of these cost factors differ in individual countries, and other external factors, such as alternative energy sources, will affect the competitive position of geothermal power. Nevertheless, some observations with regard to experience in the United States may be useful.

1. Energy Related Cost

In the United States, geothermal well drilling costs in 1976 dollars ranged from about \$150/m to \$300/m. The cost of geothermal heat delivered to the power plant at The Geysers (dry steam field) was about \$0.60/G.J. The cost of energy from other hydrothermal systems was projected to be in the range from \$0.60 to \$0.75/GJ. A better perspective of the relative value of geothermal energy can be gained by comparing it with oil-fired conventional plants. At $13.50/bbl for oil, the energy cost would be $2.50/GJ. Because of the difference in conversion efficiency of the conventional oil-fired plant compared with the geothermal plant, the oil-equivalent energy cost at The Geysers was about $1.60/GJ, a little more than one-half the price of oil. For the other hydrothermal systems the oil-equivalent energy cost would range from about $1.60 to $2.50/GJ.

2. *Power Plant Cost*

Hydrothermal power plant costs are expected to be about the same as those for oil-fired plants; however, the plant cost for dry steam fields are expected to be somewhat lower because there is less equipment in the power plant. Plant costs will vary, depending on the conversion cycle and resource type (e.g., flashed steam, binary-cycle, geopressured) but will probably not exceed nuclear power plant costs unless the lower temperature reservoirs are used.

3. *Operating Cost*

The annual power plant operating and maintenance cost is expected to be about 4.5% of the plant's capital costs. The operating and maintenance costs associated with reservoir operation are included in the estimated energy costs.

IV. NONELECTRIC USES

A. Present Uses

Geothermal energy applications other than for electricity generation are broadly categorized as nonelectric. Several factors make nonelectrical applications at geothermal resource sites desirable.

Electricity generation may not be economically feasible at geothermal resources with temperatures below 150°C.

For higher temperature resources where electricity generation is economical, thermal energy uses can reduce the cost of both the electricity and the nonelectrical uses.

Geothermal resources may contain economic quantities of minerals that may possibly be recovered.

In special circumstances the geothermal resource can be either an important supply of water or an important supply of low-cost energy that can be used to purify water.

The most widespread interest and application of geothermal energy for nonelectric purposes has been as a supply of thermal energy. These thermal energy uses have been at temperatures near the lower end of the scale, which are not suited for economic electricity generation.

Mineral recovery and water purification, although not presently significant on a worldwide basis, could be of very great importance in local situations.

Nonelectric applications, either now or recently in operation, cover a wide spectrum. At one end is the age-old balneological use (therapeutic baths), while at the other end is the use of geothermal energy for cooling, as presently practiced in New Zealand. Applications range from the warming of water for washing livestock stalls to providing the thermal energy requirements of a modern pulp and paper mill. Table XI lists the main nonelectric applications throughout the world.

The estimated total energy use rate represented by Table XI is about 7 GW (th). This amount of power is not a large contribution to the total energy use rate of the world. But in some cases it is not an insignificant amount, since for a number of the countries with geothermal potential, it represents more than their total electric power requirements. It is evident that nonelectric applications have been receiving widespread attention, but not the concerted attention necessary for a substantial number of large-scale developments. The degree of utilization in any given country may be related to the availability of geothermal resources and the relative cost of obtaining energy from other sources.

The processes that are most developed with respect to their overall significance as nonelectric applications are space heating, industrial applications, and agricultural applications.

1. Space Heating

As illustrated in Table XI, space heating with geothermal energy is becoming widespread throughout the cooler regions of the world, with sizable applications in a number of countries. The two main classifications of space heating systems are district and individual. Most of the significant applications are in the district heating category. Two noted exceptions, which are individual systems, are at Rotorua, New Zealand, and Klamath Falls, Oregon.

The largest known, and probably the most economical, district heating system is Reykjavik, Iceland. It supplies a total population of about 90,000 with space and domestic water heating from resources at temperatures varying from 80 to 120°C. The geothermal water is pumped directly from well to pipeline, held in storage tanks to meet demand, circulated to the consumer at about 80°C, and then wasted to the municipal sewer

TABLE XI

Present Nonelectric Applications of Geothermal Energy

Country	Heating (all purposes)	Drying	Animal Husbandry	Salt Production	Desalination	Mineral Production	Industrial Processes	Cleaning	Balneology	Total Capacity [GW (th)]
Argentina	X	X								Small
Chile					X					Small
Czechoslovakia	X								X	0.092
France	X									0.005–0.024
Germany (West)	X								X	0.0003
Hungary	X						X	X	X	0.46
Iceland	X	X	X				X			0.36
Italy	X					X				0.024
Japan	X	X	X	X		X				2.95
New Zealand		X					X	X		0.194
Nicaragua										0.15
Peru					X					Small
Philippines		X		X						0.005
Taiwan										0.0006
Turkey	X									0.0002
United States	X					X				0.017
USSR	X		X				X		X	0.155–4.8
Yugoslavia										0.005
										4.4–9.1

system or reinjected. The city is divided into a number of districts, each served by its own pumping station. Water meters are used to determine individual consumer use for billing purposes. The present capacity of the geothermal system is 0.35 GW, with peaking energy demand met from a 0.035-GW oil-fired heating plant. The system is presently being expanded by about 25% to serve an additional 26,000 people. For consumers being supplied by this system, the average cost of heating is about 30% of what it would be if supplied from an oil-fired heating plant.

A number of unique systems exist in one or more of the many geothermal space heating applications. For example, in Russia, France, and the United States, electrically powered heat pumps are used in conjunction with geothermal resources to provide either base or peak heating. In Rotorua, New Zealand, cooling with geothermal energy is accomplished using an absorption system.

2. *Industrial Applications*

The pulp, paper, and wood processing plant of Tasman Pulp and Paper Company, located in Kawerau, New Zealand, was the first major industrial development to utilize geothermal energy for heating purposes. The plant site was chosen because of the availability of geothermal energy. The geothermal energy was first used for timber drying in kilns and for wood preparation in 1957, then for the pulp and paper operation in 1962. Steam at pressures of 6.9×10^5 and 1.4×10^6 Pa (100 and 200 psi, respectively) is used for the various heating requirements of the plant. This system has been operating quite satisfactorily and has a power output of about 0.1–0.125 GW. A unique feature of the plant is the standby 0.01-GW noncondensing turbo-alternator, which is given priority for the geothermal steam in the event of a failure in the external electric power supply.

The production of diatomaceous earth at Namafjall, Iceland, is a significant development for geothermal energy in industrial applications. It is not only a large-scale application, it is also an example of the way in which geothermal energy can make a process economical, when it could not otherwise be justified.

Following the discovery of rich deposits of high-grade diatomite on the bottom of Lake Myvatn, technical and economic studies indicated that only by the use of potentially cheap geothermal energy from the nearby Namafjall high-temperature geothermal field could the recovery and drying of the diatomite be competitive with conventional diatomite production from comparatively dry land.

In late 1967 operation of the diatomaceous earth plant began with a production rate of 1.09×10^7 kg/yr. In 1970 the plant expanded and production increased to 2.18×10^7 kg/yr. The geothermal fluid, obtained from the wellhead at 250°C and a pressure of at least 3.9×10^6 Pa, is flashed to provide saturated steam at 1.0×10^6 Pa pressure that is transmitted to the plant. In the plant, the energy is used for drying, slurry heating, space heating, and deicing storage reservoirs during winter. The total consumption during the winter amounts to about 4.5×10^4 kg/h of the 10×10^6-Pa steam.

Another use that is indicative of the wide variety of applications for geothermal energy is the present use of relatively low-temperature waters to thaw large areas of ground to allow mining in certain regions of the Soviet Union.

3. *Agricultural Applications*

The three primary agricultural applications of geothermal energy are greenhouses, animal husbandry, and aquaculture. By far the most extensive use of geothermal fluids is in greenhouses. This application is most common in regions where growing seasons are short and greenhouses are necessary to meet the local demand for vegetables. In Iceland most of the tomatoes, lettuce, cucumbers, and other fresh vegetables are grown with geothermal energy supplying the heat. Similarly, the Soviet Union and Hungary have extensive greenhouse applications that use geothermal energy. The animal husbandry and aquaculture applications are much more restricted than are the greenhouses. Hungary has the only large application of geothermal energy for animal husbandry, and it is expected that this will not be a significant, widespread use for geothermal energy in the future. There are only a relatively few small aquaculture applications that use geothermal energy at this time.

B. Potential Uses

Many other processes may be adapted to directly utilize geothermal energy. Because geothermal energy is at a relatively low temperature (compared with that available from fossil fuels) and its direct use is limited to an area of several tens of kilometers surrounding the geothermal site, there are two main factors regarding the energy use in the particular process that must be considered when evaluating potential industrial uses. The first of these is the minimum temperature range of the geothermal fluid that can satisfy the potential use. The second is the energy-cost intensiveness of the potential application (i.e., the energy cost to produce the product relative to the value of the product itself).

If the required temperature and the resource temperatures are compatible and the process is energy-cost intensive, then there is a good chance that the use of geothermal energy in the process will be economical. The

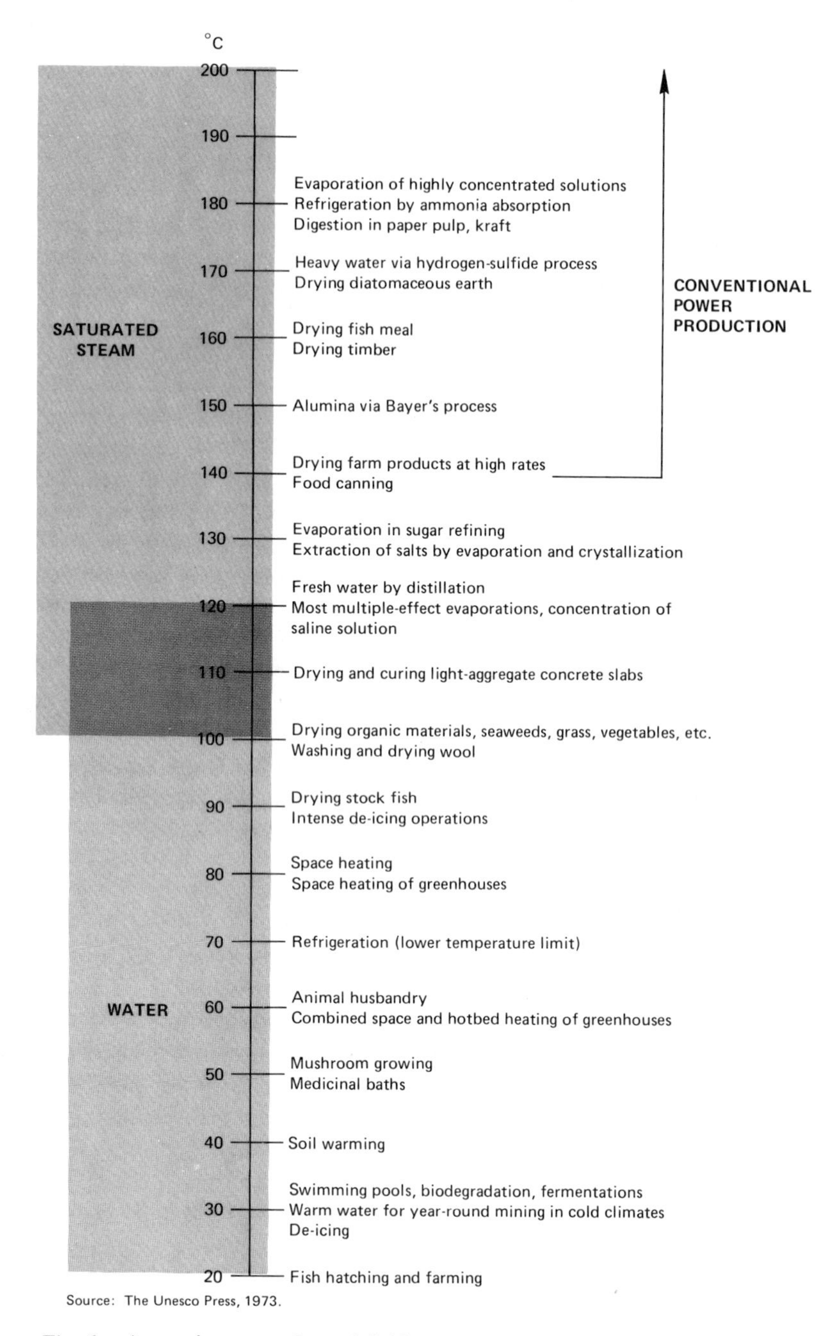

Fig. 6. Approximate geothermal fluid temperature for nonelectric applications.

location of the raw materials with respect to the geothermal resource is also an important factor. If the process is not energy-cost intensive, it probably will not be a good candidate for geothermal energy use, regardless of the temperature, unless the energy consumption in the particular industry is very high and there is an excellent match in the location of the process, the raw materials, and the geothermal resource.

Figure 6 illustrates the approximate minimum required temperatures of geothermal fluids for a variety of applications. The figure shows the temperature requirements of both the present applications and a number of potential applications. In general, as the geothermal fluid temperature increases, the application heat transfer equipment costs decrease and the operation becomes more economical. Exceptions to this occur when the increased temperature is accomqanied by increased corrosion and scaling in the system and when the increased temperature causes more complicated design problems, such as pressurization of the entire system.

Table XII lists the energy intensiveness of a variety of industrial process. Production of heavy water and fresh water appear to be very good potential uses of geothermal energy, the first depending on the need, and the latter on the coincidence of the need and the geothermal resources. A number of the other energy-intensive uses in Table XII have been examined for their potential for geothermal energy use. For many, the potential for geothermal use appears good. However, for a number of processes the electricity requirements are also quite large and the steam requirements could be generated quite cheaply in a combined system with electricity generation. For these, geothermal energy may not be competitive unless the resource is also adequate for electricity generation.

C. Problems Associated with Nonelectric Uses

Although nonelectric uses are potentially feasible for a wide variety of resource qualities, these uses are not without problems. Some of the problems are due to the use itself, while others are due to the type of resource available. One of the most readily recognized problems with direct applications, and the one that most significantly separates nonelectric from electric applications, is the fact that direct uses are necessarily restricted to relatively small areas centering on the resource. Electricity generated from geothermal resources, remote from load centers, can be tied into a grid system. But in direct applications, the thermal energy cannot be economically transported more than a few tens of kilometers, depending on the nature and size of the energy requirement.

In nonelectric applications a very significant factor, in addition to the size of the thermal energy demand and the distance between the resource

TABLE XII

Steam Requirements for Various Processes

Process	Steam Requirements (kg Steam/kg Product)
Heavy water by hydrogen sulfide process	10,000
Ascorbic acid	250
Viscose rayon	70
Lactose	40
Acetic acid from wood via Suida process	35
Ethyl alcohol from sulfite liquor	22
Ethyl alcohol from wood waste	19
Ethylene glycol via chlorohydrin	13
Casein	13
Ethylene oxide	11
Basic Mg carbonate	9
35% hydrogen peroxide	9
Solid caustic soda via diaphragm cells	8
85% hydrogen peroxide from 35% H_2O_2	4.75
Acetic acid from wood via solvent extraction	7.50
Alumina via Bayers process	7.00
Ethyl alcohol from molasses	7.00
Beet sugar	5.75
Sodium chlorate	5.50
Kraft pulp	4.20
Dissolving pulp	4.20
Sulfite pulp	3.50
Aluminum sulfate	3.50
Synthetic ethyl alcohol	3.00
Calcium Hypochloride, high test	3.50
Acetic acid from wood via Othmer process	2.75
Ammonium chloride	2.75
Boric acid	2.25
Soda ash via Solvay process	2.00
Cotton seed oil	2.00
Natural sodium sulfate	1.80
Cane sugar refining	1.67
Ammonium nitrate	1.50
Ammonium sulfate	0.17
Fresh water from sea water by distillation	0.08

and the market, is the annual load factor. Nonelectrical geothermal systems are characterized by having substantially greater initial capital costs and lower operating costs than most conventional systems; consequently, when the annual load factor is low and the large capital investment is effectively used for only a small part of the year, the desirability of such a system markedly decreases. Space heating, one of the most potentially significant nonelectric uses, has quite low annual load factors in many of the heavily populated regions of the world. Iceland, which has enjoyed much success with geothermal space heating, requires space heating throughout the year and has a relatively high annual load factor of about 0.50–0.60. In comparison, a typical load factor for space heating systems operating in New York is about 0.2–0.3. In addition, since geothermal energy requires large capital investments, it is very sensitive to interest-rate changes.

One technological problem may also substantially limit nonelectric applications at higher temperatures: as the geothermal fluid temperature increases, its tendency to cause corrosion and scaling also increases significantly. The usual surface heat exchangers that would normally be used in such applications are quite sensitive to corrosion and scaling, and an economical solution other than wasteful steam flashing has not yet been found.

Because nonelectric uses will almost exclusively utilize geothermal resources that are water dominated and at the lower end of the temperature spectrum, the overall enthalpy drop per unit mass flow of geothermal fluid can be relatively small and the total flow rate of fluid can be very large. Often this fluid is not compatible with surface disposal and reinjection must be considered. Widespread experience with reinjection of the magnitude required in such applications does not exist, and substantial study and experimentation are required before the design for such disposal systems will be routine.

D. Economics

The economic feasibility of a nonelectric geothermal system depends on a number of factors. Hypothetical economic evaluations are open to question until all factors have been examined and practical experience provides realistic data. Economic data in the literature are most prevalent for space heating systems and show that geothermal systems can be economical relative to conventional systems under the right circumstances. Such circumstances have been reported for heating in Reykjavik, Iceland, and for heating a college campus at Klamath Falls, Oregon, where geothermal energy is reported to be about one-tenth of that for conventional fuel; however, these costs may be on the low side, since the cost of capital was not included. The apparent inconsistency of greater economic advantage of geothermal energy in Klamath Falls compared with Reykjavik results from the college campus being a single concentrated load with small distribution charges, while about 60% of the Reykjavik heating cost results from distribution.

E. Outlook for the Future

The total amount of geothermal energy contained in the geothermal resource base of the world was shown to be large. How rapidly this large resource will be developed throughout the various countries is not really evident at this time. However, recent price increases of fossil fuels in world markets should stimulate such development.

Since many of the nonelectric applications can use low-temperature resources, the total resource base given in Table III may be used for such applications. A large fraction of the energy requirements of most of the world's present societies could conceivably be satisfied with thermal energy from geothermal resources. There are, however, many factors that must be evaluated to determine the feasibility of application at each resource site. Although space here does not permit a full discussion of these details, there are a number of factors that generally indicate the desirability of applications:

Space Heating. The area to be heated must be close to the resource site; the higher the annual load factor the better; the institutional barriers of having district heating systems must be removed.

Industry. The energy use and/or energy intensiveness of the process must be high; the raw materials source and resource location must be compatible; if large amounts of electricity are required in addition to thermal energy, the geothermal resource should also be able to produce electricity.

Agriculture. The main applications are greenhouse heating and aquaculture. Both of these appear to be growing in significance and have very good potential applications in countries at higher latitudes. The main limitation to such applications will come from competition with the huge quantities of waste heat available from large fossil- or nuclear-fueled electricity generating plants. The geothermal systems might have a location advantage, and in many small countries they would not have the power stations to compete with.

Drilling depth required to recover the resource is also a very important factor. The maximum economically allowable depth is dependent on many variables, one of which is the resource temperature. In general, the maximum economically allowable depth increases as the resource temperature increases, and although the resource base for all classes was considered to extend to 3 km, only Classes 3 and 4 could be expected to be economically recoverable at depths approaching 3 km.

F. Calculated Estimates of World Potential for Nonelectric Uses

An estimate of the worldwide nonelectric use potential can be made in a manner similar to that for electricity generation in Section II. The estimate has an additional degree of uncertainty in that different processes may extract different amounts of energy from the same geothermal fluid.

1. Analytical Approach

Conversion factors similar to those developed in Section III are used in the estimates. However, in this case recoverable energy rather than thermodynamically available energy is of interest. With the nomenclature of Section II, the different factors are estimated as follows and denoted by (′) to distinguish them from those used in Section III for analysis of the electric power potential.

Available Energy Fraction F'_a. The heat recovery fraction is the energy that would be obtained by cooling the resource from T_1 to T_2 relative to the resource base.

$$F'_a = E_{1-2}/Q = (T_1 - T_2)/(T_1 - T_0)$$

Wellhead Fraction F'_w. The wellhead fraction is estimated to be the same for both electric and nonelectric applications, $F'_w = F_w$.

Heating Plant Input Fraction F'_{hpi}. The pressure drops that occur in the piping and flashing units represent very small losses for most direct thermal processes, and F'_{hpi} is estimated at 0.90–0.95.

Heating Plant Conversion Fraction F'_{hpc}. The heating plant conversion efficiency depends very much on the type of plant. For an extensive district heating system with a large amount of distribution piping, F'_{hpc} may be slightly less than 0.8, while for a compact processing plant it might be 0.95.

With these factors and the data presented in Table III, the geothermal nonelectrical potential for the world is estimated at 4.18×10^{23} J $= 1.26 \times 10^{11}$ GW h (th). This estimate is based on the following:

1. Only Classes 1 and 2 of the resource base are considered for nonelectric uses (although some Class 3 resources are now being used for such).
2. Only 20% of the total resource base will be considered as being convertible by present technology.
3. On the average, $T_1 = 340°K$ (67°C) and $T_2 = 330°K$ (57°C) for the Class 1 resource; for the Class 2 resource, $T_1 = 400°K$ (127°C) and $T_2 = 360°K$ (87°C). (The T_2 values are different for the two resources because of differences in the processes that would use the resources.)

With T_0 = 288 K (15°C), these yield F'_a = 0.19 and 0.36 for Classes 1 and 2 respectively. The nonelectric analog of Eq. (3.1) is $E_{th} = F'_a F'_w F'_{hpi} Q$, and the analog of Eq. (3.10) is $\eta_0 = E_{th}/Q$.

4. Mean values of $\eta_0 = (0.19)(0.4)(0.9)(0.9) = 0.05$ and $(0.36)(0.4)(0.9)(0.9) = 0.12$ will be used as averages for resource Classes 1 and 2, respectively.

The estimates of worldwide potential for installed nonelectric capacity for the years 1985, 2000, and 2020 become 2.2×10^{10}, 7.0×10^{10}, and 1.0×10^{11} GW (th), respectively. These are shown in Table XIII.

G. National Estimates of Nonelectric Potential

In comparison, Table XIV illustrates the projections of nonelectric utilization of geothermal energy by various countries in the years 1985, 2000, and 2020. Using the above methodology to estimate the potential for just the countries listed in Table XIV gives totals of 6.1×10^9, 1.7×10^{10}, and 2.6×10^{10} GW (th) for 1985, 2000, and 2020, respectively. These values are two or three hundredfold larger than the estimates listed in Table XIV, where the totals are about 1.7×10^7, 5.2×10^{10}, and 1.2×10^8 GW (th) for the same years. The large difference between the estimates here and the estimates by different country is expected and reflects the following: (1) much of the Class 1 and Class 2 resource base may not be economically recoverable because of its occurrence at the greater depths (2–3 km) considered in the resource base, which is particularly true of the Class 1 resource; (2) limits on the anticipated domestic market for thermal energy near the resource; (3) availability of inexpensive low-temperature thermal energy (e.g., waste thermal energy from electric power plants and solar energy; and (4) site-specific difficulties such as mountainous terrain, land-use restrictions, and recreational use of the resource.

The estimates of the potential and projected uses reveal that a number of countries are planning substantial contributions from geothermal energy through nonelectric applications, but these contributions represent less than 1% of the total potential. The use estimates indicate roughly a two- to fourfold increase by 1985, a six- to tenfold increase by 2000, and up to a twentyfold increase by 2020, compared with the present use as illustrated in Table XIV. It is evident that in the next several decades there will be significant progress in this area.

V. ADEQUACY OF CURRENT TECHNOLOGY

Exploration and development of high-temperature hydrothermal geothermal systems for commercial applications are now practical, using pres-

TABLE XIII

Projected Nonelectrical Utilization of Geothermal Energy (Calculated)

Country	Nonelectrical Energy Potential, Classes 1 and 2 Resource Base [GW h (th)]	Nonelectrical Capacity Potential [GW (th)]		
		1985	2000	2020
USSR	1.9×10^{10}	3500	12,000	20,000
United States	1.5×10^{10}	3000	8,500	15,000
China	1.0×10^{10}	2000	6,000	11,000
Canada	1.0×10^{10}	2000	6,000	11,000
Brazil	7.0×10^{9}	1500	4,000	7,500
Pakistan	7.0×10^{9}	1500	4,000	7,500
Australia	6.1×10^{9}	1000	3,500	6,000
Mexico	2.6×10^{9}	500	1,500	3,000
Peru	2.6×10^{9}	500	1,500	3,000
India	2.6×10^{9}	500	1,500	3,000
Argentina	2.2×10^{9}	400	1,300	2,400
Sudan	2.2×10^{9}	400	1,300	2,400
Ethiopia	1.7×10^{9}	350	1,100	2,000
Zaire	1.7×10^{9}	350	1,100	2,000
Algeria	1.7×10^{9}	350	1,100	2,000
Saudi Arabia	1.7×10^{9}	350	1,100	2,000
Iran	1.7×10^{9}	350	1,100	2,000
Greenland	1.7×10^{9}	350	1,100	2,000
Chile	1.7×10^{9}	350	1,100	2,000
Indonesia	1.3×10^{9}	250	800	1,500
Columbia	1.3×10^{9}	250	800	1,500
Bolivia	1.3×10^{9}	250	800	1,500
Chad	1.3×10^{9}	250	800	1,500
Turkey	1.0×10^{9}	200	600	1,100
Angola	9.6×10^{8}	180	550	1,000
South Africa	9.6×10^{8}	180	550	1,000
Equador	9.6×10^{8}	180	550	1,000
Kenya	8.7×10^{8}	170	500	900
Tanzania	7.8×10^{8}	150	450	800
Japan	7.8×10^{8}	150	450	800
Egypt	7.8×10^{8}	140	400	700
Nigeria	7.0×10^{8}	140	400	700
Zambia	7.0×10^{8}	140	400	700
Venezuela	7.0×10^{8}	140	400	700
Philippines	6.1×10^{8}	120	350	650
France	4.4×10^{8}	80	250	450
New Zealand	3.5×10^{8}	70	220	390
Iceland	1.7×10^{8}	40	120	220
Hungary	8.7×10^{7}	15	40	75

TABLE XIV

Projected Nonelectrical Utilization of Geothermal Energy—National Estimates

Country	Temperature Range of Expected Use (°C)	Projected Nonelectrical Use In Each Year [GW (th)]		
		1985	2000	2020
Canada	50–100	0.0097		
	20–50	0.0005		
England	Total	Small	<3.4	
France	50–100	0.48	4.8	
	20–50		Small	
Hungary	100–150		0–0.48	0–0.97
	50–100	0.0032–0.0048	0.73–1.45	1.3–3.4
	20–50	0.18–0.29	0.35–0.58	0.69–0.97
Iceland	150–200	0.068	0.20	0.30
	100–150	0.145	0.34	0.58
	50–100	0.339	0.58	0.78
	20–50	0.012	0.023	0.047
Italy	Total	0.039		
Japan	Total	3.1		
Korea	50–100	Small		
	20–50	Small		
Mexico	Total	0.69		
New Zealand	Total	0.39	0.58	
Nicaragua	150–200	<4.8	4.8	4.8
	100–150	14.5	29.1	48.5
	50–100	<4.8	4.8	4.8
	20–50	0	0	0
Spain	Total	0.01	0.0485	
Sweden	50–100	0	0.2423–0.4846	0.7269
	20–50	0	0.2423	0.4846
Taiwan	150–200	0.0019	0.0048	0.0097
	100–150	0.0010	0.0024	0.0097
	50–100	0.0015	0.0048	0.0145
	20–50	0.0005	0.0024	0.0048
Yugoslavia	Total	0.0582	0.2423	0.4846
United States	Total	3.4	14.5–33.9	96.9–193.8

ent technology. There are two exceptions to this generalization. Current technology is deficient in the prevention and removal of scale deposition by hypersaline geothermal brines, and exploration techniques are not yet reliable in locating sealed hydrothermal deposits, deposits that have no apparent surface manifestations.

Hot dry rock resources, including magma deposits, do not yield to present technology for a number of reasons. The absence of water or steam as a convenient working fluid, or other mechanisms for extracting energy, is the primary reason. Exploration techniques are unreliable, and both the high temperatures and formation structural integrity pose formidable drilling problems in very high temperature regions. Although some encouraging progress has been made in preliminary tests in one of the hot dry rock formations in the United States, much innovative work will be required to perfect such techniques to the point where development of these resource types is commercially feasible.

The capabilities and limitations of current technology are reviewed in this section in the same general sequence as would be encountered during the course of any geothermal resource development program.

A. Geothermal Exploration

Many exploration techniques are available and may be used to aid in the search for geothermal resources. All of them may not be necessary in each particular case; however, because there is usually more than one possible interpretation of the data, several techniques are likely to be used in combination for confirmation. The various techniques are usually grouped according to visual, geological, geochemical, geophysical, and sometimes hydrological surveys. Modeling of the presumed reservoir from the available data, of course, is not an exploration tool, but it is an essential and powerful interpretive technique.

Since surface manifestations of the presence of subsurface geothermal energy are frequently conspicuous, visual surveys are the most obvious starting point in any exploration effort and are the least expensive. Surface manifestations include evidence of volcanism, geysers, fumeroles, hot springs, and snow melt patterns in regions of high heat flow in temperate or arctic zones. Remote optical sensing such as satellite imagery and airborne infrared may be useful in detecting areas with anomolously high heat flow that may not be readily detected by other methods. Visual surveys are very useful but may be insufficient to justify drilling, in which case other techniques must be used to map and model the resource.

Geological surveys are important in understanding the structure of a geothermal prospect. Geological reconnaissance includes identification of rock types, fault systems, and geochronology. Color photography and microwave radometry may be useful for the purpose of enhancing field survey data where necessary. Structural mapping and geological analogy are also useful.

In general, geochemical methods have limited use until after the first well has been drilled. There are a few exceptions, however. Where hydrothermally altered rocks and hot springs are present, geochemical surveys can add significantly to the body of site-specific data prior to drilling in that geochemical thermometry can be used to infer reservoir temperature if hot springs are present. Mineralogy, petrology, rock chemistry, and age dating may be useful in correlating prospective areas to past volcanic activity and hydrothermal rock alteration, and chemical analysis of geothermal fluids may help to infer the geothermal history of the fluids.

Of all the exploration techniques in current use, geophysical methods are the most important. Properly interpreted data from these techniques allows the geophysicist to model such important parameters as temperature profiles, fluid phase, formation porosity, fluid conduit, and volumetric dimensions associated with the heat deposit. The more frequently used techniques are electrical resistivity, seismics, and heat flow measurements. Both active and passive seismic methods are employed. Passive methods include monitoring and interpreting seismic noise and micro earthquakes. It is interesting to note that natural seismic activity is usually associated with geothermal deposits. Other methods include gravity measurement, magnetics, electromagnetics, and telluric and magnetotelluric methods. The number of techniques used at any site will depend on the clarity or, conversely, the ambiguity of the data from preceding measurements.

The purpose of these exploration techniques is to select the best site for drilling and thereby minimize exploratory drilling costs. With the proper application and interpretation, the various exploration techniques can be reasonably effective on hydrothermal systems. Additional work is needed to better adapt them to other geothermal resource types.

B. Reservoir Development

Present drilling techniques are adequate to reach depths of 10 km or so; therefore, depth is not yet a limiting factor. Since the size of a bore hole decreases with depth and both the porosity and permeability may decrease, producibility at greater depths is more likely to be a limiting factor, except where the reservoirs are overpressured. Geothermal reservoirs pose certain problems that are not generally encountered in the oil and gas operations from which the technology has been adapted. Many geothermal deposits are located in hard rock systems, where drilling is more difficult and more costly. The use of water or air as drilling fluids has been successfully adapted to geothermal fields as a means of cooling the drill bit and preventing formation plugging that may occur if conventional muds are used where the formation temperature is high. Some reservoirs exceed 315°C, the approximate temperature at which presently available cements used in well completion show accelerated deterioration; therefore, present cements are not well suited to high-temperature applications. Calcite scaling in wells at or near the flash point can be a problem if present in significant quantities in the geothermal brine and acidic brines may increase corrosion of well casing. Although the present art is being pushed in several areas, as indicated by the problems mentioned, present technology appears to be adequate for development of most identified hydrothermal systems.

Reservoir assessment techniques routinely used in oil and gas operations have been adapted to geothermal development with reasonable success. Conventional well logging, well test and computational methods are routinely used. Although logging instrumentation performance is limited by temperature, these methods have been reasonably successful at dry steam fractured rock reservoirs and sedimentary water-dominated reservoirs, while the reliability of analytical techniques for fracture controlled water-dominated reservoirs has not yet been fully demonstrated.

Higher temperature hydrothermal wells can be produced by artesian flow. For lower temperature reservoirs it may be necessary to pump the wells. In the latter case, high-temperature brine pump technology has lagged, but significant development efforts are underway.

C. Power Generation

In current practice geothermal steam is expanded through steam turbines to generate electric power. The steam comes directly from the bore hole in the case of dry steam fields and in the case of hydrothermal fields is either separated or flashed from the hot water. One or more stages of flash are possible; however, except in special situations, no more than two stages are generally used. The main items of equipment, except for condensors and cooling towers, are the steam separators, flash units, possibly steam scrubbers, and turbines. Present steam separator technology is certainly workable and does not inhibit geothermal development in any way, but more efficient separators are desirable. Some recent work with rotary separators appears promising. Steam flash technology has also proven satisfactory. Scale deposition is the most severe problem in the steam flash units but is manageable with brines low in total dissolved solids. Steam scrubbers are mostly experimental in geothermal applications. They are used to clean the steam before it enters the turbine. Noncondensible gases, if present in appreciable quantities, affect turbine performance. This has not been a problem at most geothermal fields. Steam turbine technology has been around for many years and there are no major problems. To extract maximum energy, low-pressure or double-entry turbines may be used. Turbine efficiencies of about 72% are typical. Steam turbine technology has proven economical where the reservoir temperatures are above approximately 210°C and the fluids are relatively clean. At lower temperatures the economics appear to be less favorable and alternative conversion options may be considered.

At present, alternative conversion options are limited to binary loop systems in which a secondary working fluid, instead of steam, is used to drive the turbine. Experimental work on total flow turbine technology is in progress and such machines may become available in the future. Geothermal binary loop systems are not yet in commercial use, but the first commercial-size power plant is scheduled to go on line in the United States in 1980. This technology is not new. It is closely related to that used in the petrochemical industry but must be adapted for geothermal applications. Secondary working fluids that have been considered include the refrigerants and organic fluids. The choice of fluid will depend on the temperature range of the geothermal fluid; however, most attention at present is focused on isobutane, isopentane, propane, or mixtures of these fluids for the temperature range 150–210°C. Organic fluid turbine efficiencies are expected to be in the range of 84–88%.

One of the disadvantages of the binary loop system is that heat must be transferred from the geothermal brine to the secondary working fluid and heat exchangers are prone to scaling, are large, and are significant items of cost. According to results of current study, conventional heat exchangers with binary loop technology can be used to increase the commercial viability of geothermal systems down to reservoir temperatures of about 160°C. Current work on advanced heat exchanger concepts, such as direct contact, may improve the economics and lower the economical operating temperature even further.

D. Miscellaneous Equipment

The noise created by open flowing wells and steam line vents has stimulated the development of mufflers, or silencers, and this technology is reasonably mature. Steam lines pose no particular problem except heat loss. Brine lines are prone to scaling where the rate of scaling is dependent on the chemical composition of the brine. Severe corrosion of pipe runs in certain hypersaline fields has been observed. Because of the tendency of geothermal brines to deposit scale, the maintenance of valves, pumps, and instrumentation will be higher than for fossil fuel plants. Improvements in scale-resistant brine pumps and in-process instrumentation are desirable.

In summary, hydrothermal geothermal development can proceed with today's technology. Several improvements are desired and if made will have two beneficial effects, the first being reduced cost of power, and the second being expansion of that portion of the resource base that can be developed. Development of other geothermal resource types such as hot dry rock and magma must await further technological developments.

VI. ENVIRONMENTAL CONSIDERATIONS

Geothermal energy promises to be less detrimental to the environment than energy derived from most other sources; however, it is not environmentally benign and special consideration must be given to certain unique sets of possible problems. The amount of environmental control necessary will vary from country to country and region to region, depending on local air and water quality standards, on the characteristics of each geothermal reservoir, and to some extent on the power conversion process, whether it be flashed steam or some form of closed loop with reinjection.

Currently producing geothermal fields were developed with only the most rudimentary environmental controls, and several areas of concern have been identified as society has grown more environmentally aware. Not all fields will pose the same problems. A few of the more important considerations are as follows:

Waste Brine Disposal. If the geothermal fluid is partially or totally brine, the waste brine must be disposed of in a manner that will protect groundwater supplies needed for agricultural, domestic, or industrial uses. The methods of disposal thus far have consisted of the surface streams, ponding and evaporation, and reinjection of the brine back into the reservoir. In some cases reinjection may be the preferred method, if there is a need to minimize ground subsidence or to replace the water in the reservoir to enhance production.

Noncondensible Gases. Hydrogen sulfide and other noncondensible gases may affect air quality if the concentrations are sufficiently high and if they are vented to the atmosphere. Hydrogen sulfide appears to be the major offender, and present research is focusing on a solution to this problem. In some cases, the hydrogen sulfide concentration is so low that special controls may not be necessary. In others, stack disposal may suffice; and in some, sophisticated technology may be needed, depending on local factors.

Waste Heat Rejection. Geothermal power plants are less efficient than fossil fuel plants; roughly two to three times more heat must be rejected per unit of power. Where fresh water is in abundant supply, this may not be a problem, but in arid regions cooling water may be at a premium and could slow geothermal development unless suitable alternatives are available. One solution, if the plant is operated as a flashed steam facility, is simply to use the geothermal steam condensate as makeup water in the cooling tower. This approach could pose problems if the temperature of the resource is low and if binary cycles are used in which no steam is produced. It could also pose a problem where reinjection of all the fluids is desired. Some combination of alternative approaches may prove to be the best solution in the long run.

Subsidence. Withdrawal of fluids may result in subsidence if measures are not taken to minimize the effect. Some areas, such as low coastal zones, populated areas, and agricultural irrigation systems, may be especially sensitive to subsidence.

Noise. Vented wells or steam lines can create excessive noise. Muffler systems may be needed in noise-sensitive areas.

Other Considerations. Toxic trace elements (such as arsenic and mercury), land use, cooling tower effluents, and seismicity may require consideration when assessing the environmental impact. However, these aspects appear to pose only minor problems.

Present environmental control technology will probably serve the geothermal industry's needs in the near term, and work is progressing on improved technology, particularly for the control of hydrogen sulfide emissions.

VII. INTERNATIONAL DEVELOPMENTS

This section discusses the status of geothermal energy development in countries for which information is available. The sources of the estimates included in this section are the open literature and national responses to a worldwide questionnaire. The countries are grouped by major geographic region in the following order: North America, Central America, South America, Western Europe, Eastern Europe, Asia, East Africa, and Oceania.

A. North America

1. Canada

Geothermal development in Canada is in the preliminary stage and therefore does not provide sufficient information for a detailed assessment of the resource. Suffice it to say that the western part of Canada is in a geothermal region and it is therefore likely that intensive geothermal exploration will discover significant resources. At present, geothermal electric power capacity is estimated to reach 0.01 GW (e) by 1985. Direct heat utilization would probably constitute the most important use of the resource in the near future, reaching perhaps 0.01 GW (th) by 1985. The estimated total resource base of geothermal energy to a depth of 3 km and above 15°C is 1.7×10^{24} J.

2. *Mexico*

A successful geothermal power plant of 0.075 GW (e) capacity has been in operation at Cerro Prieto since November 1973. The plant is presently being expanded through additional drilling and construction work to achieve a capacity of 0.15 GW (e) in the near future. Informal estimates suggest that the Cerro Prieto field will support 0.5–1.0 GW (e) capacity. Temperatures above 340°C have been recorded in some of the holes drilled most recently.

Extensive exploration is taking place in many other areas of Mexico, especially at Ixtlan, Los Negritos, and Los Azufres. It is estimated by the Comision Federal de Electricidad that 0.4–1.4 GW (e) of geothermal electricity would be on line by 1985 and 1.5–20 GW (e) by the year 2000. The ultimate geothermal potential of Mexico is variously estimated at 4–20 GW (e).

3. *United States*

The largest geothermal power installations in the world at present are in the United States. By the end of 1976, 0.502 GW (e) of installed electrical capacity were in existence in the United States, all in The Geysers region of California. In that region a total geothermal power capacity of 2-4 GW (e) is estimated to exist. The Geysers field is a high-quality vapor-dominated field that will increase in importance in years to come. The slow development of the field is primarily due to a variety of environmental and legal difficulties.

Active exploration and drilling is presently going on in the United States in more than twenty separate areas (Fig. 7). A major area of exploration is the Imperial Valley of California, just north of the successful geothermal power development at Cerro Prieto, Mexico. Four different geothermal fields have been discovered so far in the Imperial Valley: Niland, Brawley, Heber, and East Mesa. Estimates of the total geothermal capacity of the Imperial Valley for electric power production vary widely, ranging from 1.5 to 20 GW (e) capacity. Severe salinity problems have been encountered in the Niland field near the Salton Sea, where dissolved solids concentrations of more than 25% have been encountered

Fig. 7. Map showing geothermal regions in the western United States.

in a high-temperature field of large dimensions. The temperature range of the geothermal reservoirs encountered to date varies from 180°C in the Heber field up to 360°C in the Niland field.

Successful discoveries have been reported in the Valles Caldera in New Mexico, the Roosevelt field in Utah, and on the island of Hawaii. These fields are reported to be high-temperature liquid-dominated geothermal reservoirs. It is informally estimated that the Valles Caldera field will support at least 0.4 GW (e) capacity. It is unknown as yet what the ultimate capacity of the Roosevelt and Hawaii fields will be because only limited exploration has been completed. However, initial results at the Roosevelt Field have been encouraging inasmuch as one of the early exploration wells has produced a mass flow capable of supporting more than 0.015 GW (e) capacity.

It is the goal of the Department of Energy (DOE) to achieve a minimum geothermal power capacity of 3 GW (e) by 1985 and more than 20 GW (e) of electricity by the year 2000. At the same time, considerable effort is being devoted to the study of direct heat uses since a large portion of the energy used in the United States is of the low-grade-heat type. The federal government, through DOE, will spend more than $100 million/yr for the foreseeable future in an attempt to solve a variety of technical problems and stimulate private industry to explore and develop geothermal resources. As part of that effort, experiments are now being designed to exploit the geopressured resources of the Gulf Coast for the production of mechanical-pressure energy, geothermal heat, and methane gas.

Space heating is being implemented successfully in Klamath Falls, Oregon, and in Boise, Idaho, as well as in many other localities. The growing shortage of natural gas in the United States will no doubt cause an increase in the use of low-grade geothermal heat for both space heating and other direct heat uses such as greenhouses. The U.S. Geological Survey has estimated the recoverable heat in the United States to be about 3.5×10^{20} J of usable heat and up to 4×10^{7} GW h (e) of electric energy.

B. Central America

1. Costa Rica

Active geothermal exploration is currently taking place in Costa Rica where the Las Pailas Hornillas area in the Guanacaste province is being investigated. Exploratory drilling is scheduled to start by the end of 1977, and a geothermal power plant of 0.1 GW (e) capacity is to be installed in 1982–1984.

2. El Salvador

Practically all of Central America is an intensely volcanic, highly active geothermal region. A 0.06 GW (e) geothermal power plant has been in operation in the Ahuachapan field in El Salvador since 1976. That plant will soon be increased to 0.08 G (e) capacity through the development of a secondary flash system. It is the intention of the national electric authority in El Salvador (CEL) to develop at least 0.1 GW (e) additional capacity from geothermal resources by 1985. Exploration efforts are currently proceeding at three promising geothermal fields: Chinameca, Berlin, and San Vicente.

3. Guatemala

In Guatemala, the national electrical company (INDE), with the assistance of some outside consultants, is currently proceeding with geothermal exploration at the Moyuta field. No success has been achieved to date in proving geothermal steam for commercial power production capacity. However, simultaneous exploration through outside technical assistance is presently taking place at the Zunil prospect, and it is also likely that the Amatitlan geothermal area will be subjected to a closer examination in 1977. The objective of the present exploration in Guatemala is to achieve a minimum 100 MW (e) capacity by the early 1980s.

4. Honduras

Honduras has just embarked on an aggressive geothermal exploration program. The national electric authority (ENEE), intends to explore two areas initially: the Pavana area near Choluteca in the southern part of the country and the San Ignacio area northwest of Tegucigalpa, the capital. The objective of that exploration program is to achieve a minimum of 0.05 GW (e) capacity by 1982 and 0.1 GW (e) by 1984–1985.

5. Nicaragua

Geothermal exploration is currently proceeding in Nicaragua, where active drilling is taking place at the Volcan Momotombo field. Sixteen exploration and development holes have been drilled to date. At present, the productive capacity of the wells is uncertain because of conflicting reports about their success. The national electric authority (ENALUF), hopes to achieve a 0.1 GW (e) capacity from the Momotombo geothermal field by the early 1980s. It is estimated by ENALUF that 0.15–0.22 GW (e) of geothermal power will be installed by 1985, 0.3–0.4 GW (e) by 2000, and up to 0.8 GW (e) by the year 2020.

6. Panama

Only minimal geothermal exploration has been carried out in Panama. Some attractive possibilities seem to exist in the region of Cerro Pando and Caldera-Chiriqui province. It is not possible at present to make a judgment on the ultimate potential of geothermal energy development in Panama. However, because Panama is in a major geothermal belt, it is likely that economically attractive geothermal possibilities would materialize in the course of exploration. Some projections suggest a probable installed geothermal capacity of 0.6 GW (e) by 1985.

C. South America

1. Argentina

Geothermal exploration in Argentina has just been started under the auspices of the National Secretariat of Energy in cooperation with the

Argentina Commission on Geothermal Energy. Geothermal activity is concentrated along the boundaries of the Andes range in the vicinity of recent volcanic activity. The principal area of investigation at present is the Copahue area in the Neukuen province. Detailed exploration will indicate the nature and quality of the geothermal resources in that area. Other areas, such as the Bahia Blanca and Rosario de la Frontera, are known to have hot water systems. It is hoped that 0.02 GW (e) of geothermal power will be installed in the Copahue area by 1985. Only limited information about the ultimate potential of geothermal energy in Argentina is available.

2. *Brazil*

Geothermal exploration in Brazil has also just begun. At present there are at least 23 known hot springs, primarily in the state of Goias. There, two areas of hot springs—the Serra das Caldas, a well-known hot spring and resort area, and the Lagoa dos Peixes hot springs—indicate some potential and are presently being subjected to detailed investigation. Exploration is also planned for the regions of Bahia and Minas Gerais.

3. *Chile*

Exploration efforts in Chile, partially supported by the United Nations, have resulted in the discovery of the El Tatio geothermal field with a proven productive capacity of over 0.016 GW (e) at present. Other regions of Chile show similar potential, notably the Puchuldiza and Pollo Quere in northern Chile.

4. *Peru and Bolivia*

No significant amount of geothermal exploration has been carried out as yet in Peru and Bolivia, but both countries are located in a region of active volcanism. Numerous hot springs and fumaroles are known to exist in these two countries. It is reported that discussions are under way between national authorities and some international organizations to assist in the exploration and development of geothermal resources.

D. Western Europe

1. *France*

On the mainland, geothermal energy in France is likely to be used for direct heat purposes. In 1976 a geothermal heat installation near Paris provided heating for 10,000 apartments, resulting in an overall savings equivalent to 70,000 bbl of petroleum. Other parts of France with the potential for geothermal heat applications include the Aquitaine Basin, Bresse, Limagne, and Alsace.

Extensive exploration of geothermal resources by the French government in Bouillante, Guadeloupe, have resulted in the discovery of a geothermal reservoir; it is expected that a small geothermal power plant of perhaps 0.01 GW (e) will be installed on that island in the near future. It is expected that by 1985 the geothermal power plant at Bouillante might reach a 0.025 GW (e) capacity. On the other hand, the direct heat utilization capacity in mainland France might reach 0.6–1.6 GW (th) by 1985.

2. *Greece*

Geothermal exploration has been conducted in different parts of Greece since 1970. The most promising areas are the Ionian Islands, with the most outstanding potential existing on the islands of Milos, Santorini, and Nisiros. Some potential appears to exist also on the island of Lesbos, and there are warm springs in the regions of the Sperchios Graben, Sousaki, Methana, and others. Geothermal drilling began on the island of Milos in 1972, and deep test drilling was done there in 1975–1976. Two deep test holes encountered temperatures higher than 250°C; however, both wells are reported to be characterized by relatively low flow rates.

3. *Iceland*

The chief use of geothermal energy in Iceland at present is for direct heating of homes in Reykjavik, the capital of Iceland. More than 0.34 GW (th) is presently being utilized for space heating most of the homes in Reykjavik and neighboring towns. A small geothermal power plant is in operation at Namafjall, producing 0.0025 GW (e). A 0.05-GW (e) geothermal power plant is under construction at Krafla in an active volcanic zone. Subsequent to the start of construction the volcano became active, so the future of the power plant is somewhat uncertain at this moment.

It has been estimated that the high-temperature areas of Iceland have a production potential of 3.2 GW (th) for 50 years and that the heat content of recoverable low-temperature resources may amount to the equivalent of 5.6×10^{10} bbl of petroleum. The estimated total capacity of geothermal electrical power in Iceland is expected to reach 0.15 GW (e) by 1985, 0.5 GW (e) by 2000, and perhaps as much as 0.8 GW (e) by the year 2020. The chief use of geothermal energy in Iceland, however, is currently in direct heating. At present the total installed nonelectric capacity is 3160 GW h/yr. The level of use is expected to increase to 10,000 GW h/yr (th) by the year 2000 to perhaps 15,000 GW h/yr by 2020.

4. Italy

As of the first quarter of 1975, the total installed geothermal electric capacity in Italy was about 0.42 GW (e), of which 0.38 was in the Larderello region, 0.015 at Travale, and 0.022 in the Monte Amiata region (Fig. 8). Italy pioneered the development of geothermal energy in the world when the first electric power generator was established at Larderello early in the 1900s. It is likely that a considerably accelerated geothermal exploration and development program will be taking place in Italy within the next few years. The total direct heat utilization (nonelectric) in 1976 was estimated at 0.02 GW (th). It has been estimated that the total geothermal power capacity in Italy might increase to perhaps 0.8 GW (e) by 1985 and that direct heat utilization might increase by perhaps 60% over what it is at present. The total stored heat capacity in reservoirs of less than 100°C was estimated at 6.7×10^{21} J, and an approximately equivalent amount has been estimated for the reservoirs at a temperature range of 100–150°C.

5. Portugal

The chief exploration activity now taking place in Portugal is on the island of San Miguel in the Azores. There, geothermal exploration by the Institute of Geosciences, supported by an international contractor, is yielding initially promising results. Temperature gradients as high as

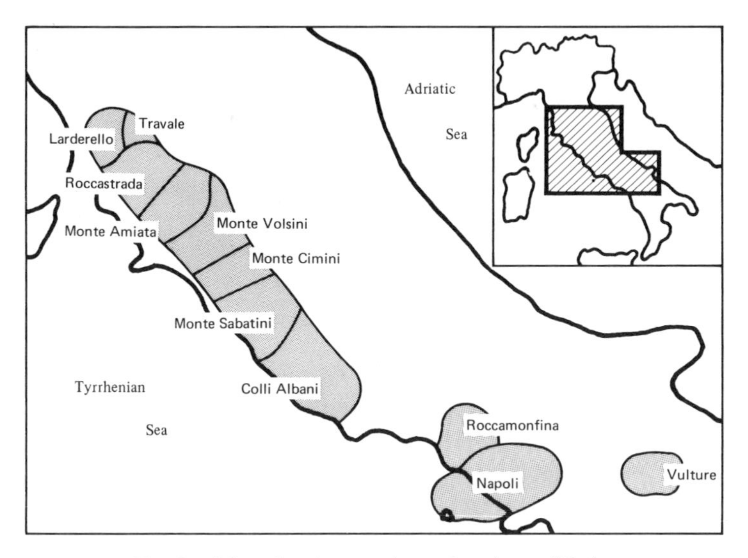

Fig. 8. Map showing geothermal regions of Italy.

0.8°C/m have been encountered in shallow test holes drilled in promising areas selected by prior geophysical investigations. The plans are to develop a 0.003–0.005-GW (e) noncondensing geothermal power plant by 1978. In case of success, the geothermal exploration and development activity would continue, with the ultimate objective of developing at least 0.03 GW (e) of geothermal electricity by the mid 1980s. Should a large geothermal reservoir be discovered, larger scale development for industrial and other purposes might follow, with an ultimate objective of 0.1 GW (e) or more by the year 2000. Exploration in San Miguel is due to be followed in short order of exploration in the other islands, especially Terceira and Fayal. Exploration on the mainland of Portugal is under way, with the objective of developing hot water resources for nonelectric uses, mainly in the northern part of the country.

6. *Spain*

The main geothermal regions of Spain are located in the Canary Islands, especially at Lanzarote, where very high temperatures have been encountered at shallow depths. Intensive exploration is currently taking place at Lanzarote, with the objective of replacing the existing diesel power. Exploration in the mainland regions will revolve around the development of lower grade geothermal resources for direct heating purposes. The objective of electrical power development in Spain is to achieve a capacity of at least 0.025 GW (e) by 1985 and perhaps 0.2 GW (e) by the year 2000. While it is difficult to estimate the nonelectric demand, it may range from 0.01 GW (th) by 1985 up to 0.05 GW (th) by the year 2000.

7. *Sweden*

At present there is no utilization of geothermal energy in Sweden. However, ongoing geothermal resource investigations suggest that some use will be made of hot water for direct heat purposes. It is estimated that by the year 2000 a total of 0.5–1 GW (th) will be yielded from geothermal water for nonelectric purposes. The estimated total stored heat to a depth of 3 km in Sweden is 4.2×10^{22} J.

8. *West Germany*

At present, the only utilization of geothermal energy in West Germany is for nonelectric purposes. At Urach and Wiesbaden the total installed capacity for nonelectric purposes is approximately 0.5 GW (th). The estimated total energy resource base of West Germany appears to be approximately 1.4×10^{23} J to a depth of 3 km. The known hot water reservoirs in West Germany occur in the Rhine graben basin, the northern Germany basin, and the southern Germany (Molasse) basin.

9. *United Kingdom*

The United Kingdom appears to be a region of normal temperature gradient (25–30°C/km), with a few anomalous regions where temperature gradients reach 40–45°C/km (in southwest England). In the North Sea Basin, thermal gradients appear to be higher, at times exceeding 40°C/km. Electric power production from geothermal resources appears to be practicable only if technology for extracting geothermal heat from hot dry rock becomes successful (i.e., at the turn of the century or later). It is further estimated that direct heat utilization of geothermal energy may reach as much as 3.4 GW (th). The total resource base of geothermal energy to a depth of 3 km is estimated to be 5.4×10^{22} J on land and approximately 2.9×10^{22} J in the North Sea Basin.

E. Eastern Europe

1. *Hungary*

Geothermal energy harnessed for direct heat uses is probably at its most advanced stage in Hungary. Extensive utilization is presently being made of low and intermediate geothermal heat for space heating, greenhouse heating, and industrial uses. It is estimated that present total utilization capacity of geothermal heat in Hungary is approximately 0.5 GW (th). Plans for further development suggest that the geothermal heat utilization will increase to 0.5–0.8 GW (th) by 1985, 1.1–2.9 GW (th) by 2000, and will ultimately reach as much as 1.9–5.3 GW (th).

In 1975 more than 430 wells were drilled in Hungary, producing water at temperatures greater than 35°C at the wellhead, that were utilized for direct heat. These wells produced over 460 m^3/min of geothermal fluid, yielding over 1 GW (th) of thermal energy. The total usable heat in the recoverable water from the main reservoir in Hungary has been estimated to be 5×10^{19} J.

2. *Soviet Union*

Electric power is being produced on a small scale in the Soviet Union, mainly in the Kamchatka Peninsula. Two small power plants, one in Pauzhetska [producing 0.005 GW (e)] and another at Paratunka [0.0007 GW (e)] have been operating for several years. At present, geothermal energy in the Soviet Union is primarily used as direct heat, including space heating, greenhouse operation, and industrial uses. More than 28 geothermal fields are now in use. Soviet scientists have estimated the potential yield of thermal water in the temperature range of 40–250°C to be 22×10^6 m^3/day. The chief regions of geothermal exploration and development in the Soviet Union are in the Kuril Islands, Kamchatka, and the Caucasus belt. Some geothermal deposits appear to occur also in Siberia, Kazakhstan, and Middle Asia. It has been estimated that the hydrothermal deposits of Kamchatka alone have the capacity to support more than 0.35–0.5 GW (e) at present, although no specific plans for such utilization have been announced.

3. *Yugoslavia*

It is unknown whether any geothermal energy in Yugoslavia will be utilized for electric power production. However, it is assumed that geothermal heat will be used. The utilization will be of the order of 0.06 GW (th) by 1985, about four times as much by 2000, and about eight times as much by 2020. Not enough information is currently available in Yugoslavia on the potential of geothermal energy; however, reasonably steep temperature gradients of the order of 60°C/km are known to exist. The estimated reserve of geothermal energy below 100°C is 6.6×10^{20} J.

4. *Other Eastern European Countries*

Limited exploration and use have been made of geothermal resources for nonelectric uses in the rest of Eastern Europe. Geologic evidence suggests that the geothermal region of Hungary extends to Czechoslovakia and Rumania and that economically attractive temperatures may exist in regions of Austria as well. Some exploration in Southwestern Poland suggests that temperatures up to 60°C may exist in depths of less than 1 km.

F. Asia

1. *China*

Exploration in China is under way in the Himalayas, where 10 geothermal fields have been discovered to date. One of these, in the Ch'iang-T'ang region in the Tibetan highlands, will be developed to generate electric power in the near future. The recently discovered Ku-Tuë geothermal well near the Bhutan border has led investigators to conlude that the magma is very close to the surface along the Himalaya Range and that the geothermal potential is likely to be very great.

2. *India*

Geothermal exploration has been concentrated primarily in the northern regions of India. Exploration in the Puga Valley in the upper Himalayas close to the northern border of the country has found temperatures as high as 135°C at a depth of less than 100 m. Geochemical thermometry suggests that temperatures higher than 250°C may be encountered in deeper drilling in that region. Preliminary exploration in the Manikaran area suggests that reservoir temperatures of 200°C or higher may be encountered in this region. A series of intriguing hot springs runs along the entire northern and southern range of volcanic deposits in western India. At the northern end of that region, in the Cambray area, geopressured geothermal energy is likely to occur. There in, oil exploration wells, abnormally high wellhead pressures in excess of 100 kg/cm^2 have been encountered. A joint geothermal exploration program with the United Nations is now under way. It is likely to result in a rapid assessment of India's geothermal potential.

3. *Israel*

Exploration for geothermal resources in Israel was started in 1974. The Geological Survey of Israel conducted numerous field mapping and geochemical studies to identify geothermal resources in the Jordan rift valley and to compile a general heat flow map of the country. The hot springs in the Sea of Galilee region reach a temperature of 72°C. Chemical thermometry suggests that reservoir temperatures of at least 140°C may occur at depth in that region. Electrical resistivity surveys in the Jordan rift valley have detected a number of anomalously low resistivity areas that will be investigated more closely in the near future. Hydrogeological investigations of the central Negev region show that a warmer aquifer (approximately 40°C) underlies much of that region. Plans are now afoot to tap those aquifers for direct heat uses (greenhouse operations). A geopressured system has been discovered in the Ashkelon region, but its pressure and temperature characteristics are not yet fully known. Hot springs in the Dead Sea region (Zohar and Ein Yesha) have been tapped for balneological purposes, and a thriving tourist industry is being developed around the springs. The Tiberius hot springs have been used since biblical times for therapeutic purposes and still have the remains of elaborate Roman spas.

4. *Jordan*

The outstanding geothermal springs in Jordan are the Zerka Ma'in Springs on the east side of the Jordan rift valley near the Dead Sea. The surface temperature of the Zerka Ma'in thermal springs is of the order of 64°C, and chemical thermometry suggests that reservoir temperatures may be higher than 180°C. At present, a preliminary evaluation of the Jordan geothermal potential is being conducted by a United Nations expert.

5. *Turkey*

Exploration drilling in the Kizildere field has resulted in the discovery of geothermal potential with a maximum temperature of approximately 207°C. However, because of serious calcium carbonate scaling problems, the field has not yet been used for commercial production. Instead, the geothermal fluid is presently being used for a pilot greenhouse operation. Extensive exploration activity by the Mineral Research and Exploration Institute of Turkey (MTA) has identified many other geothermal regions in the country. The MTA has estimated that geothermal energy is likely to supply 10% of the total Turkish electric energy requirements by the year 2000. Exploration activities in the Seferihisar and Tuzla regions of western Turkey have discovered areas of promising potential for the production of electricity from geothermal energy. Exploration activities elsewhere have suggested that in the regions near Kizildere and Ankara significant geothermal energy potential does exist. At present the geothermal electric capacity in Turkey is only 0.0005 GW (e). However, the exploration plans are to achieve a production capacity of 0.4 GW (e) by 1985, 1 GW (e) by 2000, and perhaps 1.5 GW (e) by 2020. These are maximal estimates; other estimates are somewhat more conservative.

G. East Africa

1. *Ethiopia*

An active geothermal belt runs through the rift valley of eastern Africa, starting from the Red Sea to the north and continuing southward through the length of Ethiopia, Kenya, and Tanzania. Active volcanism is currently taking place in the Afar rift valley of northern Ethiopia, where the Erfa-'Ale volcano has been boiling at 1000°C or more for over a century. It has been estimated that the amount of continuous heat flow necessary to keep the Erfa-'Ale lava pool at its current temperature is equivalent to that of a 0.1-GW (e) power plant at full capacity.

Three separate regions have been identified as potentially promising. The most attractive region for geothermal development appears to be the Danakil depression. In that region the Tendaho area seems to be particularly attractive, with geotechnical evidence suggesting the occurrence of high-temperature, low-salinity geothermal fields.

A number of favorable locations have been identified in the northern Afars area. At Dallol, geothermal springs with temperatures as high as 114°C at the surface have been measured. However, the problem confronting geothermal energy development in the Dallol area (other than the political instability of the area) is the very high salinity of the geothermal fluids. At present most of the geothermal exploration is centered on the lakes district north of Addis Ababa. Through a United Nations survey, a number of favorable geothermal prospects have been identified near Lake Langano and farther south.

The overall potential of geothermal energy in Ethiopia has been estimated from small to perhaps as much as 20 GW. However, because of the small power demand in the country, the problem of developing high-quality geothermal resources, such as those at Tendaho, is associated with the cost of transmission from the remote prospect locations to the main grid of the country, sometimes more than 300 km away. Political factors have also caused a slowdown in developing the outstanding geothermal resources of the country. Therefore, geothermal energy utilization has been restricted to minor balneological uses. However, there is no doubt that geothermal energy will eventually provide an important source of electricity in Ethiopia.

2. *French Territory of the Afars and the Issas*

Geothermal exploration in the region near Djibouti has encountered high geothermal reservoir temperatures (over 250°C at 1050 m); but because of the high salinity of the fluids, their utilization is fraught with problems.

3. *Kenya*

The geothermal exploration and development in Kenya is farther ahead than Ethiopia's. A successful exploration program at the Olkaria field about 50 km north of Nairobi has demonstrated the existence of a high-quality, low-salinity geothermal reservoir. Temperatures of up to 287°C have been encountered in drill holes less than 1.5 km deep. Further development drilling is due to take place at Olkaria, and exploration drilling is likely to take place in the near future in the Lake Hannington area to the north. Geophysical measurements of a third prospect, the Eburu Crater, have shown promising results.

The minimum installed capacity estimated by the East African Power and Lighting Company is 0.03 GW (e) of electrical power by 1985, 0.06–0.09 GW (e) by 2000, and 0.9–1.5 GW (e) by 2020. The stored energy in the Olkaria field has been estimated at 84 GW yr (gigawatt-years), of which 5% could be converted into electric energy.

4. *Other East African Countries*

Moderate exploration of the geothermal resources of Tanzania and Uganda has been carried out. On the basis of the limited exploration to date, it is too early to determine the ultimate potential of geothermal energy in those countries. Similarly, several geothermal indicators exist in Malawi and Burundi, but very little useful data are available to assess their commercial viability.

H. Oceania

1. Fiji

A number of hot springs, some at boiling point, suggest that geothermal potential may exist on Vanua Levu, the north island of Fiji. However, chemical thermometry indicates the base temperature of the reservoir to be quite low, of the order of 140°C. There are no immediate plans for development.

2. Indonesia

Indonesia is located along a major geothermal belt and is characterized by extensive volcanism and earthquake activity, common indicators of geothermal resources in the area. Extensive investigations of geothermal energy have been carried out at Java, Bali, and some other islands by the Geological Survey of Indonesia, by the national electric authority, and by the national oil company. In addition, technical assistance for the exploration of geothermal energy has been given by the governments of the United States and New Zealand. A 1927 exploration in the Kawah Kamojang near Bandung discovered that geothermal steam may be encountered at a very shallow depth. Renewed interest in the area spurred further exploration in 1973, with the assistance of the New Zealand government. As of May 1975, four holes had been drilled to depths of 500–800 m, and at least two have indicated the occurrence of dry steam underground. A small geothermal power plant is now being considered for the area, probably 0.015 GW (e) initially.

The Dieng Plateau of central Java was investigated in the early 1970s by the Indonesia Power Research Institute with the assistance of the United States. Geophysical, geochemical, and drilling activities indicated the existence of a large, shallow geothermal system of the liquid-dominated type. Technical assistance missions by the United Nations and UNESCO have shown that many other geothermal areas in Java, such as Kawah Cibureum and Cisolok, are likely to be of interest. Informal estimates suggest that perhaps 20 GW (e) of electric energy capacity occur in Java and as much as 60 GW (e) in all of Indonesia.

The estimated minimum installed geothermal energy capacity in Indonesia will be 0.03–0.1 GW (e) in 1985 and 0.5–6 GW (e) by the year 2000.

3. *Japan*

Japan witnessed a relatively slow growth of geothermal energy utilization for electric power production until the 1973 energy crisis. Prior to that only two electric generating plants using geothermal energy were in operation, in Matsukawa [0.022 GW (e)] and Otake [0.013 GW (e)]. In 1976 the total installed geothermal power capacity in Japan was 0.068 GW (e). The energy crisis has stimulated the expansion of geothermal energy exploration. As a result, it is anticipated that by the end of 1977 the installed geothermal power capacity will approach 0.15 GW (e) (based upon the completion of a 0.05-GW (e) power plant at Takinoue and the completion of a 0.05-GW (e) plant at Hatchobaru). The current objective is to achieve an installed geothermal electric capacity of 1.0 GW (e) or more by 1982 and about 2 GW (e) by 1985. In addition, ambitious geothermal investigations are currently taking place in Japan, with the goal of establishing as much as 50 GW (e) electric capacity by the year 2000.

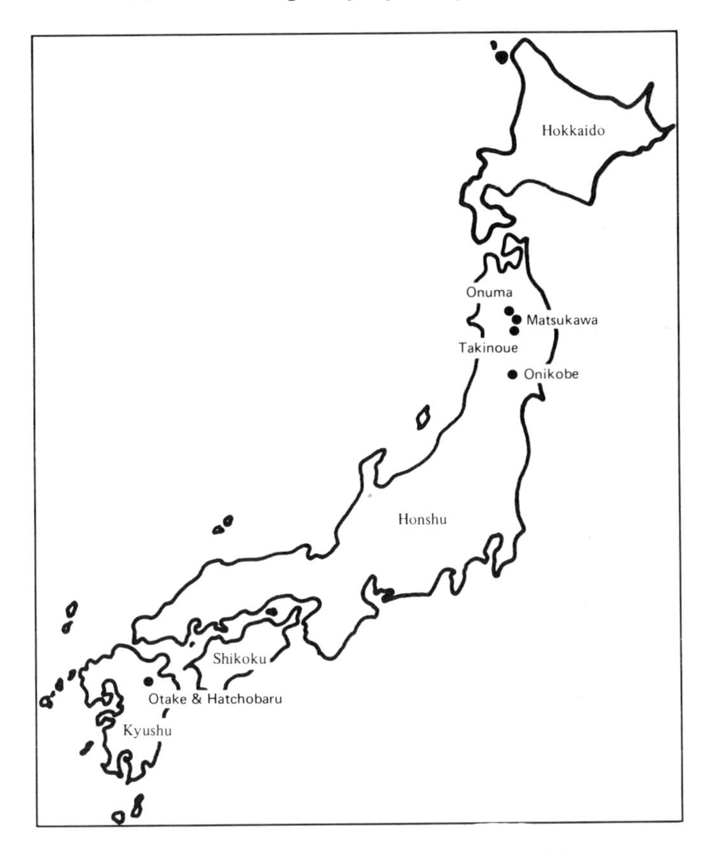

Fig. 9. Map showing geothermal regions of Japan.

More than 10,000 hot springs are known to exist in Japan. In the past they have been used primarily for balneological and religious purposes, with some minor uses such as fish breeding. Figure 9 shows the major regions of geothermal exploration in Japan at present. Approximately 29 GW (th) are now being used for direct heat (nonelectric). The total geothermal energy resource of Japan, to a depth of 3 km, is estimated at 1.2×10^{21} J.

4. *New Zealand*

The presently installed geothermal plant at Wairakei has a capacity of about 0.19 GW (e). At Kawerau, geothermal energy is being used for the combined power production and direct heat needs of the Tasman Pulp and Paper Mill. In addition, geothermal energy is extensively used at Rotorua for space heating and, in one case, for air conditioning. A second geothermal field at Broadlands, to be commissioned in the early 1980s, is likely to be developed for power production, with an estimated capacity of approximately 0.15 GW (e). The total estimated geothermal power capacity of New Zealand is rated at approximately 2 GW (e). It is estimated that 0.4 GW (e) of geothermal electricity will be on line by 1985, 1.4 GW (e) by the year 2000, and 2 GW (e) by the year 2020. The extent of direct heat utilization in New Zealand is expected to rise from 0.4 GW (th) in 1985 to about 0.6 GW (th) by the year 2000. The total resource base of geothermal energy in the explored areas of New Zealand is estimated to be 1.3×10^{20} J.

5. *The Philippines*

Extensive geothermal exploration has been taking place in the Philippines at the Tiwi, Leyte, and Los Banos areas. Successful exploration at Tiwi in southeastern Luzon has proven that sufficient geothermal steam exists for generating 0.1 GW (e) electric power. The first 0.1-GW (e) electric plant will be installed in 1977, to be followed shortly by another 0.1-GW (e) plant in the same area. Promising exploration in the Los Banos area suggests that geothermal power may be developed there in a relatively short time. Plans in 1976 called for the installation of 0.3 GW (e) of geothermal electricity by 1985. These plans are presently under review with the intent of increasing the number.

6. *Taiwan*

No geothermal power plants are in operation in Taiwan, though significant geothermal potential does exist in the country. High-temperature reservoirs have been encountered in both the Tatun and Tuchang areas. In the Tatun area, temperatures are as high as 293°C, but because of the exceptionally high acidity of the water, no development was considered to be possible in that field. Current plans call for the development of 0.05 GW (e) geothermal capacity by 1985, 0.2 GW (e) by 2000, and perhaps 0.5 GW (e) by 2020. The expected total capacity of direct heat utilization in Taiwan is 0.005 GW (th) by 1985, three times as much by 2000, and five times as much by 2020.

ACKNOWLEDGMENTS

This chapter was made possible as a result of efforts by many individuals and organizations, including the World Energy Conference, the Electric Power Research Institute, universities, private industry, and over 80 nations. One valuable source of information was the international response to a questionnaire distributed by the Electric Power Research Institute seeking current information on worldwide geothermal prospects. Sincere appreciation is extended to the many nations, individuals, and organizations that responded to the questionnaire. Special recognition goes to Dr. Tsvi Meidav and Dr. Subir Sanyal of Geonomics, Inc. for their contribution to the sections on Geothermal Resources and International Development; to Professor A. Louis London at Stanford University for his contribution to the section on Electric Power Potential; and to Professor Gordon Reistad at Oregon State University for his contribution to the section on Nonelectric Uses. Appreciation also goes to Professor Paul Kruger at Stanford University for his advice and reviews. Finally, credit goes to the World Energy Conference for having motivated this work and their assistance in making the appropriate contacts in the world community.

REFERENCES

Resources

Barnes, J. (1971). "Multipurpose Exploration and Development of Geothermal Resources," Natural Resources Forum 1. United Nations, New York.

"Index to the Geothermal Fields of Japan" (1976). Agency Nat. Resourc. Energy Geol. Surv. Jpn., Min. Int. Trade Ind., Tokyo.

Muffler, L. J. P. (1973). Geothermal resources. U.S. Mineral Resources. *U.S. Geol. Surv., Pap.* No. 820.

Rex, R. W. (1968). "Investigation of the Geothermal Potential of the Lower Colorado River Basin, Phase 1—the Imperial Valley Project." Inst. Geophys. Planet. Phys., Univ. of California, Riverside.

Rex, R. W. (1971). Geothermal resources in the Imperial Valley. *In* "California Water—A Study in Resource Management" (D. Seckler, ed.). Univ. of California Press, Berkeley.

Wells, M. (1971). Early development of western geothermal resources. *J. West.* **X**, No. 1 (1971).

White, D. F., and Williams, D. L. (1975). Assessment of geothermal resources of the United States—1975 *U.S. Geol. Surv., Circ.* No. 726.

Utilization for Electric Power

"Geothermal Energy Conversion and Economics—Case Studies." (1976). EPRI ER-301. Electr. Power Res. Inst., Palo Alto, California.

Goguel, J. (1976). "Geothermics." McGraw-Hill, New York.

Mazzoni, A. (1948). "The Steam Vents of Tuscany and the Larderello Plant." Amonina Arts Grafiche, Bologna. (In Engl.)

Milora, S. L., and Tester, J. W. (1976). "Geothermal Energy as a Source of Electric Power." MIT Press, Cambridge, Massachusetts.

"Utilization of U.S. Geothermal Resources." (1976). EPRI ER-382. Electr. Power Res. Inst., Palo Alto, California.

Nonelectric Uses

Bodvarsson, G. (1964). Utilization of geothermal energy for heating purposes and combined schemes involving power generation, heating and/or by-products. *Proc. U.N. Conf. New Sources Energy, Rome, 1961* Pap. GR/5(G). United Nations, New York.

Boldizsar, T. (1974). Geothermal energy use in Hungary. *Proc. Int. Conf. Geotherm. Energy Ind., Agric. Commer.-Residential Uses,* Ore. Inst. Technol., Klamath Falls.

Einarsson, S. S. (1973). Geothermal district heating. *In* "Geothermal Energy: Review of Research and Development." UNESCO, Paris.

Head, J. (1970). *Ore Bin* **32**(9), 182–183.

Howard, J. H., ed. (1975). "Present Status and Future Prospects for Non-Electrical Uses of Geothermal Resources," UCRL-61926. Lawrence Livermore Lab., Livermore, California.

Kerr, R. N., Bangma, R., Cooke, W. L., Furness, F. G., and Vamos, G. (1964). Recent developments in New Zealand in the utilization of geothermal energy for heating purposes. *Proc. U.N. Conf. New Sources Energy, Rome, 1961* Pap. G/52, United Nations, New York.

Lenzi, D. (1964). Utilization de l'energie geothermique pour la production de l'acide boriquet et des sous-produits contenus dans les "souffione." *Proc. UN. Conf. New Sources Energy, Rome, 1961* Pap. G/39, United Nations, New York.

Lindal, B. (1964a). Geothermal heating for industrial purposes in Iceland. *Proc. U.N. Conf. New Sources Energy, Rome, 1961* Pap. G/59, United Nations, New York.

Lindal, B. (1964b). The extraction of salt from seawater by multiple effect evaporators using natural steam. *Proc. U.N. Conf. New Sources Energy, Rome, 1961* Pap. A 3/G, United Nations, New York.

Lindal, B. (1973). Industrial and other applications of geothermal energy. *In* "Geothermal Energy: Review of Research and Development," LC No. 72-97138. UNESCO, Paris.

Ludviksson, V. (1970). "The Application of Natural Heat," Rep. 70-2. Nat. Res. Counc. Iceland, Reykjavik. (In Icel.)

Lund, J., Culver, G., and Svanevik, L. (1974). Utilization of geothermal energy in Klamath Falls. *Proc. Int. Conf. Geotherm. Energy Ind., Agric. Commer.-Residential Uses,* Ore. Inst. Technol., Klamath Falls.

Matthiasson, M. (1970). Beneficial uses of heat in Iceland. *Proc. Conf. Beneficial Uses Therm. Discharges,* New York State Dep. Environ. Conserv., Albany.

U.S. Energy Research and Development Administration (1976). "A National Plan for Energy Research, Development and Demonstration: Creating Energy Choices for the Future," Vols. 1 and 2, ERDA 76-1. U.S. Gov. Print. Off., Washington, D.C.

Valfells, A., Lindal, B., and Ludviksson, V. (1969). "An Estimate of Heavy Water Manufacturing Costs in Iceland." Iceland Natl. Res. Counc., Reykjavik.
Zoega, J. (1974). The Reykjavik municipal heating system. *Proc. Int. Conf. Geotherm. Energy Ind., Agric. Commer.-Residential Uses,* Ore. Inst. Technol., Klamath Falls.

Broad Treatment of Geothermal

Armstead, H. C., ed. (1973). "Geothermal Energy: Review of Research and Development," Earth Science Series, No. 12. UNESCO, Paris.
Burrows, W. (1974). Utilization of geothermal energy in Roturua, Ne Zealand. *Proc. Int. Conf. Geotherm. Energy Ind., Agric. Commer.-Residential Uses,* Ore. Inst. Technol., Klamath Falls.
Kruger, P., and Otte, C., eds. (1973). "Geothermal Energy: Resources, Production, Stimulation." Stanford Univ. Press, Stanford, California.
"The Times Atlas of the World" (1974). Comprehensive Ed. The Times of London and Bartholemew, Edinburgh.
United Nations Education, Scientific and Cultural Office (1970). *Geothermics, U.N. Symp. Dev. Util. Geotherm. Resour., Pisa* Spec. Issue No. 2, Vol. 2.
Wahl, E. F. (1977). "Geothermal Energy Utilization." Wiley, New York.

Clean Fuels from Coal

Harry Perry

Resources for the Future
National Economic Research Associates
Washington, D.C.

ISBN 0-12-01490-X

I. THE NEED FOR COAL CONVERSION TECHNOLOGY

A. United States Energy Demand

United States energy demand has had a 50-year average growth rate of 2.5% from 1923 to 1973, when the Oil Producing and Exporting Countries (OPEC) oil embargo and the quadrupling of prices combined with a United States and worldwide economic recession all acted to reduce this growth rate. Consumption in 1976 in the United States was approximately 75 Q (quads), about the same as 1973. Whether the change in growth rate since 1973 represents a permanent shift in the rate of increase in energy use, and by how much, has been the subject of a number of recent studies. A number of the latest projections of energy demand for 1985 and 2000 have been much lower than those made as recently as 1972, as shown in Table I.

TABLE I

Comparison of Projections of United States Energy Demand; 1985 and 2000 (10^{15} Btu)

	1985	2000
National Petroleum Council (1972)	112–130	—
Energy Policy Project (1974)	91–116	124–187
Department of the Interior (1975)	104	110–163
DRI—Brookhaven (1976)	93–100	118–156
Institute for Energy Analysis (1976)	82–88	101–125
FEA—National Energy Outlook (1977)	70–91	78–102

The reason for the lower projections in recent years is the result of numerous factors. The higher cost of energy of all kinds—not just oil—is expected to result in lower consumption of energy. At higher costs for energy, capital and labor will be substituted for energy, and efforts will be intensified to reduce waste and to increase the efficiency of energy use. Numerous studies have been made that conclude that the use of energy can be greatly reduced from that originally projected for future years without adversely affecting the quality of life.

Since early 1977 the consensus of many energy projections for the year 2000 have tended to center about a value of 100 Q, but there are some estimates as low as 75–80 Q and other more numerous ones in the range of 125–135 Q. The exact value cannot be known at this time, but prudence suggests that it would be well to plan to supply the higher estimated quantities; they may turn out to be needed to support continued economic growth.

The growth in energy demand at a nearly uninterruptible rate has been accompanied by marked changes in the form in which energy is used. In the primary fuel forms, excluding hydroelectricity and nuclear, sharp inroads were made in coal's share of the energy market, first by oil and then by natural gas. In 1925 coal supplied over 72.9% of the United States energy demand, while oil supplied 21.2% and natural gas only 5.9%. By 1950 coal's share had declined to 39.9% while oil's had increased to 41.7% and gas, to 18.4%. By 1973 coal had declined to only 18.7% of total demand, oil had increased to 49.4%, but gas had climbed to 31.9%.

In addition to changes in the types of primary fuels used, the use of energy in the form of electricity grew very rapidly. Growth rates for electricity have averaged (until about 1973) 7% per year and the share of energy used to produce electricity grew from 6% in 1925 to 27% in 1973.

There are many reasons for the shifts in use of the primary fuel forms and for the greatly increased use of electricity. The shift to petroleum first occurred because of its availability in regions where coal was in short supply, its low transport costs, and the rapid growth in the transport market; automobiles, trucks, railroads, and airplanes. Except for the railroads, the other transport modes are almost entirely captive to petroleum products for their fuel supply. In the railroad sector coal lost the market to petroleum products after World War II because the efficiency of use of diesel fuel was approximately 20%, compared to coal's 5%. This higher efficiency and the lower operating and maintenance costs of diesels combined to make the use of liquid fuels in railroad use lower in cost than coal.

In more recent years the trend away from coal to both oil and gas was the result of their much greater convenience in use in some markets, e.g., the residential and commercial sectors, and their greater cleanliness when used by the consumer. With the rise of the environmental movement the shift away from coal accelerated even more rapidly in order to meet the ever more stringent environmental regulations imposed at the Federal and State levels.

The shift to gas occurred because new technology was developed that permitted gas pipelines to be constructed more economically so that the vast reserves of natural gas in the southwestern part of the United States could be transported at reasonable cost to the large energy markets in the densely populated eastern United States. Natural gas was low cost, convenient to use, and clean, and in more recent years its price has been regulated at artificially low levels, thus increasing demand for this premium fuel by all the energy consuming sectors—even those that could use alternate fuels conveniently.

The more rapid growth in electricity use compared to that of the primary fuels was a matter of convenience and cleanliness, as well as the growth in energy uses where electricity served unique needs—lighting, electric motors, refrigerators, etc.

Projections of the forms in which fuel is expected to be used in the future have also been the subject of numerous studies. Basically, the trend is estimated to be an increased use of electricity because it is clean and convenient to use by the consumer, although it is, overall, an inefficient way to use primary energy units and it creates difficult air and water pollution problems where the electricity is generated. Domestic supplies of oil and natural gas are anticipated to decline. Renewable resources are not expected to supply any important part of energy supply for at least 25 years or more. This would leave coal and nuclear fuels as the major sources of supply of energy during this period. Nuclear fuels will probably only be used to generate electricity, while coal can be used in a variety of ways. These will be described in Sections II and III.

B. United States Energy Supplies

Several estimates of world reserves[1] and resources[2] of conventional fuels (coal, oil, gas, and uranium) are shown in Table II. Coal reserves represent nearly 75% of the total of fossil fuels in the world while oil and

[1] Reserves are that portion of total resources that is known and economically recoverable on the basis of present prices and present technology.

[2] Resources are the sum of "reserves" and "prospective reserves." Prospective reserves are that portion of the resources that is likely to be added to the reserves at some time in the future.

TABLE II

Comparison of Estimates of World Nonrenewable Energy Reserves and Resources (10^{18} Btu)[a]

	RFF	NAS	ERDA	Hubbert	WEC
Reserves					
Coal	24.0	—	14.3	47.0–178.0	14.3
Petroleum	3.5	3.6	4.1	10.1[e]	4.1
Natural gas	2.0	2.1	2.4	—	1.9
Oil shale	2.0[c]	72.0	3.0	1.7	10.4
Tar sands		4.5[d]	—	2.2	
U_3O_8 [b]	1.3	—	2.1	—	0.6
Resources					
Coal	172.5	220	120.0–160.0	—	237
Petroleum	12.1[f]	10.4	7.8–11.3	—	17.3
Natural gas	11.1	7.0	3.1–6.1	—	9.8
Oil shale	15.8[g]	205.2	6.8–9.6	—	—
Tar sands		—	—	—	—
U_3O_8 [b]	3.3–13.4	—	3.0–3.9	—	1.2

[a] Sources: Ridker and Watson (1978); National Academy of Sciences (1975); Energy Research and Development Administration (1976b); Hubbert (1974); World Energy Conference (1974); World Power Conference (1968); Linden (1974); Senate Committee on Interior and Insular Affairs (1974); Gardner (1973); Auldridge (1976); Nuclear Regulatory Commission (1976); Nuclear Energy Policy Study Group (1977).

[b] U_3O_8 up to \$50/lb (forward cost) and with 1% efficiency of utilization in lightwater reactors (37 million kWh/ton).

[c] Known deposits recoverable under 1965 conditions and containing 10–100 gal per ton of shale.

[d] Incomplete reserves of tar sands.

[e] Estimates made in 1972 by Richard L. Jodry, Sun Oil Company.

[f] This estimate assumes a continuation of current prices (in real terms) and a recovery rate of 40%.

[g] The value is for what are now marginal and submarginal known deposits containing 10–100 gal per ton of shale. Total oil shale resources of grades from 5–100 gal per ton in undiscovered and unappraised deposits could be as large as $13{,}000 \times 10^{18}$ Btu. Only 144×10^{18} Btu of this total is in the known or possible expansion of known deposits.

gas represent less than 10%. For the United States the coal share of total energy resources is even higher. The same general picture emerges when resources (rather than reserves) of these fuels are compared—coal is about 75% of the total, while oil and gas together represent only about 10% of the total. Without a breeder reactor uranium reserves and resources are comparable in size to those of oil and gas. If the breeder reactor becomes commercially available the uranium supplies are of the same order of magnitude as that of coal.

As noted above in discussing United States energy demand, the growth in fuel use has been greatest for oil and gas with a decline in the absolute amount of coal being used—in direct contrast to both the reserve and resource supply situation for these fuels. The same shift away from coal has also characterized the rest of the world energy supply situation, but for the world the growth has been much greater for oil than for gas because of the lower costs associated with oil transport.

In the United States the intensive drilling for oil and gas on shore in the lower 48 states over the first 60 years of this century has led to the conclusion that large deposits of new oil and gas will no longer be found in these provinces. This, combined with the regulated price of gas and the easy to find oil of the Middle East that can be produced at low cost, caused a steep decline in drilling activity in the United States starting in the mid 1950s. As would be expected under these conditions, reserves declined sharply (except for the years of the Alaskan oil and gas discoveries) starting about 1969 (see Table III). The declining reserves were reflected in changes in domestic production rates with production of gas having apparently peaked in 1973 and oil in 1972. The greatly increased world oil price and rapid rise allowed in the regulated price of natural gas in the past several years has resulted in an increase in the drilling rate for oil and gas during this period, but it is still unclear whether this will

TABLE III

Proved Reserves of Oil and Gas in the United States, Selected Years[a,b]

Year	Oil (10^6 barrels)	Gas (10^9 ft^3)
1950	24,649	179,401
1955	29,560	210,560
1960	31,790	261,170
1965	30,990	281,251
1968	31,376	292,907
1969	30,707	287,350
1970	29,631	275,108
1971	29,401	264,746
1972	28,463	252,806
1973	26,739	240,085
1974	25,700	223,950

[a] Excluding Alaska.

[b] Source: American Gas Association *et al.* (1976). Reserves measured at start of year.

reverse or slow down the declining domestic production rates of oil and gas.

There still remain opportunities for developing new oil and gas reserves in the United States from the resources that have yet to be discovered. The large resource potentials that remain, however, are believed to be in the offshore provinces (Atlantic, Gulf of Alaska) and onshore in Alaska. These areas present more difficult drilling and transport problems, and the oil and gas found there can be expected to command premium prices. Intensive leasing of these lands (largely Federally owned) followed by systematic drilling could result in some additional supplies of both domestic oil and gas over the next 10–20 years, but will probably not offer any long-term answer to the domestic liquid and gaseous supply problem.

Published estimates of the supply situation for these fuels for the balance of the world are contradictory to the extent that they predict differing times in the future when world oil and gas production will peak. Some estimates of when this will occur are as early as the late 1980s (Worskhop on Alternative Energy Strategies, 1977; Central Intelligence Agency, 1977) while others (Organization for Economic Development and Cooperation, 1974; Jensen, 1970) are for the year 2000 or beyond. No matter which date turns out to be correct, there remains little time left to make a transition to other liquid or gaseous sources or to other fuel forms.

In addition to accelerating the exploration for supplies of hydrocarbons in new conventional reservoirs, there remains the potential of recovering additional oil from existing reservoirs by the development of improved techniques or entirely new approaches to the secondary and tertiary recovery of oil. Average recovery rates from oil reservoirs are less than 40%, and for each additional 1% recovery rate about 4×10^9 barrels of oil could be produced—or about 10% of our proved reserves.

In the case of gas many reservoirs have been identified in which the gas is so tightly bound that it cannot be produced economically using conventional gas production methods. Methods to release this gas are under development (fracturing by various methods) and should one or more be successful, gas reserves could be increased two or more times.

There also remains the potential for producing hydrocarbons from a variety of resources that are known to exist but have been little used because of their higher costs compared to hydrocarbons produced from conventional reservoirs. However, the price of conventional hydrocarbons has increased by a factor of 4 and can be expected to go even higher as production peaks in the rest of the world. At these higher prices, the heavy oils found in large quantities in, for example, Venezuela and California, the tar sands deposits in Canada and Utah, and the oil shale deposits widely distributed around the world, could become a source of liquid (or gaseous) hydrocarbons.

Unconventional gas deposits in the United States are found in the Devonian shales, in coal beds, and in the geopressured zones around the Gulf of Mexico. There has been a small amount of production from the Devonian shales over the years and from experimental test wells drilled in coal beds from which methane has been extracted.

For some of these unconventional hydrocarbons the resource base is known to be very large. For others only scattered and incomplete resource data are available, but the resource base for these is also believed to be large. For example, preliminary estimates by one geologist project geopressured gas resources to be as large as 15×10^{15} standard cubic feet (SCF) or 70 times the proven reserves of natural gas.

Just as important as the resource base from the standpoint of the future commercial prospects of coal conversion, is the cost at which these resources could be produced. For some types of reservoirs and for some methods of stimulating increased production of oil or gas, fairly reliable cost estimates can be made, since it is basically a case of extending existing technology. For the unconventional resources, estimates of the cost of production are far less reliable because, in most cases, new technologies for their production will have to be developed. More current estimates of the costs of production should be viewed as being optimistic, since they are generally made by those trying to commercialize either the new sources or some new technology to produce them. All that can be said with certainty is that the costs will be higher than hydrocarbons produced from currently producing fields and that the costs will vary over a wide range because of differences in geologic conditions and producing methods that are suitable for use in the different deposits. The cost comparison between these resources and coal conversion is discussed more fully below.

C. Imports as a Means of Overcoming Shortfalls of Liquid and Gaseous Fuels

The United States was a net exporter of oil until 1948 and a net exporter of gas until 1958. After that, oil imports increased steadily, although they were controlled to some extent during the latter part of the 1950s and early 1960s by an import quota program. When the quotas were relaxed, imports grew at a more rapid rate, and by 1973, the year of the OPEC embargo, net oil imports represented 33% of total petroleum supply.

Gas imports (mainly from Canada but with small amounts from Mexico) increased steadily until 1972 when they totaled about 10^{12} SCF/year, or about 5% of domestic consumption. Unlike oil, gas imports have not provided a significant portion of domestic energy supply and they have been imported overland by pipeline from secure sources of supply.

Oil imports not only increased in volume but, between 1960 and 1970, the major suppliers shifted from Canada and Venezuela, both in the Western Hemisphere, to an ever larger percentage being imported from the Middle East and African countries. This shift was believed to have resulted in a less reliable supply base and these fears were confirmed by the OPEC embargo and the sharp increase in prices overnight at the end of 1973.

Immediately following the embargo, the nation adopted a policy of greater independence in energy supplies and the program was given the title of "Project Independence." The reasons for adopting this policy were to decrease reliance on energy supplies that were less secure and to assist in ameliorating the growing balance of payments problem that was exacerbated by the greatly increased volume of imports and the price increases for oil.

The original objectives of Project Independence have now been greatly modified, but a national policy to keep imports at a manageable level remains. In spite of this, by the middle of 1977 oil was being imported at a rate equal to 50% of demand, with the increasing quantities being produced in the Middle East.

Plans are also underway to import natural gas in a liquid form (LNG) to supplement declining United States gas production. While the quantities involved will be much smaller than for oil, it will increase the country's reliance on energy imports and increase the size of the balance of payments deficit.

D. The Potential of Alternative Fuels to Oil and Gas

As was discussed in Section I,B, the United States has ample supplies of coal to provide its energy requirements and with the development of the breeder reactor, the ability to generate electricity from an almost infinite energy source. With existing technologies, only part of the markets for oil and gas can be supplied by these fuel forms. Liquid fuels are needed for the transport sector, the very clean heat from gas is needed for food processing and other special applications, and both oil and gas have very large economic advantages over coal for the production of the nonfuel uses of energy.

In the case of gas the very large investments in gas transmission and distribution systems and in the stock of gas utilizing devices is a major economic reason for continuing to keep gas consumption at or near current levels.

Finally, oil and, even to a greater extent, gas, result in greatly reduced environmental, occupational health and safety, and public health impacts in all phases of the fuel cycle—exploration, production, transportation, upgrading, conversion, and utilization—when compared to coal.

Even with greater emphasis on conservation, the limited oil and gas resources will be depleted in a relatively short time and a transition to other fuels will have to occur in some markets. For the nonspecialized uses the choice in the near term will be between electricity (generated by coal or nuclear energy) and oil and gas produced from either unconventional resources or by the conversion of coal to either (or both) oil and gas. In the longer term the use of renewable resources—e.g., solar geothermal, biomass—will grow and provide an increasing share of energy supplies and replace these shorter term fuel options.

The first choice to be made, where the markets are such that the alternative fuel forms can compete, is between electricity, gas, and liquids. If electricity use appears to be preferred then the choice must be made as to which fuel is to be used to generate it—coal or a nuclear fuel. If gas is preferred, then a selection must be made between using unconventional gas sources or gas produced from coal. In the case of gas from coal, three different types of gas can be produced and the type selected would be that with the lowest overall delivered cost that still meets market requirements. In the case of liquids, unconventional petroleum sources, oil shale or hydrocarbons or methanol made from coal, would be the alternatives that could be used.

The selection of which of these transition energy options will be used by consumers will depend on overall economics, including the environmental and health and safety costs for all parts of the fuel cycle for each fuel. This in turn will be greatly affected by regional and local factors, such as nearness to the different energy resources, the rate at which technology develops to produce, transport, convert, and use the fuels, socioeconomic considerations, and regulatory and institutional factors. Another important factor that should effect the choice of which fuel to use is the life cycle costs of the alternatives. Complicating such estimates are the prediction of how the relative price of the different fuels will change over time, what future requirements will later be imposed on the different options (more nuclear safety, greatly reduced sulfur oxide emissions), and whether solutions will be found to difficult technologic and policy problems (the effect of carbon dioxide on the climate and nuclear proliferation).

II. COAL CONVERSION TECHNOLOGY[3]

A. Gasification

1. Historical

Gas manufactured from coal was first produced in the late eighteenth century by heating (distilling) coal in the absence of air. The first coal-to-gas companies were formed in the early nineteenth century and distributed a gas that had a heating value of about 500 Btu/SCF. Since the same type of gas is produced when coal is carbonized to make coke, this gas, too, in the early twentieth century was used to supplement the coal treated solely to manufacture gas.

As much as 70% of the coal is left as a carbon residue when heated to produce coal gas. To increase gas production and to utilize the carbon residue it was gasified to produce a mixture of carbon monoxide and hydrogen. This gasification product had a lower heating value than the distillation gases, but when enriched by gases produced from cracking liquid hydrocarbons and mixed with the distillation gases it provided a satisfactory gas for commercial distribution.

The processes used to gasify the carbon residue or coke produced a gas known as "blue gas" or "water gas." The carbon residue (or coal) was fed into a gasifier and heated to a high temperature by burning with air. After a hot bed of carbon was created by combustion of part of the carbon the air was shut off and steam introduced. The hot carbon reacted in the gasifier with the steam to produce the product gas. The chemical reactions involved are:

First stage

$$\text{C (coke or coal)} + \text{air } (O_2) \rightarrow CO_2 + \text{C (hot)} \qquad \Delta H = -94.45 \text{ kg cal/g}$$

Second stage

$$\text{C (hot)} + H_2O \text{ (steam)} \rightarrow CO + H_2 \qquad \Delta H = +41.4 \text{ kg cal/g}$$

The production of manufactured gas in this manner was expensive and created severe environmental problems. A large number of mechanical improvements were introduced over the years but it remained a difficult and expensive technology because of the cyclic nature of the operations and its operation at atmospheric pressure.

A second type of gas, "producer gas," that could be used at industrial operations was also made from coal in commercial quantities. In circum-

[3] For bibliography, see Appendix.

stances where gas was the preferred fuel for industrial operations a low heating value gas could be produced from coal by reacting it with a mixture of air and steam. The heating value was low because it contained all of the nitrogen in the air used to heat the coal, but it could be produced at lower cost than "blue gas" because the process was continuous instead of cyclic. However, because of its lower heating value it could not be transported economically any distance from where it was produced.

With the introduction in the United States of long distance transmission lines for natural gas after World War II the manufacture of gas from coal declined quickly and by the mid 1950s little or no gas was being produced from coal. In Europe, where coal remained the dominant fuel much longer, the technology of coal gasification was further improved utilizing technical advances made in other fields. These included development of relatively low cost, large scale oxygen plants, improved materials of construction, and newer methods of reacting solids and gases.

Even in Europe, however, Middle East oil soon began to replace coal and gas derived from coal so that few new coal gasification plants were constructed. The improved gasification processes that were used were the Lurgi fixed bed process, the Koppers–Totzek entrained process, and the fluid bed Winkler or Davy Power process. The product gases were mainly used to make a feed for the production of ammonia or other chemicals rather than as a means of supplying gas for residential, commercial, or industrial use.

2. *Current State of Technology*

Although the European-developed gasification processes were an improvement over the cyclic processes used earlier, none meet the qualifications of an ideal gasification scheme. An ideal process would be a single-stage continuous operation, employing air as an oxidizing medium, that could convert any type of coal at pressure into a gas low in inert constituents. None of the more recently developed processes meet all the requirements of an ideal gasification process. For example, the Lurgi process requires a sized noncoking coal and the Koppers–Totzek and Winkler processes operate at atmospheric pressure. In addition, none of these processes produces a product that could be used as a substitute natural gas since the gases produced do not have the heating value or chemical composition required to be used as a natural gas substitute. The gas from these processes is basically a mixture of carbon monoxide and hydrogen and a certain amount of methane distilled from the coal during heating.

On the other hand, these processes do make a raw synthesis gas that (after purification) could be upgraded by methanation to make a suitable substitute for natural gas. In addition, they are continuous rather than cyclic processes, which makes their operation more efficient and lower cost. However, the new processes that have been installed cannot make a product that can compete economically with the regulated price of natural gas. As a result, more advanced or "second generation" gasification processes are now under intensive development.

3. Coal Gasification Process Steps

Figure 1 is a flow diagram of the various process steps common to nearly all different types of coal gasification processes. Coal is first prepared by crushing or grinding to the size needed for the gasification process that is to be used. If the coal is a coking coal, for certain gasification processes it is necessary to pretreat the coal to destroy its coking properties. This is generally done by heating the coal in a mixture of air or oxygen and steam. The time and method of pretreatment will vary for different coals.

The pretreater can use a fixed, free fall, or fluidized bed. In a fixed bed the temperature is about 800°F and pressure about 325 psi. In a free fall pretreater the temperature is 1100°F and the pressure 300 psi. In a fluidized bed reactor the temperature is approximately 700–745°F. Since

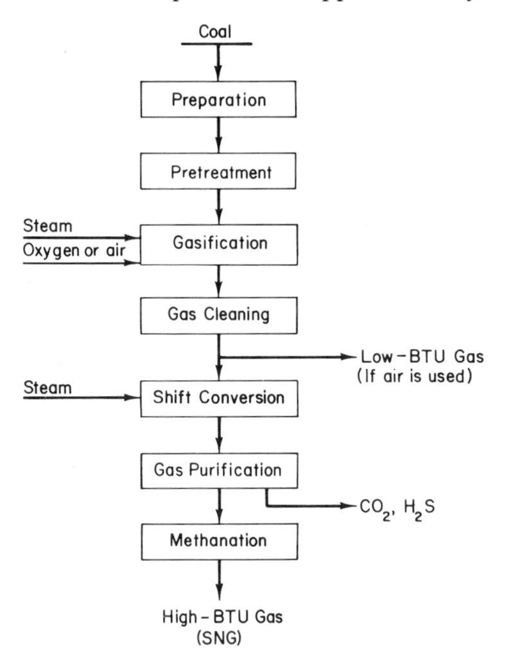

Fig. 1. Flow diagram for coal gasification.

valuable volatile matter is lost during pretreatment the amount of pretreatment is kept to a minimum and the product gases are used if at all possible.

The coal is then fed to a gasifier where it is gasified with air or oxygen. The steam carbon gasification reaction is favored at temperatures over 1700°F. The gas is then cleaned to remove the impurities. Depending on the type of gas to be made it may be treated in a shift converter to obtain the desired carbon monoxide and hydrogen ratio. This is followed by a second purification to remove carbon dioxide and hydrogen sulfide. Finally, if a substitute natural gas is to be made, the gas is passed through a methanation step where a gas that is essentially a substitute natural gas is produced.

Coal can be gasified by several different methods—pyrolysis, hydrogasification, and synthesis gas production. In pyrolysis the coal is heated in the absence of air. Some gas and liquids are produced but the major product formed is a char or coke residue. The gas that is produced contains methane, higher molecular weight hydrocarbons, carbon monoxide, and hydrogen. The amount of gas and liquids produced depends on the type of coal used and the temperature at which the pyrolysis is carried out.

In hydrogasification, coal, coke, or char is reacted with hydrogen (made from coal) to form methane. In synthesis gas production the coal or char is reacted with steam and oxygen. Part of the coal burns with the oxygen and provides the heat for the steam—carbon reaction that produces a mixture of hydrogen and carbon monoxide. The gas has a heating value of 300–500 Btu/SCF. If air is used in place of oxygen, a low-Btu gas (100–300 Btu/SCF) is produced. To make a substitute for natural gas medium-Btu gas is cleaned and further reacted by methanation (see Fig. 1).

Substitute natural gas can be mixed with natural gas and transported by pipelines over very long distances and still not increase the cost of the delivered gas excessively. Medium-Btu gas cannot be transported long distances, but the quality of the gas is such that a central medium-Btu gasification plant can transport gas to industrial users in special pipelines for distances up to 50 miles. The advantage of this system instead of direct use of coal is that the gas can be cleaned before use, the transport costs are smaller than they would be for coal, and, for some uses such as food processing, a clean fuel is needed.

Low-Btu gas must be used on site, where it is produced, since transport costs of the low-quality gas would be excessive and the sensible heat in the gas (which can be a significant portion of the total heat) would be lost. Low-Btu gas is expected to be used where gas is the required fuel form or where a clean fuel is needed. Low-Btu gas has been proposed as an alternative to stack scrubbing for the control of sulfur oxides from coal combustion. It is also being considered for use in advanced power cycles that use a combination of gas turbines and steam turbines to increase the efficiency of converting fuel to electricity. Coal could not be used directly because the ash in the gas would erode the gas turbine blades rapidly.

4. *Chemistry of Coal Gasification*

a. Pyrolysis

(1) coal + heat → char + tars + gas (CH_4, CO, H_2, others)

b. Hydrogasification

(2) coal + hydrogen → CH_4 + heat

c. Synthesis gas

(3) $C + O_2 \rightarrow CO_2 + \text{heat}$ $\qquad \Delta H = -94.45$ kg cal/g

(4) $C + H_2O + \text{heat} \rightarrow CO + H_2$ $\qquad \Delta H = +41.4$ kg cal/g

(5) $CO + 3H_2 \xrightarrow[\text{(nickel)}]{\text{catalyst}} CH_4 + H_2O + \text{heat}$ $\qquad \Delta H = -59.8$ kg cal/g

Processes that give higher yields of methane according to reaction (2) are more efficient since if the methane is made in a separate step outside the gasifier [reaction (5)] most of the heat released is lost and is not available for the endothermic steam–carbon reaction.

5. *Methods of Classifying Gasifiers*[4]

a. By the method of supplying the heat required by the gasification reactions:
- (1) Internal heating
  - Autothermic
  - Cyclic heat carrying fluids or solids
- (2) External heating
  - Heat transferred through walls of reaction vessel

b. By the method of contacting reactants:
- Fixed bed
- Fluidized bed
- Suspension of particles in gasifying medium (entrained)

c. By the flow of reactants:
- Concurrent
- Countercurrent

[4] See Elliott and Von Fredersdorff (1963).

d. By the gasifying medium:
Steam with oxygen or air or oxygen enriched air
Hydrogen
e. By the condition of residue removed:
Dry ash in nonslagging operation
Slag in slagging operation

6. *Commercial Reactors Producing Medium-Btu Gas*

A very large number of gasifiers of different types have been proposed and many have been tested experimentally over the past 20 years.

a. Fixed Bed Gasifier (Lurgi). In a fixed bed gasifier sized coal is supported on a grate which is located near the bottom of the gasifier. The grate serves to distribute the incoming gases and to remove the ash. A typical continuous fixed bed (Lurgi) gasifier is shown in Fig. 2. Raw sized coal is fed into a lock hopper at the top of the gasifier and then is fed, as required, into the gasifier itself, which is pressurized. Steam and oxygen are fed into the bottom of the gasifier through the rotating grate on which a bed of ash is maintained to protect the grate from overheating. The coal reacts with the oxygen and the hot gases produced flow up the gasifier. The coal is spread evenly over the gasifier by a distributor and the ash is collected in a lock hopper from which it is removed periodically. Raw gases leave the top of the gasifier at about 850°F and the raw gas is cooled and scrubbed before further treatment.

The fixed bed process, with the countercurrent flow of gas and coal, permits the efficient use of heat released in the gasifier. By transferring the heat in the hot gases to the incoming coal less oxygen is required and a larger volume of methane is found in the product gas. The Lurgi process, the best known of the fixed bed processes, operates at elevated pressure, and this also results in some economies since the gas is almost always distributed under pressure and does not have to be compressed for distribution.

The fixed bed processes have three disadvantages—they require a sized coal so that the fines produced during mining cannot be used without briquetting, they can only gasify a noncoking or weakly coking coal, and they are low throughout devices that require a large number of gasifiers.

About 60 Lurgi gasifiers have been used commercially and most of the commercial high-Btu gasification plants that have been proposed for construction plan to use the process. The typical Lurgi gasifier has a diameter of 12 ft and produces a gas with the composition given in Table IV.

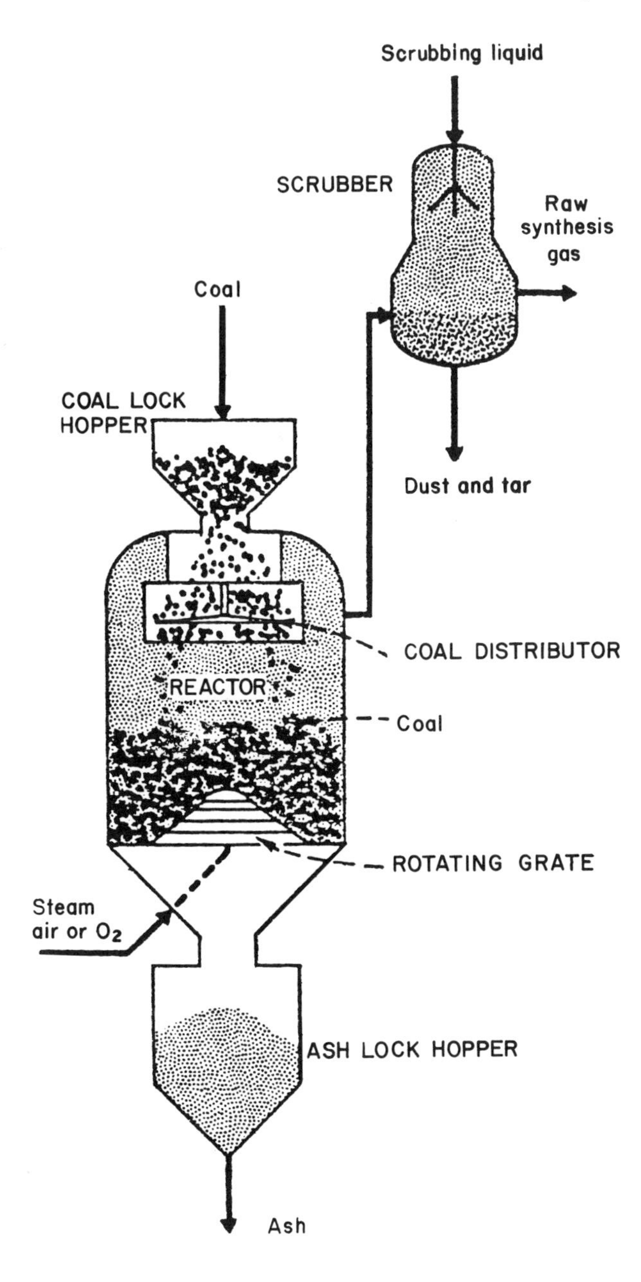

Fig. 2. Lurgi gasifier.

TABLE IV

Composition of Gas Produced in Lurgi Gasifier

	Volume Percent, Dry
Hydrogen	37.4
Oxygen	nil
Carbon monoxide	26.0
Carbon dioxide	26.1
Methane	9.1
Other hydrocarbons	0.6
Nitrogen	0.8
Heating value: 310 Btu/SCF	

b. Entrained Gasifier (Koppers–Totzek). As shown in Fig. 3, the entrained gasifier reacts a mixture of powdered coal, steam, and oxygen in an entrained state. Most of the ash leaves with the raw product gas but a portion is collected as slag. The Koppers–Totzek process using coal and operating at atmospheric pressure has been used in a number of commercial plants abroad. Commercial gasifiers use either two or four opposing burners. Because of this mode of operation the product gases leave at temperatures as high as 3300°F, so that oxygen consumption per unit of

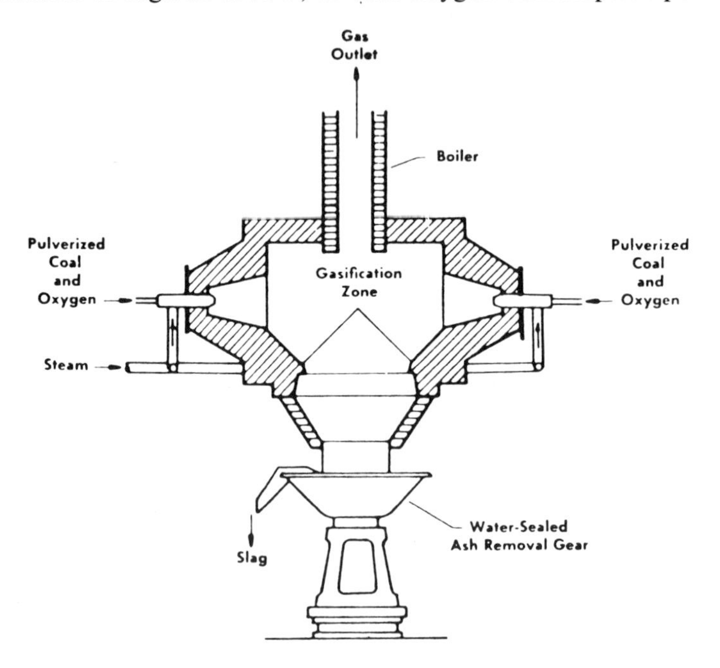

Fig. 3. Koppers–Totzek gasifier.

TABLE V

Gas Compositions from a Koppers–Totzek Gasifier with Two Types of Bituminous Coal

	Volume Percent, Dry	
	Coal A	Coal B
Hydrogen	34.0	31.1
Oxygen	0.1	4.1
Carbon monoxide	51.1	55.3
Carbon dioxide	12.6	11.9
Methane	0.1	0.1
Other hydrocarbons	N.1	N.1
Nitrogen	1.9	1.3
Hydrogen sulfide	0.2	0.3
Heating value: 275 Btu/SCF		

product gas is higher than for fixed bed gasifiers. On the other hand, the Koppers–Totzek gasifier can handle any rank of coal and can use all of the coal—even the fines. It has one further advantage in that the high temperatures destroy all the tars, phenols, and light oils, which makes for fewer environmental control requirements.

The product gas compositions produced from two different types of bituminous coal are given in Table V.

c. Fluidized Bed Gasifier (Winkler and Davey Power Gas). The Winkler process, the only commercial fluid bed process that has been constructed, operates at atmospheric pressure with a steam oxygen gasifying medium (Fig. 4). The process uses a noncoking subbituminous coal or coke and is operated at a temperature of 1600–1800°F. Most of the oxygen and steam is injected at the bottom of the bed but there is a disengaging space above the bed to react the carryover fine particles with a second steam–oxygen stream.

Thirty-six generators were built at 16 plants in a number of different countries. The process is able to handle a wide range of coal sizes. Oxygen consumption is intermediate between the fixed bed and entrained bed processes. The heavier ash particles settle to the bottom of the bed, although most of the ash is carried over with the product gas. The product gas composition for a typical dry brown coal is given in Table VI.

d. Other Commercial Gasifiers. A number of other gasifiers have been used, but only a few have been tested to produce medium-Btu gas. Nor-

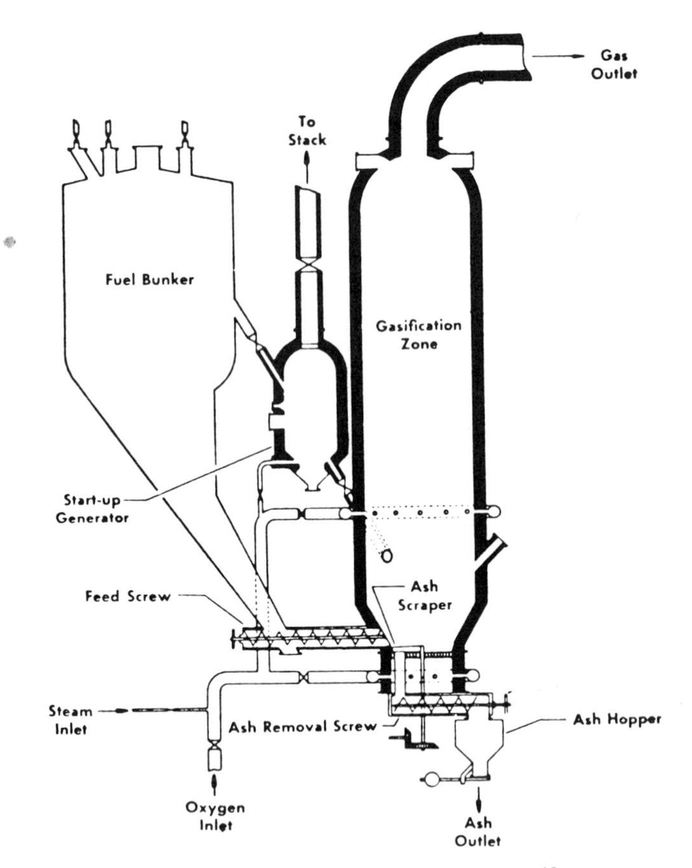

Fig. 4. Typical Winkler fluidized bed gasifier.

TABLE VI

Gas Composition Produced in a Fluid Bed Gasifier with a Typical Dry Brown Coal

	Volume Percent, Dry
Hydrogen	36.0
Oxygen	nil
Carbon monoxide	44.4
Carbon dioxide	15.7
Methane	1.6
Nitrogen	0.8
Hydrogen Sulfide	1.5
Heating value: 270 Btu/SCF	

mally the commercial Wellman Galusha generators used steam and air to gasify the coal to produce a low-Btu gas. However, tests have been conducted on a Wellman Galusha commercial sized gasifier that was modified to enable a medium-Btu gas to be produced using a steam–oxygen gasifying medium. This is a fixed bed gasifier with a bottom grate that operates at atmospheric pressure and requires the use of a noncoking feed (coke, anthracite, subbituminous coal).

The unit tested had a 10-ft internal diameter and a gasification capacity of just under 3 million cubic feet per day. The gas had a heating value of 260 Btu/SCF.

During the 1940s and into the 1950s a number of other industrial scale gasifiers were tested for producing medium-Btu gas. These included the B&W du Pont gasifier in West Virginia, the Würth gasifier at Leuna, Germany, and the Rummel gasifier at Wesseling, Germany. None of these were later installed in any large numbers.

7. *Commercial Gasifiers Producing Low-Btu Gas*

Many commercial gasifiers that were operated during the period when coal gasification was used extensively produced a low-Btu gas because large scale oxygen plants had not yet become available commercially. A wide variety of processes were developed in order to improve the efficiency, operability, and costs of producing the low-Btu gas. With a shift away from gas made from coal there has been little interest in recent years to develop new low-Btu gas processes.

One exception to this generalization was the development of an air blown fixed bed pressurized gasifier for use in advanced power cycles. At a STEAG plant in Leuna, Germany five Lurgi gasifiers produce 160 million SCF per day of low-Btu gas that is burned in a combined cycle generating plant. A pressurized boiler with a pressure recovery turbine is used.

8. *Pilot Gasifiers Producing Medium- and High-Btu Gas*

a. Improved Fixed Bed Lurgi Gasifiers. Three new developments in fixed bed Lurgi gasifiers have received attention over the past several years—modifications in the generator to permit the use of coking coals, large-diameter gasifiers, and removal of ash in the form of slag rather than in a dry state. Tests using coking coals were carried out in a Lurgi gasifier at Westfield, Scotland. It was reported (Energy Research and Devlopment Administration, 1975) that a modified stirrer was used and that successful tests were made on Illinois No. 5 and Illinois No. 6 bituminous coals and on Montana Rosebud subbituminous coal.

A second improvement in the conventional Lurgi process is the development of a new 13-ft diameter unit that is being tested at Sasolburg,

South Africa. An even larger unit (16 ft in diameter) is also to be tested at Sasolburg (National Research Council, 1977b).

The third potential improvement in Lurgi design is the use of a slagging generator. In this design, in place of a grate the fixed bed of coal rests on a slag layer which is supported on a hearth. The temperature of the ash must be maintained high enough so that the slag will flow freely from the base of the generator. The two major advantages of the slagging operation are the use of smaller quantities of steam (used in the conventional Lurgi to keep the grate cool) and a much larger throughput for the same diameter gasifier. In tests (American Gas Association *et al.,* 1975) conducted on a modified Lurgi using slagging operation only one-eighth as much steam was required, compared to a dry ash system, and throughputs were more than three times as great.

b. Hygas Gasifier. The Hygas process has been operated by the Institute of Gas Technology in a pilot plant at Chicago, Illinois that has a rated capacity of 3 tons/h. Coal is fed in an oil slurry into a vessel pressurized to 1000–1500 psi. The oil is vaporized and recovered for reuse. The coal flows downward countercurrently to the flow of gases, but coking coals must be pretreated before use. There are two fluid beds in series within the gasifier. In the upper stage the dried coal is gasified and reacted at high temperatures in the presence of hydrogen. It then flows to a lower fluid bed operating at 1700–1800°F where it is further reacted with hydrogen. The unreacted carbon from this stage reacts with oxygen and steam to produce the hydrogen used in the upper stages of the reactor. Other methods to produce the hydrogen required, such as the steam–iron process, are under development.

The Hygas process is designed to make a substitute natural gas, so that production of the maximum amount of methane in the gasifier is desired. After cleaning and shifting to get the desired carbon monoxide to hydrogen ratio, the gas is then upgraded by reacting it over a methanation catalyst.

The composition of the gas leaving the gasifier in a recent Hygas test (54) is given in Table VII.

c. Synthane Gasifier. The Synthane process employs a fluidized bed gasifier in the range 600–1000 psi and temperatures in the range of 1500–1800°F (Fig. 5). Coal is fed in a dry condition in a size range of 20 mesh. In the early stages of operating the 75 ton per day pilot plant, subbituminous coal was allowed to fall freely into the space above the fluid bed. In more recent experiments the coal is introduced deep in the fluid bed with better operation of the gasifier by eliminating the tars, oils, and phenols. This in turn permitted the gas to be cooled in ordinary heat transfer equipment

TABLE VII

Composition of Gas Produced in Hygas Gasifier

	Volume percent, dry
Hydrogen	26.1
Carbon monoxide	16.0
Carbon dioxide	27.6
Methane	28.1
Ethane	0.1
Nitrogen	7.0
Hydrogen sulfide	0.7

and thus generate large amounts of steam not possible with the other mode of operation. A pretreating section is provided when coking coals are used.

Steam and oxygen are distributed through nozzles at the bottom of the gasifier, which is fitted with an internal cyclone to reduce fine carry over. About 65–80% of the carbon is gasified and the balance is discharged as a char. The char is used within the plant or sold.

When using a Rosebud coal and deep fluid bed injection the gas had the composition given in Table VIII.

d. CO_2 Acceptor Gasifier. A 40 ton per day pilot plant was constructed and operated at Rapid City, South Dakota to process western coals into pipeline gas. Coal, after being devolatilized, is fed into the bottom of a gasifier where the char reacts with steam in a fluid bed operated at 150 psi and 1500°F. The reaction takes place in the presence of hot dolomite. The heat for the steam char reaction is provided in two ways—by the sensible

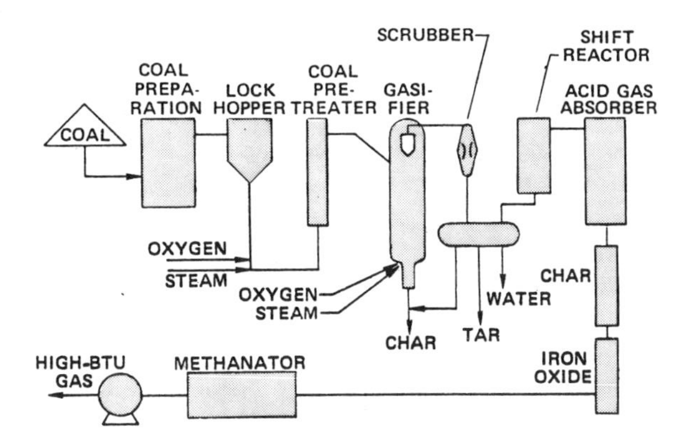

Fig. 5. Synthane process.

TABLE VIII

Composition of Gas Produced in Synthane Gasifier Using Rosebud Coal and Deep Fluid Bed Injection

	Volume Percent, Dry
Carbon dioxide	40.4
Carbon monoxide	12.2
Hydrogen	36.5
Methane	10.8
Ethane	0.1
Nitrogen	0.1
Hydrogen sulfide	0.3

heat of the hot dolomite and by the heat released when the dolomite reacts with the carbon dioxide produced during the gasification process. The unreacted char and spent dolomite flow to a regenerator where the char is burned with air and the heat released calcines the dolomite and heats it to 1900°F. Ash is removed from the generator by elutriation.

The process produces a medium-Btu gas that can be upgraded to a high-Btu gas by methanation. The advantages of the process are that it does not require the use of oxygen, carbon conversions are high, and gas clean-up requirements for carbon dioxide and hydrogen sulfide are low. The major disadvantage is that it requires a very reactive coal to be utilized successfully.

e. Cogas Gasifier. The Cogas process, like the CO_2 acceptor process, does not require the use of oxygen, but in this case the heat for the steam char reaction is provided by a recycle char stream that has previously been heated by burning part of the char fines. The gasification reaction takes place in a fluidized bed at 15–40 psi and 1600–1700°F.

The Cogas process is intended to be used in conjunction with the COED process, a multiple fluidized bed process that pyrolyses the raw coal and produces liquids, pyrolysis gas, and the char that is gasified in the Cogas process. A 100 ton per day gasifier has been operated at Leatherhead, England for several years.

f. Bigas Gasifier. This process is being tested on a 120 ton per day scale at Homer City, Pennsylvania. The gasifier is vertical and gasification is carried out in two stages. Coal is introduced in an entrained state into the upper stage where it reacts with the hot gases flowing upward from the lower stage. The coal is heated rapidly to over 1500°F and produces a char and a significant amount of volatiles. The char and product gas leaving the gasifier are separated in a hot cyclone. The gas, after purification, has a heating value of about 350–400 Btu/SCF. The char is reacted with oxygen and steam at 2800°F in the lower stage to produce a mixture of carbon dioxide, carbon monoxide, and hydrogen. The ash is removed from the lower stage as slag.

Although oxygen is being used as the gasifying medium, air could be used and a low-Btu gas would be produced.

g. Grand Forks Gasifier. This gasifier has a capacity of 25 tons per day and is a fixed bed, pressurized, slagging reactor. It was used to test various western lignite coals, using oxygen and steam as the gasifying medium. It produces a gas that, after cooling and scrubbing, had a heating value of 350 Btu/SCF.

The unit is being reactivated to study the nature and quantity of the residuals produced from a fixed bed gasifier and to determine what process alterations might reduce them.

h. Union Carbide/Battelle Gasifier. A 25 ton per day pilot plant was tested at West Jefferson, Ohio to determine if agglomerates formed from the ash in coal can be used to provide the heat for the steam–carbon reaction. This would avoid the use of oxygen or dilution of the gas quality by the nitrogen in air when the coal is gasified. The process is a pressurized two-stage fluidized bed system in which coal or char is burned in one fluidized bed and gasified with steam in a second fluidized bed. Successful operation requires that the ash in the coal is agglomerated in such a way as to form free-flowing inert solid pellets.

The gasifier is operated at 1800°F and 100 psi, while the combustor operates at 2000°F at the same pressure. The steam is fed into the bottom of the gasifier and the hot agglomerates into the top. The product gas leaves the gasifier at the top and is cleaned by cyclones and a venturi scrubbing system. The agglomerates leave the bottom and are recycled to the combustor for reheating with the hot product gas made from the combustion of part of char separated from the exit gases.

i. U Gas Gasifier. This process is under development by IGT in a 7 ton per day atmospheric pressure pilot plant that has been operated at atmospheric pressure using air, oxygen, steam, and coke breeze. When coking coal is used, it must be pretreated with air in a fluidized bed at about 750°F and 350 psi. The pretreater off gas is mixed with the product gas. The gasifier, in which coke or pretreated coal is reacted with air (to make low-Btu gas) or oxygen (to make a medium-Btu gas), is operated under conditions (1900°F) in which the ash in the coal is agglomerated. The design of the grid in the reactor and the method of injecting the gas are such that agglomerates are formed and grow in size until they are heavy enough to fall and be separated from the bed. The dust in the raw gas is separated by both internal and external cyclones, and the raw product gas has a heating value of either 300 Btu/SCF, when oxygen is used, or 150 Btu/SCF, when air is used.

Plans have been made to operate a modified U gas process using raw coal and to operate the gasifier under pressures of 90 psi.

9. Pilot Gasifiers Producing Low-Btu Gas

A number of the gasifiers for producing medium-Btu gas, described previously, can also produce a low-Btu gas when air is the oxidizing medium. These are the Bigas process, the Grand Forks fixed bed gasifier, and the U gas process. In addition, a number of pilot plants are being operated which are designed primarily to produce low-Btu gas. These are:

a. Morgantown Energy Research Center Gasifier. This is a fixed bed gasifier, 42 inches in diameter, operating at pressures up to 300 psi at feed rates of 25 tons per day. As with other fixed bed gasifiers the coal is supported on a grate, which is also used to remove the ash. The unique feature of this gasifier is a water cooled stirrer which has three arms and is capable of stirring and of agitating the bed over its entire depth since it not only rotates but moves in a vertical direction as well. The coal is fed from lock hoppers into the top of the gasifier and a steam air mixture is fed below the grate. Temperatures in the gasifier range from 2400°F above the grate to between 800 and 1200°F at the gas exit. The product gas has a heating value of 130–170 Btu/SCF.

The gasifier has been successfully operated using a strongly coking coal with air as the oxidizing medium. The product gas is cleaned and passed through a cascade section of turbine blading for corrosion/erosion tests. Modifications in the gasifier train are being made and will include methods to test for the presence of various residuals which may be of environmental concern.

b. Westinghouse Gasifier. This gasifier is a fluidized bed designed to recirculate the coal within the bed where devolatilization, desulfurization (with added lime), and hydrogasification occur at 1300–1700°F and at 10–20 atm (Fig. 6). The coal is fed into the central draft tube where it is diluted with large volumes of recirculating solids consisting of char and lime adsorbant. Char is removed from the top of the gasifier and spent sorbent is removed from the bottom. The char removed with the spent sorbant is separated and the spent sorbant is regenerated with steam and

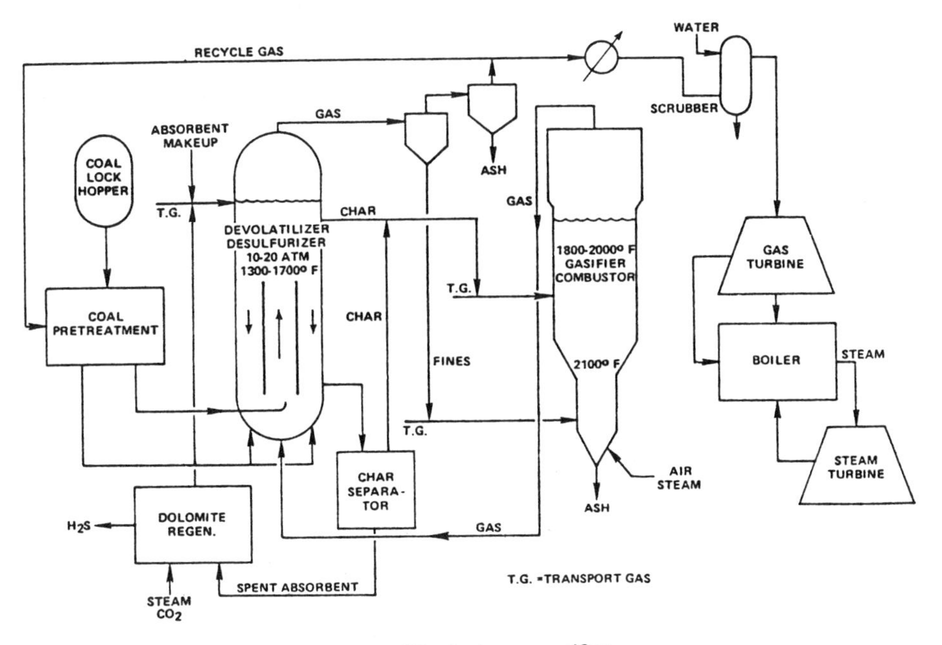

Fig. 6. Westinghouse gasifier.

carbon dioxide. The dry low-sulfur char is gasified in a second fluidized bed in which the heat for the gasification is provided by burning part of the char in a lower leg of the gasifier. Coal ash is removed as agglomerates from this combustion section. The gasification reaction occurs at 1800–2000°F.

To date, a 15 ton per day process development unit has been operated, but only certain parts of the integrated process have been tested.

c. Combustion Engineering Gasifier. The Combustion Engineering gasifier is an entrained flow atmospheric pressure gasifier that generates a low-Btu gas in two chambers, each enclosed by refractory-covered water-cooled walls. In the lower section char or coal is burned with preheated air and ash is removed as slag. Steam and pulverized coal are fed into the upper section, where the coal is heated by the rising hot combustion gases and devolatilized and gasified by reaction with steam at 1600°F. The product gas is cooled and cleaned and has a heating value of approximately 130 Btu/SCF.

The gasifier can use all coals without pretreatment, is expected to have a high carbon conversion, and does not require special facilities for feeding coal into a pressurized vessel. Although low-Btu gas at high pressure (e.g., the Westinghouse process) is preferred for use in combined cycles for generating electricity, the ability to use conventional and commercially tested and available equipment makes the use of atmospheric pressure attractive. A 120 ton per day unit is now undergoing shakedown tests.

d. Foster Wheeler Gasifier. Another entrained air blown gasifier is being investigated by Foster Wheeler, but this process will be operated under pressure (30 atm). Pulverized coal is fed into the upper stage of a two-stage gasifier, where it is heated to 1800°F by contacting with hot combustion gases produced in the lower stage. Part of the char produced is reacted in the lower stage with a mixture of air and steam, where it is gasified and produces hot gas for use in the upper stage. The lower stage is operated so that the coal ash is removed as slag. The gas is expected to have a heating value of 165 Btu/SCF.

10. Bench Scale Experiments—Medium- and High-Btu Gas

The processes described in Sections II,A,8 and II,A,9 are generally referred to as "second generation" gasifiers. Experimental work is already underway on a smaller scale on "third generation" gasifiers, which it is hoped will further increase efficiency and reduce costs.

a. Hydrane Gasifier. The Hydrane process is designed to produce a high-Btu gas that can be made with a minimum amount of methanation as the final step in making a substitute natural gas. The gasifier is operated with two zones at 1000 psi and 1650°F. The coal falls freely into the upper zone in a dilute phase through a gas, produced in the lower zone, that contains hydrogen and some methane. This devolatilizes the coal and produces methane and char. In the lower zone the char is reacted with hydrogen, made by reacting the residue from the two-stage gasifier with oxygen and steam, in a fluidized bed external to the hydrane gasifier.

b. Exxon Gasifier. Exxon Research and Engineering Company is studying a process to make a medium-Btu gas in a small (0.5 ton per day) fluidized bed gasifier. It has been reported that the heat for the endothermic steam–carbon reaction is supplied by burning part of the char with air in a separate vessel and returning the hot unburned part of the char to the gasifier.

Exxon has also reported that it has developed a catalytic fluidized bed gasification process which operates at 1200–1400°F and at pressures of 500–1000 psi. The catalyst is an alkali carbonate used at high concentrations (up to 20%). The carbonate is recovered by water washing of the ash and char from the process.

11. Bench Scale Experiments—Low-Btu Gas

a. Gegas Process. General Electric is testing a small (0.5 ton per day) fixed bed, air blown, pressurized (300 psi) gasifier to be used in a combined cycle power plant. The coal is ground and mixed with a pasting agent and extruded into a pressurized chamber where the extruded stream is cut to desired length. The gasifier is fitted with two agitators—the upper one fractures any coke that is formed and the lower one breaks up clinkers.

If the process is successful it should: (1) reduce the cost of coal crushing, (2) permit the use of coal fines in a fixed bed, and (3) allow more economical feeding of coal under pressure. General Electric also claims that steam requirements are lower than for other processes.

b. BCR Three-Stage Pressurized Gasifier. This process consists of three fluidized beds that operate at pressure and can gasify any type of coal and produce no tars or oils as by-products. A 1.2 ton per day unit is now being tested. Coal is fed by lock hopper into a first-stage pretreatment step, where it is devolatilized at a temperature of 1200°F with gases produced in the third stage. The pretreated coal, and entrained tars and oils are gasified in the second stage with steam and air at 2000°F. The residual char is gasified with steam and air in the third stage.

12. Other Gasification Processes

A large number of other gasification processes have been proposed, tested, investigated, or used on both a large and small scale. However, except for those that use the "molten bath" principle, all of the gasifier "types" are described in the above sections.

A molten bath can serve to provide good heat transfer characteristics and to remove sulfur from the product gas. The molten baths that have been studied include iron, salts, and coal slag. If the molten bath is circulated between two vessels it is possible to produce a medium-Btu gas using air. If a single vessel is used, oxygen is required to make a medium-Btu gas.

Two types of molten bath gasifiers were tested or used on an industrial scale—the Rummel (single shaft) process and the Otto Rummel (double shaft) process. At present the only significant research effort on molten bath gasifiers is that being carried out by Rockwell International.

Tests have been carried out in a 4-ft-diam, 10-ft-high reactor in which coal is injected at rates of 6 ton/day into a molten pool of sodium carbonate, sodium sulfite, and sodium sulfate through which air is blown. The molten bath operates at 1800°F and at 5–10 atm. The product gas has a heating value of 100–150 Btu/SCF and is relatively free of ash and sulfur, which are trapped in the molten bath. The process can use either coking or noncoking coals and gasification rates are much higher in the slag than in processes where gas–solid reactions take place. Part of the slag must be withdrawn continuously so that ash and sulfur can be removed, after which the slag can be recycled back to the process.

Construction of a larger unit (120 tons per day) is currently underway.

13. Underground Gasification

Underground gasification of coal has been a subject of extensive investigations in the Soviet Union since the early 1920s and a number of industrial scale projects have been developed. Although it has been studied in the Soviet Union for over 50 years, the technology still only provides a very small amount of the energy supply of that country.

Following World War II, a large number of underground gasification experiments were initiated in a number of countries having indigenous coal supplies. These included underground gasification installations in the United Kingdom, Belgium, Morocco, and the United States. A wide variety of different types of technology for producing gas were tested, but none were used commercially outside of the Soviet Union. After about 10 years of investigation, interest in development of the technology abated outside the Soviet Union and very little research was done until interest was revived in the early 1970s.

Interest in developing commercial underground gasification techniques has been based largely on the expectation that successful development of the technology would improve the economics of removing energy from the coal in the ground and reduce adverse health, safety and environmental impacts by not requiring that miners go underground to extract the coal. In addition, it would permit recovery of coal from reserves that cannot be mined economically by other means.

New experiments have been undertaken in the United States, starting in 1973. Those with government support are: the linked vertical well (LVW) process in Western subbituminous coal being tested at Hanna, Wyoming; the packed bed process (PBP) to be used in thick Western subbituminous coals, which is now under investigation by the Lawrence Livermore Laboratory; the Longwall generator (LG) process being studied for use with an Eastern bituminous coal seam; and a test on a steeply dipping bed (SDB) that is to be studied and carried out jointly by the government and industry. In addition, a private underground gasification experiment has been reported to be under study by Texas Utilities and Williams Brothers using licensed Soviet technology.

In the LVW process, vertical wells are drilled from the surface to the coal seam and the boreholes linked by reverse combustion or high-pressure air injection. Air is injected in one series of boreholes and product gas, formed by the partial combustion of the coal seam between the boreholes, removed from the second series of boreholes. When linkage is achieved between the boreholes there is an abrupt pressure drop and a change in gas composition; from a gas high in methane and low in carbon monoxide to one high in carbon monoxide and low in methane.

To be successful the method requires that the flow be controlled and that the coal be utilized efficiently. To date the LVW experiment has resulted in steady gas production for 6 months, producing a gas with a heating value of about 125 Btu/SCF. A second test gave a gas with an even higher heating value, and other tests are planned.

In the PBP, a thick western subbituminous coal will be fractured using chemical explosives to create a permeable reaction zone. Collection wells will be drilled to the bottom of the coal seam and air will be injected in wells drilled to the top of the thick seam. In essence, an underground packed bed reactor would be created. Six sites have been investigated for possible use and tests are to be initiated.

The LG method of underground gasification is proposed to be used in eastern bituminous coals that occur in thin seams and low dip. Deviated wells are drilled from the surface and produce long boreholes in rows in the coal seam (Fig. 7). The coal is burned between the parallel horizontal boreholes and it is expected that the need for fracturing may be elimi-

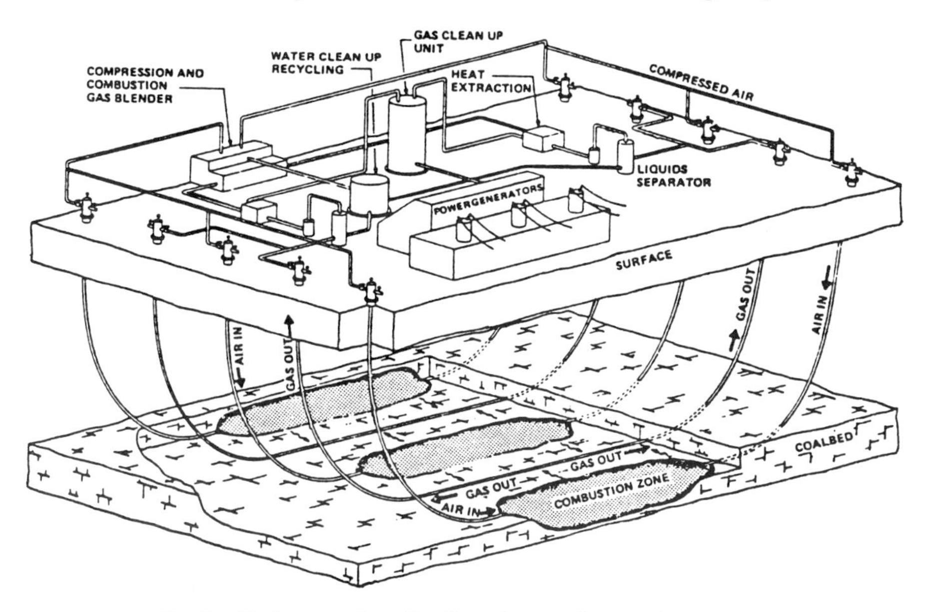

Fig. 7. Underground gasification—Longwall generator process.

nated. Deviated wells are now being drilled and the concept is ready for field testing.

Experiments on SDB were started during 1977 and field work will be carried out in 1978.

B. Coal Liquefaction

1. Historical

Unlike coal gasification, conversion of coal to liquids has never been accomplished commercially in the United States. However, during World War II, Germany produced liquids from coal in industrial amounts, using two different coal liquefaction processes. Coal-to-oil plants were constructed in a number of other countries, but only one plant in South Africa is still producing liquids from coal.

Liquid production from coal by the Bergius hydrogenation process involved the reaction of hydrogen with a mixture of coal and oil (made in the process) at high temperatures and pressures. In the Fischer–Tropsch process a synthesis gas (a mixture of carbon monoxide and hydrogen) is passed over a catalyst at elevated temperatures and moderate pressures to form paraffinic oils and waxes. In both processes the gas is made from coal in one of the gasification processes described above. A more complete description of these two liquefaction processes is given below.

The German production of liquids from coal was never very large—12,000 barrels/day of gasoline, with the largest plant turning out 4000 barrels/day. (Current United States gasoline consumption is approximately 8 million barrels/day.) Thus coal liquefaction has never been a major commercial industry. For a commercial industry to develop, the liquids from coal must be able to compete with oil produced and transported from conventional petroleum reservoirs, from oil extracted by secondary and tertiary recovery and from unconventional sources.

2. Current State of Technology

Improvements on the German-developed liquefaction processes were made following World War II on a pilot plant scale but, except for the processes installed in South Africa, none of these have been used commercially. The fixed bed catalytic reactor of the type used by the Germans was used in South Africa, but a second type of reactor, using an entrained catalyst, was also installed. Both are still in commercial operation.

At the pilot plant level improvements were made with the types of catalysts used so that they were of lower cost and more selective in the products made. In addition, several engineering modifications were tested that could remove more efficiently the large amounts of heat released during the catalytic reaction and thus improve the throughputs of the reactors and further reduce costs. One of these modifications involves submerging the catalyst in oil and circulating the oil from the reactor to a heat exchanger to remove the heat. The cooled oil is then returned to the reactor. In another modification the heat of reaction is removed by recycling large volumes of gas with fresh feed and removing the heat from the recycled gas in a heat exchanger.

Improvements were also made in the direct hydrogenation (Bergius) process on a pilot plant scale. In one modification the catalyst reacts with the coal-oil mixture in an ebullating bed reactor. In several others the reaction takes place over a fixed bed of catalyst. In still a third process dissolution of the coal and catalytic hydrodesulfurization take place at the same time and produce a liquid similar to No. 6 oil. The status of these and other processes still under development are discussed below, since no commercial application of these processes has been made.

3. Coal Liquefaction Process Types

Coal liquefaction processes can be classified into four different types. These are (1) pyrolysis, (2) solvent extraction, (3) direct hydrogenation, and (4) indirect hydrogenation.

a. Pyrolysis. When coal is heated in the absence of air, crude tar and light oils are produced. These products have been produced from coal for over 150 years as by-products of the coking operation. The crude oil and light oils are usually refined to give either chemicals or pure compounds. The amount of liquid fuels produced in this manner has always been small. The pyrolysis processes now being studied are aimed at the production of larger amounts of liquids and gases and lesser amounts of solids. This can be accomplished in several ways, including using hydrogenation as part of the pyrolysis process.

b. Solvent Extraction. Solvent extraction of coal can be used to produce liquids. In the extraction process coal is dissolved in a liquid solvent at elevated temperatures (900°F) and pressures of 275 psi. In one type of solvent extraction process, the dissolved coal is separated from the undissolved coal and ash and the solvent is recovered for reuse. In another type of solvent extraction process, the solvent also acts as a hydrogen donor to the coal under reactor conditions. If hydrogen is introduced into the process at some point the donor solvent can be hydrogenated, which can then transfer more hydrogen to the coal.

An extraction process that uses a supercritical gas as the solvent reacts with the more volatile coal species and after recovery of the solvent dissolved coal mixture leaves a reactive carbonaceous residue.

c. **Direct Hydrogenation.** This classification includes all processes in which hydrogen is added to the coal in the presence of a catalyst that is in direct contact with the coal, thus eliminating the need for a special reactor in which the donor solvent is hydrogenated. The reaction takes place at about 900°F and at pressures of from 1500 to 10,000 psi. A large number of different catalysts have been tested and several different reactor types have been tried to obtain good contact between the coal and the catalyst.

d. **Indirect Hydrogenation.** In indirect hydrogenation the coal is first gasified to make a synthesis gas. This is then passed over a catalyst to produce alcohols (methanol) or paraffinic hydrocarbons. The reaction takes place at 500–600°F and at pressures of 375 psi.

4. *Chemistry of Coal Liquefaction*

Coal differs from oil in that it has a deficiency of hydrogen compared to oil, or a higher carbon-to-hydrogen ratio. To convert coal to oil either the hydrogen-rich part of the coal must be separated from the coal and leave a residue much higher in carbon content than was in the original coal, or hydrogen must be added to the coal. In addition, if an ash-free liquid is desired the ash must be separated at some stage of processing. Pyrolysis and extraction represent the first type of process to make liquids; direct and indirect hydrogenation, the second.

The chemical structure of coals will determine the type of chemical reactions that will take place during liquefaction by pyrolysis, extraction, or direct hydrogenation. The chemical structure is much less important in indirect hydrogenation, since the coal is completely gasified before it is recombined into a liquid.

The chemical structure of coal varies with rank, and the aromatic character increases with higher rank coals. Liquefaction behavior is also strongly influenced by the petrographic constituents found in the coal. These constituents are the layers of coal that have different visual properties.

When coal is heated to liquefy it the physical bonds are broken, chemical bonds are broken to form free radicals, and the free radicals are stabilized in different ways. In pyrolysis the heating of the coal breaks off hydrogen-rich free radicals from the coal to form a liquid product. When a solvent is used the free radicals formed on heating react with hydrogen from the donor solvent, forming additional liquids. In direct hydrogenation the hydrogen is chemisorbed on the catalyst, which then can transfer the hydrogen to the coal radicals and to the larger chemical clusters making up the coal structure.

The major factors affecting liquefaction are (1) the reactivity of the coal, (2) the rate of heating, (3) the liquefaction temperature, (4) the catalyst, (5) pressure, and (6) contact time.

In the Fischer–Tropsch indirect hydrogenation process, the chemical reactions are:

$$nCO + 2nH_2 \xrightarrow{\text{catalyst}} (CH_2)_n + nH_2O \qquad \Delta H = -3.99 \text{ kg cal/g}$$

$$2nCO + nH_2 \xrightarrow{\text{catalyst}} (CH_2)_n + nCO_2$$

In the methanol process the reactions are:

$$CO + 2H_2 \xrightarrow{\text{catalyst}} CH_3OH \qquad \Delta H = -21.684 \text{ kg cal/g}$$

$$CO_2 + 3H_2 \xrightarrow{\text{catalyst}} CH_3OH + H_2O \qquad \Delta H = -0.833 \text{ kg cal/g}$$

5. *Commercial Reactors Producing Liquids Fuels from Coal*

The only large commercial coal liquefaction plant in operation is in Sasol, South Africa. Two engineering variants of the Fischer–Tropsch (FT) process are being used. The plant is currently undergoing a very large expansion.

Smaller commercial plants using a low-pressure pyrolysis process (Lurgi–Ruhrgas) are still in operation. Coal, after crushing, is rapidly heated by direct contact with a hot recirculated char which had been heated by partial combustion of part of the char. The heating value of the product gas is 700–850 Btu/SCF. The liquids, which represent about 18% by weight of the coal (1 barrel per ton of coal) are condensed and hydrotreated to produce commercial fuels.

In addition to these two processes, commercial plants for making chemical grade methanol are in operation at a number of locations in the United States and abroad. Methanol, of less than chemical grade purity, could be used as a clean boiler fuel and in mixtures with gasoline to reduce the amount of liquids needed to be produced from petroleum for the transport sector. The lower quality methanol could be produced at lower cost than the chemical grade type. Some of the commercial plants abroad have produced a methanol from gas made from coal, but even the largest commercial methanol plants would be considered small when compared to commercial liquid fuel plants.

a. Fischer–Tropsch Process. In this process synthesis gas (carbon monoxide and hydrogen) made from coal is reacted in the presence of a catalyst (Fig. 8). In the first bed reactor the catalyst is contained in vertical tubes and the heat released during the synthesis is absorbed by boiling water outside the tubes. Process conditions are 430–490°F and 360 psi.

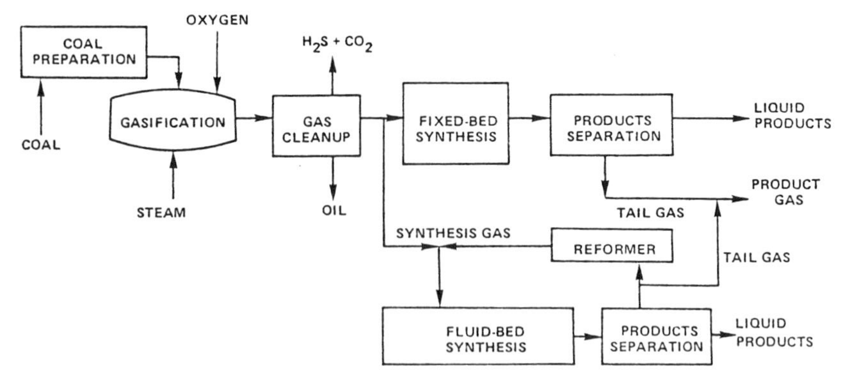

Fig. 8. Fischer–Tropsch process.

The products from this reactor are straight chain and medium boiling oils, diesel oil, LP gas, and oxygenated compounds.

In the entrained bed reactor the plant is similar to that shown in Fig. 8 but catalyst and synthesis gas (after increasing the hydrogen to carbon monoxide ratio with reformed product gas) are circulated together. Product gas and catalyst are separated in cyclones and the catalyst recycled. Process conditions are 600–625°F and 330 psi. Products from the entrained reactor are mainly low-boiling hydrocarbons and gasoline.

The commercial indirect liquefaction process as used at Sasol would have limited use in the United States. The coal gasifier could not handle United States coking coals without modification of the reactor and the product mix is not suitable for United States markets because the gasoline fraction has too low an octane rating and the medium-Btu gas could not be used in natural gas pipelines, although it could be used for industrial markets after purification.

b. Methanol Synthesis. Synthesis gas is converted to methanol in a catalytic reactor. There are a variety of processes for methanol production operating at 500°F and high pressure (4500 psi), intermediate pressure (2000–2500 psi), and low pressure (750–1500 psi). The different processes use catalysts with different chemical compositions. The crude methanol is condensed and separated from unreacted gas and is then purified to the level desired. When used as a fuel it is expected that the methanol will contain about 2% of impurities and have a heating value of 10,000 Btu/lb.

6. *Pilot Plants Producing Liquid Fuels from Coal*

a. Pyrolysis.

(1) *COED Process.* This is a four-stage multiple fluid bed pyrolysis process for making liquids and gases from coal, but the process also

produces a large amount of char. It was tested in a 36 ton per day plant at Princeton, New Jersey for a number of years, but the plant has now been dismantled. Each fluidized bed is operated at increasingly higher temperatures with the temperature in the first bed being 650°F and in the last bed 1600°F. Gas and coal flow in countercurrent directions (see Fig. 9) through the four fluidized beds.

The liquids produced are about equal in value to a low-sulfur crude oil. Tests have been made on catalytic hydrogenation of the COED oil and on methods to upgrade the product oil as a source of petrochemicals. Since the major product is a char, successful development depends on finding uses for the char either as a boiler fuel or as a gasifier and liquefaction feedstock. Unfortunately, the sulfur content per million Btu of char is about the same as the sulfur content per million Btu of the raw coal used. Moreover, since the char is low in volatile matter it is more difficult to burn than coal, so it is not a promising fuel for boiler use.

A contract has been awarded by DOE to design a large-scale demonstration plant using the COED and Cogas processes (see Section VIII,B).

(2) *Flash Pyrolysis*. The Occidental Research Corporation has been studying flash pyrolysis in a 4 ton per day plant for a number of years. This is a low-pressure pyrolysis process that is carried out in an entrained bed carbonizer. Heating to 1100°F is done rapidly using a mixture of coal and recycled char from a char heater. The products leaving the carbonizer pass through cyclones, where the char is separated from the gas. The gas stream is cooled and scrubbed to remove tars which can then be hydro-

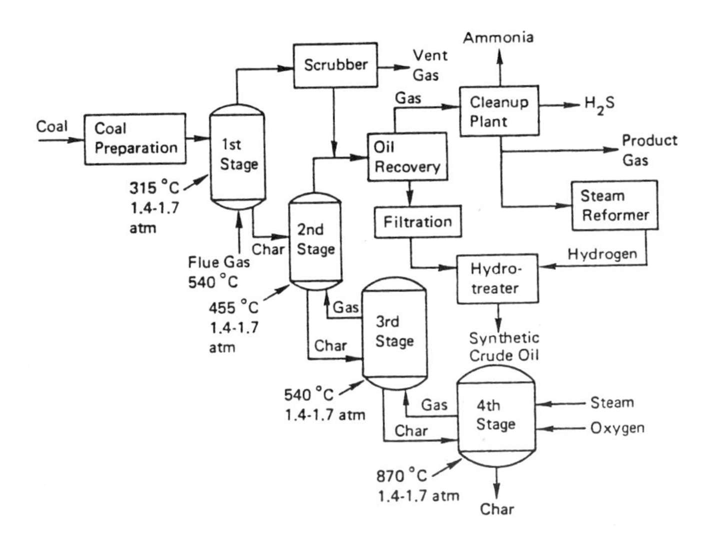

Fig. 9. COED process.

treated to obtain a synthetic oil. Part of the gas is used to produce hydrogen for the hydrotreatment, and the balance, after cleaning, can be marketed either as a medium-Btu gas or, after further upgrading, to a pipeline quality gas.

Like the COED process, the principal product is char for which a market would have to be found, although the rapid heating is said to produce more oil than other pyrolysis processes.

(3) *Toscoal.* A 25 ton per day pilot plant has been operated using subbituminous coal in a process originally designed to produce oil from shale oil. Coal is crushed and preheated in a fluid bed. It is fed to a rotating pyrolysis drum, where it is heated to carbonization temperature (about 950°F) with hot ceramic balls. The ceramic balls in the solids that are discharged are separated from the char and sent to a heater where they are reheated.

The product gas contains about one-half barrel of liquids per ton of coal and is a gas with a heating value of 500–600 Btu/SCF. The liquids are condensed, hydrotreated, and fractionated. As with other pyrolysis processes the major product is a char.

(4) *United States Steel Clean Coke Process.* This process is designed to make a metallurgical grade coke from a high-sulfur coal, but it also produces some liquid and gaseous fuels. After cleaning and sizing, part of the coal is carbonized at about 1300°F and 150 psi in a fluid bed where char, liquids, and a hydrogen-rich gas are produced. The gas is cleaned and recycled to fluidize the bed and the reaction of hydrogen with the coal removes part of the sulfur. The char is pelletized with oil derived from the process and the pellets calcined to produce a low-sulfur coke and a hydrogen-rich gas.

The other part of the coal is slurried with part of the oil produced and reacted with hydrogen at 800°F and 2000–4000 psi in the absence of a catalyst. Liquids and gases are produced and separated from the solids by flash vaporization. This part of the process has been studied in autoclaves and more recently in a 10-inch solvent extraction reactor.

(5) *Other Processes.* The Australian Commonwealth Scientific and Industrial Research Organization (CSIRO) has studied a fluidized bed process in which the pyrolysis is carried out under conditions of rapid heating. Air is used to fluidize the bed and supply the heat for the reaction. About 8% (about one-half barrel) of liquids are produced per ton of coal.

The Rocketdyne Division of Rockwell International is testing a pyrolysis that involves the reaction of coal with hydrogen, using the propellant injection technique used in rocket engines. The short residence times and rapid rates of heating in the presence of hydrogen should result in high conversions to liquids. The reaction takes place at 1830°F and 1000 psi and the products are quenched with a water spray. The process has only been tested on a bench scale.

A fluidized bed process operating at 300 psi of hydrogen and at temperatures of 1200°F has been studied by the Bureau of Mines. Part of the cleaned product gas is returned to the carbonizer after it has been mixed with hydrogen and pretreated by the gases leaving the carbonizer. A small, 10 pound per hour unit is now being tested at Oak Ridge National Laboratory.

b. Solvent Extraction.

(1) *Consol Synthetic Fuel (CSF).* Figure 10 is a schematic flow sheet of the CSF process. Coal is mixed with a recycle hydrogen donor solvent, heated, and fed to an extractor. After extraction at 750°F and 150–450 psi for 1 h, the product is separated (a variety of solid liquid separators have been tested) and the solid product carbonized to char, which is used to produce the hydrogen required for the process. The liquids are hydrogenated to produce a low-sulfur fuel oil and a solvent for use in recycle.

Typical products per ton of raw coal (14.4% moisture and 10.8% ash) are as follows:

3425 SCF gas—933 Btu/SCF	71.00 lb of sulfur
3.5 bbl of oil	213.60 lb of ash
11.00 lb of ammonia	492 lb of char

The pilot plant at Cresap, West Virginia, originally designed to produce gasoline from coal, is being reactivated to test this process.

(2) *Solvent Refined Coal.* This process is being tested in two pilot plants. One is a 50 ton per day plant located at Tacoma, Washington and the other is a 6 ton per day plant at Wilsonville, Alabama. Pulverized coal is mixed with an unhydrogenated coal derived oil and the slurry, along with gaseous hydrogen, is preheated to 850°F and pumped to an extractor–dissolver operating at pressures of 1000–2000 psi. After a flash separation, the solution containing the dissolved coal and unreacted coal is filtered to remove the solids. The solvent is recovered from the liquid product and recycled back to be used to produce a fresh coal slurry. The final product is cooled and yields a solid low in ash and sulfur.

A combustion test in a full-scale boiler with the solid solvent refined coal product was conducted successfully and new studies have been initiated on modifying the process to produce a liquid rather than a solid product.

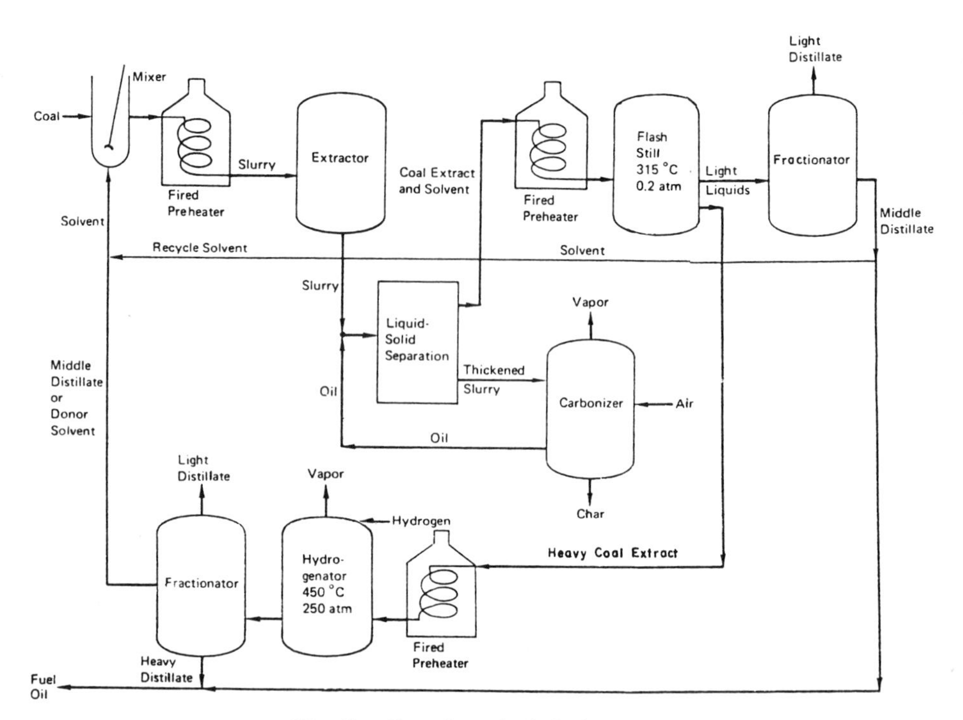

Fig. 10. Consol synthetic fuel process.

(3) *Solvent Refined Lignite Process.* This process is being studied at the University of North Dakota and uses a modification of the solvent refined coal process to produce a high heating value product from lignite. Synthesis gas is used in place of hydrogen to solvent extract lignite under pressure and recovers 70% of the feed as a clean solid along with lighter liquids and gases. The oils can be upgraded to liquid fuels or refinery feedstocks.

(4) *Exxon Hydrogen Donor Solvent.* Coal is heated with a hydroaromatic donor solvent material in the presence of hydrogen. The solids are then separated from the liquid products which are upgraded by hydroprocessing. The solids separation is accomplished by vacuum distillation and this avoids the solid–liquid separation problems that have made most solvent extraction processes difficult to engineer.

Research has been carried out in a 1 ton per day pilot plant, and construction of a 250 ton per day plant has been started.

(5) *Arthur D. Little Extractive Coal Liquefaction.* A hydrogen donor solvent oil is added to coal and the mixture is reacted in a coke drum for one hour at about 100 psi and 750°F. Hot solvent vapor is passed continuously through the drum to provide the heat for the vaporization, and the

drum is finally raised to coking temperature to separate the solids and liquids. The donor solvent is hydrotreated and recycled. The process is being tested on a bench scale.

(6) *Supercritical Fluids*. A stream of supercritical gas at a pressure of 1500 psi and a temperature of 570–750°F is passed through a bed of hot coal. The high hydrogen content volatile matter in the coal is extracted, leaving behind coal ash and char. The pressure of the gas and products are reduced in a second vessel and the extract is precipitated. The gas is repressurized and recycled. Tests have been conducted on bench scale equipment in the United Kingdom.

(7) *Other Processes*. Several different solvent extraction processes using crude oil or asphalt as the solvent have been reported to be under study in Japan. Other research on solvent refined coal processes is being conducted in Australia, Poland, South Africa, and the Soviet Union.

c. Direct Hydrogenation

(1) *H-coal Process*. This process converts coal into either a low-sulfur boiler fuel or into refinery syncrude. Coal is slurried in a recycled oil and reacted with hydrogen at 3000 psi. The mixture is pretreated and pumped into the bottom of an ebullated bed catalytic reactor. The recirculated oil maintains the catalyst in a fluidized state. Fresh catalyst is added and spent catalyst removed on a semicontinuous basis. Vapor product leaves the top of the reactor, where it is separated by cooling. Liquid–solid product, containing unconverted coal and ash, is flashed in a separator. The heavy oil and solids are treated in hydroclones, in a liquid solid separator and a vacuum still. Yields of 3.5–4.0 barrels per ton of coal fed to the reactor are obtained.

Design of a 600 ton per day pilot plant has been completed and the construction phase should be over in the middle of 1978.

(2) *Gulf Catalytic Coal Liquids Process*. The process is similar to the Synthoil since it involves hydrogenation of a coal slurry using a fixed bed catalyst. In this case the catalyst is held in special radially placed baskets. The catalyst is claimed to have a high resistance to carbon deposition, to maintain its activity for long periods, and to tolerate the metallic compounds in the coal ash without being deactivated. The reactor operates at 750°F and 2000 psi.

(3) *Synthoil Process*. This process converts coal into a low-ash low-sulfur fuel oil. A mixture of coal and a recycled oil is mixed with hydrogen and heated in a preheater. The slurry then enters a fixed bed catalytic reactor. The catalyst is cobalt molybdate on silica-promoted alumina.

Liquids and unreacted solids leave the reactor and are centrifuged to separate the solids, which are then pyrolyzed. The solids that remain are gasified to make the hydrogen required for the process. The rapid flow of the slurry through the reactor prevents plugging by coal mineral matter and increases mass and heat transfer in the slurry. A 10 ton per day pilot plant has been designed, but not yet constructed.

(4) *Costeam.* The process is designed to convert low-rank coals such as lignite into low-sulfur fuel oil by the noncatalytic reaction of a coal oil slurry with synthesis gas (or CO) and steam. A coal–oil slurry made using a recycled oil is reacted at 4000 psi and 800°F in a stirred reactor. The water in the lignite reacts with the synthesis gas to produce hydrogen by the water gas shift reaction. The gases are separated and solids are removed from the liquid product.

(5) *Disposable Catalyst Hydrogenation.* This process uses inexpensive catalysts so that they do not have to be recovered and reused. Coal is slurried with a recycled oil and preheated in the presence of the catalyst. The reactor is kept in a highly turbulent condition. Process conditions are 850°F and 2000–4000 psi pressure. Screening of catalysts to obtain the lowest cost reactive catalyst is continuing.

(6) *Other Processes.* A number of other processes are at various stages of development. These include the clean fuel from coal process of the C. E. Lummus Company, the University of Utah two-stage process, the zinc chloride process of Consolidation Coal Company, and the Universal Oil Products Company processes.

d. Indirect Hydrogenation.

(1) *Fischer–Tropsch.* In addition to the commercial FT plant in South Africa, a small-scale FT reactor is being tested at the Pittsburgh Energy Research Center. It differs from the Sasol plant in that two different types of reactors will be used to remove the reaction heat. One system uses a hot gas recycle of product gas, while the other removes the heat in a tube wall reactor.

(2) *Methanol Production.* A large number of economic studies have been made on the cost of making methanol from gases produced by the gasification of coal. A number of different coal gasification reactors have been examined to estimate the costs of producing methanol in large (5000–13,000 tons per day) methanol plants, and product costs are relatively expensive.

C. Auxiliary Gasification and Liquefaction Process Requirements

The description of the various coal gasification and liquefaction processes above has been mainly concerned with reactor vessel configurations. In both coal gasification and liquefaction these represent only a small part of the total cost of the conversion system, although reactor characteristics are critical in determining the size and cost of the auxiliary equipment that is required. In most coal gasification processes the reactor represents only 10–15% of capital investment, but an improved conversion reactor design could result in reduced costs in the "balance of plant."

In addition to the reactor itself, coal conversion plants must be able to: (1) receive, store and handle coal, (2) crush and grind coal to the size required, (3) feed coal of the needed size in some way into atmospheric or elevated pressure vessels, (4) pretreat coal as required, (5) remove dust and solids from gas streams, (6) recover sensible heat, (7) carry out shift conversion, (8) purify gases from a variety of impurities, (9) operate air and water pollution control equipment, (10) recover a variety of byproducts, (11) manufacture the steam, oxygen, and electricity required, and (12) provide the other plant facilities that are needed.

Improvements in these processes could reduce costs significantly, but reduction in the cost of these operations will be difficult to accomplish. Coal handling and preparation, solid–gas separation, heat recovery, shift conversion, and other process steps have been the subject of extensive research and development for many years, since they are widely used commercially, and dramatic improvements in costs of these operations are unlikely. For other unit processes involved there may be important opportunities for cost improvement. These include coal feeding under pressure, gas purification, air and water pollution control technologies, liquid–solid separation techniques for coal liquefaction processes, and new and improved coal gas methanation processes for producing a substitute natural gas.

The two most important areas for further investigation in coal conversion plants are gas clean up and liquid–solid separation. Processes exist to perform both of these operations, but they are costly and, in the case of coal–liquid solids separation, unreliable. The coal gasification plants that have been proposed for construction have selected a variety of gas clean-up systems. This is probably the result of the uncertainty with respect to operability, cleaning efficiency and costs of the various processes that are commercially available.

Many of the coal liquefaction processes that are being studied require a liquid–solid separation, but no entirely satisfactory method has been devised for making this separation at the operating conditions that are encountered.

III. ENERGY EFFICIENCY

Although the energy efficiency of conversion processes is not as important as capital costs per unit of product, it does have some impact on fuel

TABLE IX

Gasification Efficiencies for a Lurgi Gasifier

	Percent
Crude gas + tar + oil + naphtha	
Gasification coal	89.0
Crude gas	
Gasification coal + steam-raising coal	68.3
Crude gas	
Gasification coal + steam-raising coal + + power for oxygen production	62.6

TABLE X

Coal Conversion Efficiency

Process Type	Thermal Efficiency (%)
Gasification	
High-Btu gas	56–68
Low-Btu gas	70–80
Liquefaction	
Direct hydrogenation	60–70
Fischer–Tropsch	40–45
Methanol	60–67

products costs and is critical in determining how much primary energy is required to supply a useful unit of energy to the consumer. However, it is difficult to estimate accurately values for coal conversion efficiencies.

Efficiencies reported in the literature often do not indicate whether the value is for a plant that purchases all, part, or none of the supplemental energy needed in the conversion. These purchased energy forms include electricity, steam, and other utilities. Moreover, efficiency can be defined in a variety of ways—as shown in Table IX.

Because of the difficulties in estimating efficiencies a range for various products is given in Table X.

IV. ENVIRONMENTAL, OCCUPATIONAL HEALTH AND SAFETY, GENERAL PUBLIC HEALTH AND SAFETY

A. General

In considering the potential adverse effects of coal conversion facilities, the entire fuel cycle must be examined—from exploration for coal to the air pollution emissions resulting from the use of the converted coal. Some of these effects are common to both coal gasification and liquifaction, but the size of the impact is dependent on the overall efficiency of the coal conversion plant and the efficiency of utilization of the converted coal. The lower the combined efficiency of conversion and utilization, the larger the amount of coal that will have to be produced, upgraded, and transported to the conversion plant to obtain the same amount of useful energy. There are also differences between gasification and liquefaction with respect to the environmental effects at the conversion plants.

The data base for evaluating environmental and health and safety effects (on occupational workers and on the general public) is very limited for part of the coal cycle, and the reported information is frequently contradictory. The adverse health impacts expected at coal conversion plants are particularly difficult to estimate. No large-scale plants for converting coal to either liquids or gases are in operation in the United States and the only available data on the emissions and effluents that are released are estimated from limited information collected at pilot plant operations. Since these plants are almost always designed to determine if a new process is operable and has economic potential, generally little attention is paid to residuals that are produced and how they would be controlled. Moreover, the nature of pilot plant operations is such that, even when data on residuals are collected and attempts are made to estimate the amount of effluents and emissions that result, they may not be representative of what would be produced in commercial plants.

Some potential environmental impacts are very site specific. These include factors such as the population profile at a particular site, including the number of people that would be exposed to any new adverse impact and their age distribution. The greater the population that may be exposed, the greater will be the total adverse impacts of pollutants that are emitted. Children and older people are generally more susceptible to health impacts from environmental pollutants, so that an area with a high concentration of these age groups will be more severely impacted than areas where the age distribution of the population includes a higher percentage of people in the young and middle-aged population group.

The quantity of emissions and effluents that will be permitted from a coal conversion plant will be dependent on the state of the degradation of the environment from other sources. This means that even a plant emitting very minimal amounts of pollutants may be precluded from being located at a particular site for environmental reasons.

Finally, plant locations are determined in part by the type of ecosystems with which they must coexist. A plant located in an area where the surrounding flora and fauna are fragile should be required to meet standards for emissions that are more stringent than for plants in areas where the ecosystem is more hardy.

B. Coal Exploration, Mining, Upgrading, and Transportation to Conversion Plants

In the exploration and mining operations the largest share of mining wastes (both from underground mining and coal beneficiation plants) is in the eastern part of the United States. As much as 25% of the raw coal is rejected at coal preparation plants and the waste piles are a source of both water and air pollution. Leaching of the piles from exposure to rainfall can result in contamination of streams from silt and dissolved salts. Air pollution can occur from windage losses and from air pollutants released when the waste pile burns, if it ignites as a result of spontaneous combustion.

The amount of land that must be reclaimed after strip mining is a function of the thickness of the coal bed being mined. With the thick coal deposits of the West, less land is disturbed when a ton of coal is extracted. However, the most difficult strip mining conditions to reclaim are in the arid regions of the Southwest and Northern Great Plains. Rainfall may be insufficient in some areas to revegetate the disturbed land.

Acid mine drainage is a difficult problem in the Appalachian region, and large volumes of acid waters reach the streams from drainage of mines that are above the water table.

Occupational health and safety problems are also greater in the East because of the high percentage of production from underground mines. With the lower productivities that exist in these mines, the number of accidents and fatalities to produce a ton of coal is much higher in the East, compared to the West. In addition, underground miners are exposed to a much greater extent to the danger of developing coal miner's pneumoconiosis (CMP). The magnitude of this health related problem is still in doubt, but there is little question that it is a major problem at least as far as past damages are concerned.

Transportation of coal results in emissions of relatively large amounts of particulates. The emissions occur during loading and unloading of the coal and from losses of fines from the railroad cars during transport.

Transportation of coal by rail involves considerable risk to both employees and the public because of deaths at rail crossings. In fact, transportation of coal has been projected to cause about four times as many deaths (occupational and general public) as coal mining in the year 2000. Whether these deaths should be attributed to the use of coal is still a matter of debate, but transporting the coal to where it is needed is part of the energy supply–delivery problem.

Coal storage piles are also the source of contaminated water due to leaching and can create air pollution problems from particulates from windage losses. Coal storage piles, if not properly compacted, can ignite from spontaneous combustion and emit air pollutants such as sulfur oxides, nitrogen oxides, carbon monoxide, hydrocarbons of various kinds, and particulates.

Both coal gasification and coal liquefaction plants must crush and size the coal prior to use so very similar types of fugitive dusts, particulate matter contaminates, water, and complex liquid and gaseous compounds are produced at both. The quantities produced will depend on the overall efficiency, from production of coal to providing useful energy to the consumer. However, most of these types of emissions can be controlled by commercially available equipment and provisions will have to be made to maintain satisfactory levels of dust emissions.

C. Coal Gasification

About 10% of the coal consumed at a coal gasification plant is used to generate electricity and steam. Combustion of coal for these purposes creates the same type of air emissions that arise at a coal fired electric generating plant. These air pollutants are sulfur oxides, nitrogen oxides, particulates, and small amounts of trace metals and hydrocarbons. The same methods of pollution control can be used at the gasification plant as are used at a power plant.

The annual controlled air emissions from a unit high-Btu gasification plant (0.25×10^{12} Btu/day) is shown in Table XI. There are significant differences in the quantity of air pollution emissions, depending on which type of coal is used. The amount of sulfur oxides emitted varies by as much as a factor of 4 for different coals, but other air emissions are less dependent on coal type.

Waste water pollution that must be controlled results partly from cooling tower water and ash quenching operations and partly from the waste streams created at the gasifier. The amount of water that has to be treated from the gasifier depends on the type of reactor used. With fixed bed reactors relatively large amounts of oils, tars, phenols, and other residuals are formed. With entrained gasifiers that have high exit gas temperatures the amount of contaminants are either greatly reduced or are nonexistent. Fluidized bed gasifiers usually produce an intermediate amount of contaminated water, but in a recent test in the Synthane pilot plant, when deep bed injection of raw coal feed was used only negligible amounts of residuals were found in the product stream.

Recycling of contaminants (tars, oils, phenols), after removal from the

TABLE XI

Annual Controlled Air Emissions from a Unit High-Btu Fixed Bed Gasification Plant (250×10^6 SCF/day)[a,b]

Region/Coal	Total Partic-ulates (tons)	Sulfur Oxides (tons)	Nitrogen Oxides (tons)	Hydro-carbons (tons)	Carbon Monoxide (tons)	Alde-hydes (tons)
Appalachia						
Monongalia Co., West Virginia	450	7,350	4580	100	340	1.7
Belmont Co., Ohio	500	10,860	4870	110	360	1.8
Jefferson Co., Alabama	520	2,130	4650	105	345	1.7
Eastern Interior						
Fulton Co., Illinois	460	8,530	5730	130	425	2.1
Henderson Co., Kentucky	770	11,910	5340	120	395	2.0
Powder River						
Campbell Co., Wyoming	530	2,600	7570	170	560	2.8
Fort Union						
Mercer Co., North Dakota	800	4,190	9140	200	680	3.4
Four Corners						
San Juan Co., New Mexico	1020	3,820	6690	150	495	2.5

[a] Source: Synthetic Fuels Commercialization Program, ERDA (1976a).
[b] 328 operating days a year (90% stream factor).

waste streams, back to the reactor allows the trace amounts of hydrocarbons that remain to be treated with commercial biological waste treatment processes. By evaporating any contaminated water to dryness, zero waste water discharge from these coal conversion plants is possible and is planned for the new Sasol plant in South Africa. This, however, creates a solid waste disposal problem.

The solids discharged from the plant come largely from the ash in the coal and can be handled by either landfill or mine burial. Because of differences in the chemical composition of the ash, care must be taken to prevent leaching when burial locations are selected. Other sources of solid waste for disposal are the sludge from limestone scrubbing of the sulfur oxides and residues from evaporated liquid condensates. If these later wastes are soluble they will require special treatment.

Occupational health problems at coal gasification plants should be less severe than those at installations handling similar process streams, such as coke oven plants, steel mills, and coal washing plants. This is expected because some coal gasification processes operate at pressure and all produce large amounts of poisonous gases, so that leaks and fugitive emissions will have to be controlled. Experience with petroleum refining provides the basis on which to achieve good hygienic practices, since most of the harmful health hazards expected at coal gasification plants are similar to the industrial streams at petroleum refineries. However, plants must be operated so as to ensure safe health conditions for employees.

Injury and fatality rates to be expected at coal gasification plants should also be similar to those at petroleum refineries.

The effect on the general public's health and safety of a coal gasification plant should be minimal because no significant amount of the residuals produced should be emitted during routine operations.

D. Coal Liquefaction

Waste water, air, and solid waste management in a coal liquefaction plant is very similar to those described above for coal gasification. Zero waste water discharge could be achieved if contaminated water that is produced is evaporated to dryness. In addition, gaseous effluents from coal liquefaction plants may contain significant amounts of volatile materials stripped from the cooling water. It is believed that it will be possible to control pollutants that have been identified as being generated at coal liquefaction plants, but complete data on all the emissions have not been obtained.

The occupational health impacts of coal liquefaction plants are greater than those at coal gasification plants. Much larger amounts of liquids are produced and the oil that is formed from the direct hydrogenation of coal has been found potent in generating skin cancers. On the other hand, Fischer–Tropsch oils did not produce cancers in National Cancer Institute tests. Lung cancer should not be a problem because of the continuous nature of the process, so that uncontrolled emissions, such as those at coke oven plants, should occur only when accidents occur.

Little data are available on the carcinogenicity of materials formed by the newer, more advanced liquefaction processes. Medical programs developed at Union Carbide during the 1950s and, more recently, by Sasol, appear to show that avoidance of prolonged exposure to carcinogenic types of material will provide adequate protection to employees. However, additional data to confirm these findings are needed.

Coal liquefaction plants should have no greater impact on the community health and safety than coal gasification plants, and this is minimal during normal operations. However, since consumers could come in contact with liquid products, precautions should be taken to assure that they can avoid exposure to distillate fuel oils and heavy residual fuels made by coal hydrogenation. Lighter products, after refining, and liquid fuels made by the Fischer–Tropsch process should not represent a hazard to consumers.

E. Environmental Regulatory Considerations at Coal Conversion Plants

This section covers the major Federal environmental regulations that must be met but will not attempt to enumerate the State and local regulations that must also be satisfied. Each state and local community has important responsibilities for environmental matters but the regulations are so numerous and diverse that no short discussion of them is possible. However, their overall impact is to delay development and increase costs, although they may be necessary to protect health and safety and amenities in a community.

Any coal conversion facility will be required to file an environmental impact statement (EIS) under Section 102(c) of the National Environmental Policy Act. Based on experience in other industries, obtaining approval of an EIS for a coal conversion plant will be time consuming and may be subject to litigation with further delays. Since these are new types of commercial plants there will be great uncertainties about what the EIS should contain. All of the delays caused by the need to satisfy the EIS procedures will result in higher plant costs as a result of the rapid rate of inflation in the construction industry.

The Clean Air Act of 1977, just recently enacted, will result in even stricter air emission standards for coal conversion plants than those contained in the 1970 Clean Air Act. The exact nature of these standards is not yet known since the regulations have not been promulgated. The new source performance standards of the Clean Air Act could affect both the synthetic fuel plants that manufacture liquids or gases from coal and the facilities that use the synthetic fuels that are produced.

The hazardous air pollutant provisions of the Clean Air Act could eventually affect synthetic fuels plants, although none of the hazardous pollutants designated (or those expected to be designated shortly) are emitted by these plants. Another section of the Clean Air Act that could affect the siting of coal conversion plants is the "nonsignificant deterioration" provision.

In addition to the Clean Air Act, the Federal Water Pollution Control Act could affect the siting and cost of production of synthetic fuels in three ways:

1. The National Pollutant Discharge and Elimination System (NPDES). This section of the Act will require any new water discharge source to use the best available control technology.
2. Effluent guidelines and standards. These guidelines were developed by the states to comply with the FWPCA Act. These are now under review.
3. Pretreatment standards. These are now under review.

The Safe Drinking Water Act of 1974 could affect synthetic fuel plants if surface storage of fuels were being leached by water. The Occupational Health and Safety Act contains a large number of specific standards that will apply to coal conversion plants such as noise, heat, and machinery hazards. Restrictions may even be imposed on hazardous substances that occur in process streams.

The Resources Recovery Act of 1970 requires that plants comply with the guidelines for solid waste disposal established by the Secretary of Health, Education, and Welfare. Coal conversion plants will be subject to these guidelines.

The above discussion covers the major Federal environmental regulations that will affect the siting and the cost of producing synthetic fuels. As noted earlier, individual states also have a number of environmental provisions that will have to be met in order to obtain permission to construct a new coal conversion plant.

The leasing and health and safety regulations that must be met during coal mining are discussed in Section V.

V. RESOURCE REQUIREMENTS AND CONSTRAINTS

A. Coal

1. Size and Location of Deposits

The coal reserves and resources of the United States are very large. With the tonnages physically available under favorable geologic conditions there should be no problems with respect to coal availability in the orderly development of a synthetic fuels industry. Even with greatly increased production of coal for direct utilization, reserves and mining capacity for producing the coal needed for a commercial coal conversion industry should be easily provided.

The United States coal deposits are widely spread geographically. Bituminous coals are found in the Appalachian Region, Illinois Basin, and the Western Interior Region. Subbituminous coals are found in Wyoming, Montana, Utah, Colorado, New Mexico, and Arizona. Lignite deposits occur in North Dakota, Montana, and Texas.

On a calorific value basis, the coals are about equally distributed between those east and west of the Mississippi, but, because of their lower heating value, on a tonnage basis, the Western deposits are larger. Besides the rank of coal and differences in other chemical properties, two other characteristics distinguish the Eastern and Western deposits. Most of the Western deposits of subbituminous coal occur in very thick seams and nearly half of these resources can be mined by stripping. In contrast, much of the Eastern coal occurs in thinner seams—2–7 feet—and at greater depth so that less than 20% is suitable for strip mining.

Western coal is also largely owned by the State and Federal government and is not available for use until it is leased, while the greatest part of the Eastern coal deposits are in private ownership. This difference in ownership patterns makes Eastern coal development much easier to undertake since there is no need to wait for Federal lands to be offered for lease.

2. *Coal Production Trends*

Coal production in the United States has historically been concentrated east of the Mississippi. This occurred because of the concentration of energy markets in the area, the relatively high cost of coal transport, and the much greater availability of oil and gas in the West compared to the East. Because of the low sulfur content of Western coals and the low cost of strip mining the thick Western deposits, the share of United States coal production is expected to grow at a rapid pace in the West.

The passage of the Coal Mine Health and Safety Act had a profound effect on coal mine productivity in Eastern underground mines. Coal productivity in these mines declined from over 16 tons per man per day to over 8 tons per man per day between 1970 and 1976. This trend, along with other factors, greatly increased the price of this coal and resulted in a dramatic shift in the percentage of coal mined by surface methods. By 1976 nearly 60% of the coal was produced in strip mines, compared to less than 40% in the late 1960s.

Coal costs are one of the factors that determine the economic feasibility of coal conversion process. The increasing cost of producing coal in the East compared to lower costs of production of Western deposits is an important element in determining where coal conversion plants will be located. For many uses of coal, transport costs are a significant cost factor and Western coal (with its low heating value) is a long distance from many major energy markets. For coal conversion plants, if they can be located near the low-cost Western coal reserves, this may not be a significant problem, since the cost of transporting the oil or gas that would be produced is much lower than that for moving coal.

Siting coal conversion plants in the sparsely settled and arid regions in the West may be more difficult than in the East. The availability and cost of mining coal in the West depends on the policies of a large number of Federal (and State) agencies. Water requirements could limit the number of Western coal conversion plants that could be constructed and the resistance of local communities to large-scale industrial development for social and infrastructure reasons could affect decisions as to where to locate commercial coal conversion plants.

3. *Regulatory and Other Constraints on Coal Production*

No attempt is made in the following discussion to compile a complete list of the regulations of the State or local governments or the complications caused by Indian rights which permits them, in some instances, to determine the conditions and rate at which coal and water resources will be developed. The variation among the states is very great (see, e.g., Section V,A,3,b below) and they are usually numerous and complex. However, development of a coal conversion industry will require that their rules and regulations be met so that the planning process is further complicated and additional delays in plant construction are inevitable.

a. Federal Coal Leasing Program. The Federal coal leasing program is mainly of interest in connection with Western coal deposits and it could be a critical factor in the rate at which coal conversion plants are constructed. Leasing of Federal coal lands has been at a virtual standstill since about 1971 because of earlier speculation on the coal lands under the terms of the Mineral Leasing Act of 1920. This speculation and the environmental controversies that developed subsequently have resulted in attempts to develop new procedures to dispose of Federal lands in ways that would prevent speculation and bring a fair market return to the government while permitting the orderly development of the Western coal resources. This has proved to be a difficult problem and new leasing procedures are still being developed. Meanwhile, although large tonnages of coal were previously leased and are available for private development, it is not certain that those who have the leases will be willing or capable of starting a commercial synthetic fuels industry. For this reason it is highly desirable to develop a method of offering Federal coal lands for development on a sounder basis than was previously used. If the lease terms and conditions are very restrictive and leases are not offered in a timely fashion, development of a coal conversion industry in Western regions could be seriously retarded.

Another factor that could complicate the use of Federally leased coal for coal conversion plants is the Diligent Development (45 C.F.R. Part 3520) provisions that became effective in June, 1976. The objective of these regulations is to reduce speculation on Federal coal leases, but they will also affect decisions on whether conversion plants will be constructed since mine production must be at a certain level by a given date or the lease is forfeited. Thus planning of the conversion plant and the coal mine associated with it has to be closely coordinated under conditions where a prescribed deadline on mining operations must be met.

b. Reclamation of Surface Mined Land. Federal coal leases (mainly Western coal) have always contained environmental stipulations with respect to restoring strip mined lands. The Surface Coal Mining Act of 1977, as far as Federal lands are concerned, will result in some changes in leasing terms with respect to reclamation provisions, remove certain lands from leasing (alluvial valley floor and prime agricultural lands), and clarify the State-Federal relationships on both leasing and reclamation. The large increase in the number of regulations and the nature of the requirements that must now be met can only result in increased costs for coal because of the very much longer planning procedures prior to mining, the large number of records which must be maintained, and the increased requirements for the restoration of the strip mined land.

On non-Federal lands the effect of the new strip mine legislation on mining procedures and cost of coal will vary from state to state. Most states have had strip mine laws but their requirements have varied widely, as has the effectiveness of enforcement. The effect of the new Federal law is to set minimum standards for reclamation and enforcement of the regulations, but it permits those states with approved plans to enforce standards above the minimum required by the Federal law. The overall impact nationwide, therefore, will be to improve strip mining reclamation practices and increase coal costs since all State regulations will have to meet the minimum Federal standards.

c. Health and Safety Regulations. The Federal Coal Mine Health and Safety Act of 1969 (modified in 1977) and the mandatory health and safety standards (30 C.F.R. 70) that were promulgated as a result were designed to improve the health and safety conditions in both strip and underground mines. The results to date have been favorable in that fatality and injury rates have been reduced. It is too early to determine if coal workers pneumoconiosis (black lung) has been reduced as a result of the new dust standards because of the long periods required for the disease to develop and the difficulties of detection.

While it appears that the major objectives of the legislation have been met, it has resulted in higher coal costs. These increased costs have been the result of two factors—first, there is the need to purchase more expensive measuring devices for detecting dangerous conditions and higher costs for equipment that is able to meet the more stringent safety standards. Second, the testing procedures for detecting dangerous conditions require time previously used to produce coal and the assignment of additional personnel that are largely concerned with ensuring that the working conditions are safe have both contributed to reduced productivity in the mines. In underground mines the average productivity was reduced to one-half its former value in less than eight years, while the historic long-term growth rate in productivity in strip mines has been greatly reduced.

B. Water Resources

1. Water Requirements and Availability for Coal Conversion Plants

All of the coal conversion processes require large volumes of water, part of which is consumed in process use (e.g., to produce hydrogen), part through evaporation, losses from cooling systems, and part is lost in the disposal of solid wastes. In most plant designs the water to be used for solid waste disposal is one of the contaminated streams that is produced in the plant. In addition, water is required during the plant and mine construction for sanitation and dust control, for strip mine area reclamation to aid in revegetation, and for increased municipal demands of the influx of population required to construct and operate the plant.

The amount of water consumed per unit heat value of the product varies to some extent with the nature of the product, with the process selected, and with plant design. In water-short areas it is possible to use air cooling instead of water cooling and the amount of water used for cooling will be dictated by overall economics. For these reasons and

TABLE XII

Water Required for Coal Conversion Plants

	Water Consumed (10^6 gal/day per 10^{15} Btu)
High-Btu gas or coal to oil plant	64–158
Low- or medium-Btu gas or coal to low-sulfur oil plant	5–10

because there is no actual commercial experience, the published values of water consumed vary widely. The table (Energy Research and Development Administration, 1976a) shows the range of values of water that have been estimated to be required for coal conversion plants.

For comparison, a coal-fired plant electric generating plant would require 120–204 million gal/day per 10^{15} Btu, while a nuclear plant would require 235–300.

Water availability (Hudson Institute, 1977c) for coal conversion plants depends on the drainage basin being considered. In the Appalachian region there should be no water problems in the coal regions of western Pennsylvania, southeastern Ohio, West Virginia, and Alabama. In the Kentucky–Tennessee–Virginia region the stream flows are less dependable and if a large number of plants were to be constructed, the water would have to be provided by a combination of impoundment, ground water, municipal waste waters, or by importation.

Water availability should not present problems in the area covered by the Eastern Interior coal region, Illinois, Indiana, and western Kentucky. However, depending on the site, the water could come from ground water, surface water, or waste water sources.

In the Northern Great Plains the water supply situation is not as satisfactory as in the eastern part of the United States and local shortages could develop if a number of coal conversion plants were constructed. Regionally there should be no physical scarcity of water for coal conversion plants for 30 years or more, but to provide water may require transporting the water from areas within the region with ample supplies to areas where coal conversion plants are to be constructed but that are water short.

The water supply needed for a large number of coal conversion plants in the Colorado area will be difficult to obtain. The situation for fresh water supply is complex because of uncertainty with respect to water rights. Saline aquifers appear to exist in the area but their size and availability are unknown.

The most difficult water supply problem is in the Four Corners area. With the rapid growth in the area and the intense competition for the limited fresh water supply, it is likely that little, if any, of this water will be available for coal conversion plants. Underground fresh water is also limited, but a number of saline water aquifers have been identified. How difficult it will be to develop these saline aquifers is unknown, but the flow rates in the formations are low and well yields may be low.

Three methods of supplying water in areas with a fresh water shortage are (1) water imports, (2) use of waste waters, and (3) use of saline groundwater. Transport of water by pipeline can be achieved at reasonable costs when large volumes are transported. At a flow rate of 30 million gal/day the cost of 1000 gal of water transported 200 miles has been estimated to be $0.80, compared to an estimated desalination cost of about $1.00.[5]

Waste water has the potential for supplying water for industrial cooling operations. In water-short areas, this could extend the water supply, but numerous institutional problems will have to be solved. Saline ground waters also represent a new source of cooling water, or if fresh water is needed, a feedstock for desalination. In addition to a lack of detailed resource information, the saline water aquifers may be difficult to utilize directly because of their potential for pollution when discharged after use.

2. *Regulatory and Other Constraints on Water Development*

There are a number of legal restrictions on the development of water resources and these are particularly difficult to resolve in the western United States. Unless water is contained in a federally regulated project, water is subject to state regulation. These regulations are very extensive and involve consideration of interstate water compacts as well as international treaties. Because of the complexity of western water rights issues it may be difficult to obtain the quantity of water required for coal conversion plants without extensive litigation and delays. Further complicating the water situation is the uncertain state of Indian water rights in areas where water supply for coal conversion facilities is already short.

To coordinate federal, state, and Indian water interests will require: (1) an inventory of federal and Indian rights, (2) new laws redrawing interstate water compacts, and (3) development of federal–state compacts.

C. Manpower Requirements and Availability

In evaluating manpower requirements a distinction must be made between the construction and the operation phases. Estimates of manpower requirements during the construction phase will vary from year to year, building up to a peak about 60% of the way through construction. The maximum number of construction workers required for a 250 million cubic foot per day high-Btu gasification plant has been estimated in one study to be 3000. (American Chemical Society, 1976). Construction skills required on site are pipe fitters, electricians, carpenters, iron workers, boiler makers, and welders. In addition, shop fabrication of the equipment needed will require a large number of machinists and steel and foundry workers. No national shortages of these kinds of skills should occur if

[5] See Hudson Institute (1977c, pp. 16, 19).

synthetic fuel development occurs in an orderly fashion. For example, one study (Hudson Institute, 1977a) has shown that even for the skill (boiler makers) in shortest supply in 1994 that the demand, under conditions of maximum synthetic fuels growth, would require only 6.1% of the total number of boiler makers available. However, local shortages of employees with certain skills could occur, particularly if a number of plants were to be constructed in a sparsely populated region. However, such shortages can be expected to be temporary.

In addition to construction and equipment producing personnel, a large number of other types of personnel will be required. This would include mining, metallurgical, mechanical, electrical and chemical engineers, and draftsmen. Although a new synthetic fuels industry should create no long-term problem with respect to these skills, there could be an initial period in which there would be shortage until engineers and others could be trained in the special skills needed for the new technologies.

The number of operating personnel required at a coal conversion plant will be much fewer than during most of the construction period. For example, in the study which estimated that 3000 employees would be needed during the peak of the construction phase it was estimated that only 800–900 would be required to operate the plant. While some of the same skills will be needed during operation as were required for construction—electricians, carpenters, pipefitters, mechanics, etc.—for maintenance, another set of skills will also be required—operators, controlmen, foremen, etc. The greatly reduced work force needed for operation compared to construction and the need for different types of skills for at least part of the work force will create difficult social problems in some areas.

As with construction personnel, however, the total manpower with the required skills should be available nationally, although local shortages could occur in sparsely populated regions. Such shortages should be temporary as the trained labor required either moves to areas where jobs exist or is trained locally to meet job needs.

D. Financial Requirements

The cost of a unit sized coal conversion plant will be over $1 billion and this will place a very large financial burden on most individual companies that can be expected to become involved in such a new industry if funds have to be raised in the equity market. For example, even the largest gas transmission companies, prime candidates for high-Btu gas–coal conversion plants, would nearly double their total capital investment in all their facilities by constructing a single coal to high-Btu gas plant.

On the other hand, while a synthetic fuels industry will be capital intensive, once the institutional and economic problems are resolved, the burden on financial markets should not be excessive. The total cost of a coal conversion industry capable of providing a useful unit of energy to the consumer should be less than that for plants producing the same amount of energy in the form of electricity. Under these circumstances there should not be an unacceptable demand placed on the financial community for investment funds for energy supply. Investment in energy facilities, including coal conversion plants, as a percentage of total new investments should not be significantly greater than that that has prevailed historically.

If one examines the cash flow of the oil industry, the prime candidates for constructing coal to oil plants, adding the financing required to meet even the maximum rate of development of a synthetic fuels industry would not change the financial situation greatly. By the year 2000 the increase in borrowing by the industry would be only about 10 billion dollars (Dickson *et al.*, 1976) per year compared to total investment in that year by the industry of nearly $180 billion.

VI. OTHER POLICY CONSIDERATIONS

A. Community Impacts

A large new facility of any type can have a major impact on the community where it is located. This impact will be greater in regions where the population is small and there have been no similar types of industrial development. In the Western regions with large coal deposits that can be mined at low cost and where large coal conversion plants may be constructed, major impacts can be expected on the social and political conditions that exist. An influx of new residents in a sparsely settled area could drastically change the political and social conditions that exist and have a major impact on local affairs.

Even during the operational phase, when the employees needed to operate the plant are much less than during construction, a single 100,000 barrel per day plant would require an operating and associated population of 15,000. In Campbell County in Wyoming (Dickson *et al.*, 1976) where the 1975 population was only 18,000 the influx of new residents could create serious local problems. These include lack of adequate housing, roads, schools, hospitals, and other infrastructure facilities needed to support a new large town.

More important to local communities in sparsely settled regions is the prospect of a "boom and bust" cycle of development. Boom towns can create serious community problems which severely diminish the quality of life; when the resource on which the new facilities are dependent is depleted, the town has no source of income and the facilities that were created are no longer required. Small western towns have experienced this phenomenon with other mineral resource developments and are reluctant to repeat the experience with large coal conversion plants where the impacts could be greater than for previous mineral resource developments.

In the long run a large new industrial facility may be beneficial for a region but in the short run, before taxes on the plants are collected community services fall behind needs. These include potable water supplies, sewage treatment, fire and police protection, hospitals and schools.

B. Regulatory and Other Factors Affecting the Development of a Synthetic Fuels Industry

In addition to the regulatory measures described above, there remain a large number of other uncertainties that restrain the development of a synthetic fuel industry. These include oil and gas price regulation, threatened legislation on divestiture (both horizontal and vertical) for major oil companies (which would preclude the companies most likely to be able to raise the funds required for constructing plants) long lead times for construction caused by delays in obtaining approvals, environmental reviews, permits, and licenses.

All of these factors increase the cost of the synthetics and increase the risk of making investments in a capital intensive industry. Only the large oil (or gas transmission) companies are likely to undertake a venture in synthetic fuels since only they can raise the funds required and have a vested interest in a source of liquid supply to keep their existing refinery pipelines and market facilities in operation. The oil companies, thus, have a much greater flexibility than others to integrate the new fuels into their existing businesses. It is for this reason that it is unlikely that methanol will be developed as an automotive fuel since it does not fit easily into the existing distribution system for fuels for transportation. Methanol as a clean boiler fuel may be developed and this could release some oil for the transport sector and some natural gas for high priority customers.

State and local regulations often duplicate or overlap Federal regulations and this places a great additional burden on those wishing to construct conversion plants. Jurisdictional questions are inevitable and lead to excessive delays and higher costs.

C. Organizational Responsibilities

A number of different groups will participate in the establishment of a synthetic fuel industry. These include the local, state, and federal governmental bodies, the business community, new workers at the plant, environmentalists, the Congress, and the courts. At the federal level the Department of Energy (DOE) will play a major role in synthetic fuel development through its support of research and development to improve the economics of coal conversion processes and through support of demonstration plants large enough to gain experience with the technology and to get on the learning curve to reduce costs. DOE will probably also be selected as the administrator of any federally funded (or partially funded) plant.

The Environmental Protection Agency (EPA) will be heavily involved in the environmental aspects of any large plant and establish emission standards to protect the general public's health from harmful effluents that may be released by the plant. The Occupational Health and Safety Administration (OSHA) will be responsible to ensure that both the occupational safety and health of the plant workers are protected.

Congress has been and will continue to be heavily involved in the development of a synthetic fuels industry. Congress must appropriate the funds for the research and development that is needed and will have to set policies for government support of the demonstration plants and to appropriate the government's share of the funds needed for them.

Inevitably the courts will become involved as the various groups that are affected by construction of a large conversion facility appeal adverse decisions of DOE, EPA, etc. to the courts. In recent years, the courts have become very much involved in energy decision making and there appears to be no change in that trend.

VII. ECONOMICS OF COAL CONVERSION

A. General

Estimates of the costs of converting coal to synthetic fuels vary widely. Since there is no actual experience with commercial-size plants, the estimates will continue to be "estimates" until full-size plants are constructed and operated. Even the gasifiers that were in operation in the past were constructed so many years ago and under such different conditions that little can be learned about costs from that experience.

The optimism of those seeking research and development funds from the government about the costs of the product from the improved technology that they are proposing to study is to be expected but it makes understanding actual costs very difficult. Each succeeding firm seeking government support must make estimates of costs that are less than the last firm's. Venders of equipment and architect–engineering firms who wish to build plants reinforce this bias in favor of low costs since they take no risks of their own with respect to either plant operability or costs.

Estimates also vary greatly depending on when they were made and what type of dollars are used—current or constant. In a period of rapid inflation in the heavy construction industry, which has occurred over the past 10 years, a very large difference can result among estimates from these two factors alone, if the same assumptions are not used. Moreover, the same plant constructed in different regions can differ in costs by as much as 25% or more and costs are also affected by characteristics of specific sites. Differing assumptions with respect to the amount of environmental controls needed, fuel and water costs, by-product credits, and time for construction also affect costs markedly.

Another factor influencing product costs is the method of financing—private or a utility-type rate of return. In addition, the debt–equity ratio selected and assumptions with respect to interest on the debt and return on equity account for part of the wide spread in the cost estimates that have been published. Until a number of plants are constructed we will have no accurate way to know what true product costs are, since the first few plants will be custom built with no industry infrastructure to supply much of the equipment or any experience with the design and construction of such plants.

Moreover, after a few plants are in operation, additional cost reductions should be possible as a result of the learning curve. Synthetic fuel prices could eventually be reduced to about half of the current estimated cost. This estimate is based on cost reductions experienced in other capital-intensive industries. Between 1939 and 1968 innovations in petroleum cracking processes resulted in cost reductions of 25% per volume of gasoline produced (expressed in constant dollars), equivalent to reductions in capital and operating costs of about 60% (Fisher, 1968). In another study of four major petroleum process innovations, capital costs per unit of capacity were halved over a 5-year period (Enos, 1958). A study of the machine tool industry found that unit costs were reduced by 20% for each doubling of cumulative output (Hirsch, 1952). And as a consequence of evolutionary innovations and economies of scale, the real cost of electricity declined approximately 70% in the thirty years between 1940 and 1970. Based on these experiences, the future cost of synthetic fuels, which has not yet been produced on a commercial scale, could be reduced from current estimates, possibly by as much as 50%.

A number of cost estimates for coal conversion plants were examined and an effort made to put them on as common a basis as possible. Unfortunately, frequently the information to do this is not given in the published cost estimates. As a result, the best that one can do is give a range of estimates after the data have been adjusted to the extent possible.

TABLE XIII

Estimates of the Cost of Low-Btu Gas

Gasifier	Estimator	Cost (dollars/10^6 Btu)	
		Utility Financing	Private Financing
1. Wellman–Galusha	Ashworth (1977)	1.99	2.66
2. FW/Stoic	Bardos (1977)	2.72[a]	3.33[a]
3. Wellman Incandescent	Bardos (1977)	4.19[a]	5.41[a]
4. Woodall–Duckman	Bardos (1977)	2.92[a]	3.74[a]
5. Lurgi Moving Bed	Kimmel *et al.* (1976)	4.61	5.77
6. Hygas	Kimmel *et al.* (1976)	2.99	3.64
7. CE Entrained Bed	Kimmel *et al.* (1976)	2.66	3.26

[a] The assumption is made that the original cost estimates by Bardos are in June 1977 dollars.

B. Coal Gasification Costs[6]

The cost of low-Btu gas estimated for seven different processes by three different estimators ranged from approximately $2.00 to $4.61 per million Btu when using utility financing, and from $2.66 to $5.77 per million Btu with private financing. All of these estimates are shown in Table XIII. The first four estimates are for plants with a capacity of 1.25 million SCF/h or less while for the last three estimates the plant size ranged from 31.3 million to 68.5 million SCF/h. If the two high estimates are not used the range of the other five is much narrower. For utility financing the range is $1.99–$2.99 per million Btu and for private financing $2.66–$3.74 per million Btu.

The estimates of costs for medium-Btu gas made by five different estimators for five different processes are shown in Table XIV. The plant sizes ranged from 17.8 to 26.35 SCF/h for estimates 1 and 5–9, and from 1.66 to 6.7 SCF/h for estimates 2–4. For utility financing, the costs range from $2.81 to $5.34 per million Btu and for private financing from $3.40 to $6.77. If the single high estimate is not used the costs range from $2.81 to $3.84 per million Btu with utility financing and from $3.40 to $4.91 for private financing.

[6] Based on an unpublished manuscript of Cobb, Chiang, and Klinzing of the University of Pittsburgh, January 1978.

TABLE XIV

Estimates of the Cost of Medium-Btu Gas

Gasifier	Estimator	Cost (dollars/10^6 Btu) Utility Financing	Cost (dollars/10^6 Btu) Private Financing
1. Winkler	Goodman and Bailey (1977)	3.35	3.93
2. Koppers–Totzek	Ashworth (1977)	3.31	4.59
3. Winkler–Type	Murray and Chamberlin (1977)	3.37[a]	3.99[a]
4. Koppers–Totzek–Type	Murray and Chamberlin (1977)	3.84[a]	4.91[a]
5. Koppers–Totzek	Mitzak *et al.* (1976)	3.03	3.83
6. Lurgi Moving Bed	Kimmel *et al.* (1976)	5.34	6.77
7. Hygas	Kimmel *et al.* (1976)	2.81	3.40
8. CE Entrained Bed	Kimmel *et al.* (1976)	2.97	3.50

[a] The assumption is made that the original cost estimates by Murray *et al.* are in June 1977 dollars.

TABLE XV

Estimates of the Cost of High-Btu Gas

Gasifier	Estimator	Cost (dollars/10^6 Btu) Utility Financing	Cost (dollars/10^6 Btu) Private Financing
1. Lurgi	Gallagher (1976)	4.53	5.85
2. Hygas	Detman (1976)	2.92	4.17
3. Hygas	Detman (1976)	4.41	6.27
4. CO_2 acceptor	Detman (1976)	3.64	4.93
5. Bi-Gas	Detman (1976)	3.77	5.25
6. Synthane	Detman (1976)	5.06	7.22
7. Synthane	Detman (1976)	4.42	6.35
8. Synthane	Detman (1976)	3.99	5.64
9. Lurgi	Detman (1976)	3.56	5.06
10. Hygas	U.S. Bureau of Mines (1976a–e)	2.28	2.80
11. Hygas	U.S. Bureau of Mines (1976a–e)	3.61	4.24
12. Bi-Gas	U.S. Bureau of Mines (1976a–e)	2.58	3.19
13. Bi-Gas	U.S. Bureau of Mines (1976a–e)	3.34	3.93
14. Lurgi	U.S. Bureau of Mines (1976a–e)	3.60	4.61
15. CO_2 acceptor	U.S. Bureau of Mines (1976a–e)	2.80	3.49
16. CO_2 acceptor	U.S. Bureau of Mines (1976a–e)	2.03	2.61
17. Synthane	U.S. Bureau of Mines (1976a–e)	3.47	4.33
18. Synthane	U.S. Bureau of Mines (1976a–e)	3.63	4.34

For high-Btu gas plants 18 estimates were examined. All of these estimates were for plants of 10.4 SCF/h. These were made by three different groups using five different processes. The estimates ranged from $2.03 to $5.06 per million Btu for utility financing and from $2.61 to $7.22 per million Btu for private financing. As shown in Table XV, the spread in costs is greater for these estimates than for low- and medium-Btu gas since eliminating the high and low estimates still leaves a wide range of costs.

Attempts to determine what caused the wide range in the estimates have been unsuccessful. For example, for a Lurgi plant the capital costs for the gasifier were estimated to be $66.2 million, $186 million, and $206.3 million by three different estimators. Total capital costs for a medium-Btu gas plant were $462.8, $490.8, and $797.3 million. For a high-Btu gas plant, they were $592.8 and $1060 million. Examination of certain individual plant components was not helpful in determining the reasons for this disparity. Even those responsible for the plant estimates believe that at present they are probably no more accurate than ±20–±40%.

For high-Btu gas plants the capital costs (in June 1977 dollars) to produce 250 million SCF/day of SNG will be in the range of $1.5 billion ± $0.4 billion, with an annual operating budget of $140 billion ± $40 million. Estimated average gas costs for high-Btu gas for the Lurgi process range from $3.56 to $5.34 per million Btu when using utility financing.

Attempting to get a "best" estimate of gas costs was made by making intraprocess comparisons, interprocess comparisons, and comparisons of different plant sizes. The best estimates are shown (as of June 1977) in Table XVI using the utility financing method.

TABLE XVI

Gas Costs (dollars/10^6 Btu)

	Small Plants	Large Plants
Low Btu	3.10 ± 1.10[a]	3.45 ± 1.65[b]
Medium Btu	3.10 ± 1.10[c]	3.45 ± 1.65[d]
High Btu (all processes)	—	3.85 ± 1.85[e]
High Btu (Lurgi)	—	4.60 ± 1.10[e]

[a] Plant size between 0.25 and 1.25 SCF/h.
[b] Plant size between 31.3 and 68.5 SCF/h.
[c] Plant size between 1.66 and 6.7 SCF/h.
[d] Plant size between 17.8 and 26.3 SCF/h.
[e] Plant size 10.4 SCF/h.

Although gas costs for low and medium Btu appear to be the same it is likely that when actual operating experience is gained that low-Btu gas will be lower cost than medium-Btu gas. High-Btu gas costs are estimated to be only slightly higher when all processes are considered, but it is more probable that the significantly higher cost of high-Btu gas when using a Lurgi generator will turn out to be nearer to actual costs, since most of the commercial experience has been with Lurgi gasifiers.

Several estimates of the cost of producing gas of various qualities from the LVW and LWG methods of underground gasification have been made. However, the experimental data base is too limited to make good estimates of what could be accomplished commercially. As a result, the estimates reflect the optimism of the investigators and gas costs are either the same or lower than those for conventional gasification. At this stage of development these estimates must be used very cautiously.

C. Coal Liquefaction Costs

For a number of reasons there has been a much greater effort devoted to converting coal to gas than there has been to producing liquids from coal. The long period of commercial use of coal gasification technologies and the continued existence of a number of coal gasification plants (compared to only one coal liquefaction plant in South Africa) has led to a greater confidence in the state of coal gasification technology compared to that for liquids. More importantly, however, coal gasification will lead to a product that will be marketed under regulated conditions, while coal liquefaction will not have this advantage. In a regulated market it is probable that funds can be raised using the more favorable utility financing rather than private financing. The higher priced gas produced from coal can be "rolled in" with lower priced natural gas, resulting in only a relatively small increase in the price to the consumer, and those producing the gas will probably be guaranteed a fair rate of return on their investment. As the institutional arrangements now exist, none of these benefits will accrue to a coal liquefaction plant operation.

Coal conversion costs for liquefaction are also now likely to be higher than for coal gasification because of the comparative status of the technologies.

In a report published in February 1976 (Hoertz, 1976), estimates of the cost of producing liquids from coal ranged from a low value of $13.90/barrel with coal at $15/ton and a DCF rate of return of 12% to a high value of $24.95/barrel with coal at $25/ton and a 20% DCF rate of return.

In a 1977 (National Research Council, 1977a) study of the NRC, capital and operating costs were estimated for the different types of coal liquefac-

TABLE XVII

Estimated Capital and Operating Costs of Coal Liquefaction Plants

Plant Type	Capital Cost (10^9 dollars)	Operating Costs[a] (10^6 dollars)
Pyrolysis	1.0–1.3	71
Solvent extraction	0.9–1.15	85.5
Catalytic liquefaction	0.8–1.05	92.4
Fischer–Tropsch	0.9–1.15	81.6
Methanol	0.85–1.20	84.2

[a] Not including coal or capital related costs.

tion plants, each with a capacity of 50×10^{12} barrels/day of oil equivalent (Table XVII).

Estimates of NRC of the cost of producing liquids from coal range from \$3.35 to \$5.00 per million Btu. The lower value is for coal at \$10/ton and the higher value for coal at \$35/ton. However, the report concludes with the following[7]:

"At the present time no process for coal liquefaction is sufficiently well developed to permit the detailed engineering required for the 50,000 or 100,000 barrels per day plants necessary for a large scale industry and it is improbable that the necessary data will be available before 1980. At this early stage of development most estimates of costs for coal liquefaction are optimistic and understate actual costs by significant amounts. Therefore, unless there is an important technologic breakthrough, costs for liquids from coal will be higher than high-Btu gas from coal for the foreseeable future."

VIII. STATUS OF COMMERCIALIZATION

A. Synthetic Fuels Commercial Demonstration Program

With the present state of technology, costs of producing either liquids or gas from coal are such that they cannot compete with natural gas or imported oil prices. This combined with the very large investments required for a single plant and the technologic risks associated with "first of a kind" plants prompted the Administration to propose a Synthetic Fuels Commercial Demonstration Program that would receive government sup-

[7] National Research Council (1977a, p. 145).

port for the pioneering plants (Energy Research and Development Administration, 1976a). The program included (1) plants to convert oil shale to synthetic petroleum, (2) coal conversion to gas and clean boiler fuels, and (3) waste conversion to gas, oil, or other fuels.

The program was designed to obtain information on the many facets involved in creating a new synthetic fuel industry—financial, environmental, economic, institutional, technical—and to promote experience in the private sector in synthetic fuels production. The program called for construction of a number of different types of plants for producing synthetics with a total capacity of 350,000 barrels of oil equivalent per day by the early 1980s.

The justification for this program was the need to provide millions of barrels of oil equivalent per day to the economy by the mid 1990s if the United States was not to be too heavily reliant on oil imports. With declining supplies of domestic oil and natural gas, imports would increase appreciably unless synthetic fuels are available. To obtain the amount of synthetic fuels needed by 1995 requires that a number of major problems be resolved if the private sector is to make the needed investments to produce these fuels. Since there remains a great uncertainty about future world oil prices the Federal government will probably become involved in reducing the private sector risks.

Without Federal assistance, industry cannot be expected to undertake synthetic fuel plant construction because of uncertainties in costs of synthetics, the possible volatility of world oil prices, and the effect a drop in oil prices could have on capital intensive synthetic fuel industries, the large capital investments required for a single plant compared to the total assets of most of the companies involved, and the risk of project delays due to obtaining the numerous permit approvals that are required.

The commercialization program was designed to provide financial incentives to the private sector, to expedite federal regulatory clearances, to give local communities assistance when required to meet socioeconomic infrastructure and planning problems, and to meet environmental protection requirements.

The Federal assistance proposed under the commercialization program included a variety of incentives tailored to the particular conversion process that was being encouraged. It included nonrecourse loans of up to 75% of the project cost for high-Btu gas, and construction grants of up to 50% of the estimated project cost for substitute fuels for use by utilities and industry. For other synthetic fuel programs (oil shale and biomass) loan guarantees, purchase agreements, price guarantees, and tax changes were proposed to be used as incentives.

The criteria used to select the incentives were (1) expected cost to the government, (2) effectiveness of obtaining the target production, (3) the degree of industry participation expected, (4) complexity of administering the incentives, and (5) the existence of the authority necessary to undertake the program so that new legislation would not be required.

Competitive bidding would be used to determine which firm would receive the Federal incentives and the size of the incentive to be offered. Offers would be solicited from interested and qualified firms and their bids evaluated on the basis of the expected net value to the government. The government, under the program proposed, would mainly guarantee against plant failure and market price uncertainty.

Under this program approximately nine plants would be constructed with a total capital investment of $8.3 billion (1975 dollars) of which the loan guarantee in the construction phase would be $6.0 billion. In addition, there would be $600 million in construction grants. In the operation phase the price guaranty provisions would lead to a financial exposure to the government of $4.5 billion.

In order to speed up the commercialization program the Senate Committee in the fall of 1975 gave ERDA the authority to provide loan guarantees to industry and provided $6 billion in authority for the loan guarantee program in the course of its authorization hearings. When this new authority and request for appropriation was presented to the House of Representatives there was concern that no hearings had been held by that body on an important new and expensive program. As a result, authorization for the program was voted down by the House. A second attempt to revive the program in 1976 also failed in the House by one vote.

The impetus to get a synthetics program underway seems to have lost its momentum and there does not appear to be any possibility that the program will be revived unless the energy supply situation changes radically, or there are major changes in attitude in the Executive and Congress about the need for the program.

One other much larger program for establishing a synthetic fuel industry was proposed involving a $100 billion Energy Independence Authority, but it never reached the stage of a formal proposal similar to that of the Commercial Demonstration Program.

B. Coal Gasification Programs

Because of the favorable institutional factors for making high-Btu gas from coal, several commercial plants have been proposed. Table XVIII is a list of high-Btu gas projects (commercial and demonstration) that have been announced, giving their location, type of process to be used, size,

TABLE XVIII. Status of Commercial and Demonstration High-Btu Coal Gasification Projects[a] (as of June 1977)

Controlling Company[b]	Site	Process	Coal Feed (tons/day)	Plant Output (10^6/ft^3/day)	Status[c]
Commercial projects					
WESCO; Texas Eastern Transmission Corp. and Pacific Lighting Corp.	Four Corners Area, New Mexico	Lurgi gasification with methanation	25,625 / 26,625	250 / Expansion 250	The plan is for a plant on the Navajo Indian Reservation near Farmington, N.M. Negotiations for site lease have not yet been completed. Utah International Corp. will supply the coal and water for the plant. Water and coal are also available for an additional plant. Estimated project cost for the first plant is $1.3 billion (1977 dollars). The start-up gas price authorized by the FPC is set at $2.50/10^6 ft^3 with a minimum-bill provision to set rates, after operation to cover basic costs. The environmental impact statement was completed in early 1976.
American Natural Resources Co., Peoples Gas Co. (North American Coal Gasification Corp.)	Beulah-Hazen Area, North Dakota	Lurgi gasification with methanation	—	275	The plant will be built in two phases. The first phase will be construction of a gasification train capable of producing 137 × 10^{12} ft^3/day. Construction is to begin in 1978, with completion scheduled for 1982. Under the second phase, a second train, also with a capacity of 137 × 10^{12} ft^3/day, would be built after 1982. The expected total output of 250 × 10^{12} ft^3/day is based on an anticipated operational factor of 91%. Cost of the initial phase is estimated at $532 million (1976 dollars). Total cost for both phases is estimated at over $1 billion. The Bureau of Reclamation has completed its draft environmental impact statement.
El Paso Natural Gas Co.	Four Corners Area, New Mexico	Lurgi gasification with methanation	28,250	288	El Paso Natural Gas Co. plans to construct and operate the Burnham Coal Gasification Complex on the Navajo Indian Reservation. Capital costs (1975 basis) will be $225 million for the mining operation and $1 billion for the gasification plant. The average unit cost of the synthetic gas over a 25-year plant life is estimated at $2.90/1000 CF. El Paso has requested deferral of the Federal Power Commission (FPC) decision.
Panhandle Eastern Pipe Line Co. (Peabody Coal Co.)	Eastern Wyoming	Lurgi gasification with methanation	27,700	270	Plant operation is anticipated in the early 1980s. Investment costs are estimated at $1.3 billion (early 1976 dollars). No filing has yet been made to the Federal Power Commission.
Natural Gas Pipeline Co. of America	Dunn County, North Dakota	Lurgi gasification with methanation	30,000	250	Plans call for construction of one (and possibly four or more) plants. The coal supply would be mined from deposits within a 110,000-acre area. Capital investments for the initial plant will be approximately $954 million (1976 dollars), and for associated facilities, including the mine, approximately $46 million (1976 dollars). The first plant is scheduled to go on-line in 1985.
Northern Natural Gas Co.	Powder River Basin, Montana	Lurgi gasification with methanation	30,000	250 Expansion 250	Project suspended.
The Columbia Gas System, Inc.	Illinois	—	—	300	Columbia Gas has agreed to exchange a 50% interest in 43,400 acres of its 300,000 acres of West Virginia coal lands for a 50% interest in 35,000 acres of Illinois coal lands held by Exxon's Carter Oil Co. The Illinois coal will be held by Columbia for coal gasification pending development of an economically and technically sound coal gasification process. Preliminary capital cost estimates are in excess of $1 billion for the plant, and the related investments in mines would approximate $100 million.
Panhandle Eastern Pipeline Co. and City of Wichita	Wichita, Kansas	—	—	250	A feasibility study is underway of a plant to be financed through the sale of municipal bonds.

Panhandle Eastern Pipe Line Co. (Peabody Coal Co.)	Southern Illinois	Lurgi gasification with methanation	—	—	A feasibility study of a plant has been completed. No plans have been announced.
Cameron Engineers	Colorado	Lurgi gasification with methanation	34,250 plus 6000 tons of solid waste per day	250	The company has filed plans with the U.S. Bureau of Land Management which is reviewing tentative coal leases. Plant operations could begin in the early 1980s.
Mono Power Co., Resources Co., and New Albion Resources Co.	Kaiparowits Plateau, Utah	Lurgi gasification with methanation	32,900	360	The three companies have completed a conceptual study for a coal gasification project which will use the coal and water rights of the abandoned 3000-MW Kaiparowits power plant. Three plants of 120×10^6 ft^3/day are envisioned with the first plan operational by 1985. Annual water use would be about 30,000 acre-ft. Total plant, mine and pipeline costs would be approximately $2 billion (1976 dollars) with gas cost about $($10^6$ ft^3). 3600 permanent employees would be necessary and 4400 indirect and supportive jobs would be created.
Exxon Corp. (Carter Oil)	Northern Wyoming	—	—	—	Project deferred.
El Paso Natural Gas Co.	Southwestern North Dakota	—	—	—	Reserves of 2 billion tons of coal are under lease.
Consolidated Natural Gas Co.	Southwest Pennsylvania	—	—	—	The company has purchased about 500 million tons of recoverable coal for gasification.
Pennsylvania Gas and Water Co.	Pennsylvania	HYGAS[a] or similar one	5000	80	The company had proposed to ERDA a plan for financing and operating a demonstration plant. The proposal was not accepted and the project is currently suspended.
Demonstration projects					
Coalcon Dept., Union Carbide Corp.	New Athens, Illinois	Union Carbide	2600	22 (plus 2900×10^9 barrels of syncrude per day)	The process design stage was completed in June 1977. Discussions are continuing for a pilot reactor facility prior to consturction, a demonstration plant. The pilot project would scale-up the existing test reactor to answer uncertainties concerning the use of caking-coal without pre-treatment. A 2000 acre site has already been selected for a demonstration project with plant costs estimated at above $200 million. The Illinois Energy Resource Council has agreed to provide up to $25 million for the project. ERDA is expected to provide about $130 million.
Conoco Coal Development Co. & ERDA	Noble County, Ohio	Slagging Lurgi with methanation	3800	60	ERDA has awarded the company a $24 million contract to design the demonstration facility. The design phase will last about 22 months followed by a construction period of 2½ years. Plant operations are expected to last for 3½ years. Total project cost is estimated at $324 million. The cost for construction and operations will be shared equally by ERDA and the company.
Illinois Coal Gasification Group & ERDA	Perry County, Illinois	CO GAS Process	2200	18 plus 2400×10^9 barrels/day of synthetic crude oil	ERDA has awarded this group of companies a $22.5 million contract to design the demonstration plants. The design phase will last about 21 months. Construction and operation phases are expected to last 34 and 42 months, respectively with costs shared equally between ERDA and the group. Total costs are estimated to be $334 million.

[a] Source: American Gas Association *et al.* (1977). [b] Mining partners in parentheses. [c] Reported costs may not reflect recent increases caused by inflation.

and status. None of these plants has started construction and only three[8] have applied to the Federal Power Commission (now FECR) for permits to construct the plants.

All three of the proposed plants would use Lurgi gasifiers followed by a methanation of the cleaned raw gas to increase its heating value. Only the WESCO plant has all the approvals that it requires from the FPC but it has said that federal assistance is essential in order to obtain the funds to construct the facilities. (The California State Public Utility Commission, where most of the WESCO gas would be sold, has supported the WESCO position concerning the need for Federal assistance.)

Several financial institutions also indicated that government assistance and changes in regulatory requirements would be needed for the regulated natural gas companies to accomplish the financing. Even with the latest, more favorable November 1975 ruling of the FPC, WESCO continues to claim that the project cannot be financed without government loan guarantees.

El Paso has asked FPC to defer any further action on its application of March 1975 and this request has not yet been withdrawn. The American Natural Resources proposal is still before the FPC. Hearings were expected to resume in late May of 1976 on new data which the company has supplied and on a new proposal to construct the first plant at only one-half the size that had been originally proposed by American Natural Resources. However, none of these proposed commercial plants are now under construction (June 1978).

Part of the difficulties faced by these projects have been the rapid escalation in capital costs noted above between the original application and the more recent estimates that have been made. This is shown in Table XIX In four and one-half years El Paso's estimates of plant costs have increased fourfold, WESCO's estimates doubled in two years, and American Natural Resources' increased 70% in one year.

In order to speed up the commercialization of coal gasification technology, ERDA, now the Department of Energy (DOE), has announced a series of Industrial Fuel Gas Demonstration Projects for producing medium- and low-Btu gas for industrial uses and three demonstration projects for high-Btu gas.

One of the low- and medium-Btu gas projects is to be constructed at Shelby, Tennessee and will use the U gas process developed by the Institute of Gas Technology. It will produce a gas with a heating value of slightly less than 300 Btu/SCF using a fluidized-bed oxygen-steam gasifi-

[8] WESCO, American Natural Resources Company, El Paso Natural Gas Company (see Table X).

TABLE XIX

High-Btu Coal Gasification Projects—Capital Cost Estimates (as of March 1977)[a]

Project Sponsor	Date When Project Completion and Cost Was Estimated (1)	Estimated Project Completion Date (2)	Estimated Project Cost (millions of dollars) (3)
El Paso Natural Gas Co.	August 1971	1976	250
	November 1972	1976	353
	October 1973	1978	491
	December 1975	*b*	1000[e]
	January 1977	*b*	1000[e]
WESCO	February 1973	1979[c]	406
	June 1974	1979	447
	January 1975	*d*	853
	January 1977	*d*	1000
American Natural Resources Co.	March 1974	1980	450
	April 1975	1980	778[f]
	January 1977	1980	800[g]

[a] Source: 1972–1975: General Accounting Office: (1976); 1977: Wahrhaftig (1977).

[b] El Paso no longer projects a specific completion date other than it would occur three to three and a half years after the date when all necessary approvals were obtained and financial arrangements completed.

[c] An earlier estimate projected that the plant could become operational in 1977.

[d] Construction was scheduled to begin in early 1976, but has not. WESCO currently claims that its project cannot proceed without government incentives.

[e] Approximately.

[f] Does not include capitalized interest during construction, which is expected to be provided on a current basis through a surcharge on gas customers. This is subject to FPC approval.

[g] Plant size reduced from 275 million cubic feet per day to 125 million cubic feet per day.

cation process at 90 psi pressure. It will produce 50 billion Btu/day of industrial gas using 2600 tons/day of Western Kentucky No. 9 coal with a sulfur content of 4.1%. Total project cost will be $180 million, with the government providing $92 million. Thermal efficiency of the process is expected to be 72%. Construction is expected to be completed by the end of fiscal year (FY) 1981 and commercial operation should start at the end of FY 1983.

A second industrial gasification process supported by DOE is one which would make a synthesis gas from coal that could be used to produce ammonia. The plant would be constructed in Western Kentucky and will use a Texaco gasifier[9] that operates at 2500 psi. The coal is ground and slurried with water and fed with oxygen into the gasifier. The gas produced has a heating value of approximately 250 Btu/SCF and the process has a thermal efficiency of 73%. Total cost of the plant (not including the ammonia synthesis and recovery unit) is $318 million, with the government share being $164 million. The coal feed rate will be 1700 ton/day of a Western Kentucky No. 9 coal and the plant will produce 1200 ton/day of ammonia.

A much smaller gasification plant making industrial gas is being supported by DOE at Hoyt Lake, Minnesota and will be used to dry and harden taconite pellets. The process is a two-stage fixed bed gasifier and five gasifiers will be used to produce 7.4 billion Btu/day of gas with a heating value of 160–180 Btu/SCF. The coal feed will be either a high-sulfur (2.5%) Eastern coal or a low-sulfur (0.2%) Montana coal. The process to be used will be selected during the conceptual design phase and thermal efficiencies are expected to be about 71%. Total cost of the project is $47 million, with the government's share being $27.8 million. Commercial production of gas is expected by the end of FY 1983.

The DOE program for high-Btu gas consists of contracts for the conceptual designs of two different processes, one of which will be selected later for detailed design and construction. In addition, Congress appropriated funds for conceptual design of a third high-Btu gas process which may be supported if the 50/50 cost sharing by government and industry can be arranged. At present, authorization has been given only for process design for this plant.

Conoco has proposed the construction of a slagging Lurgi gasifier to verify the economic and technical acceptability of the slagging gasifiers tested at the Gas Research Laboratories and at the Westfield gasification plant in Scotland. If selected for construction that plant would be located in Ohio and would produce 59 million SCF/day of high-Btu gas. The gasifier has been designed to operate at higher temperatures than the conventional Lurgi gasifier and to be able to handle coking bituminous coals. The thermal efficiency of the plant is expected to be 67% and total coal feed will be 3800 tons/day of a high-sulfur bituminous coal. Total project costs are estimated to be $371.6 million, of which the government's share is $198.7 million. If selected as the candidate plant for construction, construction should be completed in early FY 1982 followed by a shakedown and operation phase which could last to FY 1985.

The second high-Btu gas plant selected for a conceptual design study is

[9] This gasifier was first tested with coal in a pilot plant in the mid 1940s and commercial installations using oil have been constructed. There are no commercial coal gasification plants using this process in operation.

the Cogas process which was proposed for demonstration by the Illinois Coal Gasification Group (ICGC). In this process the coal is pyrolysed in four-stage fluidized bed units to produce liquids (COED process) and the remaining char is then gasified in an air blown two-component gasifier. If selected for construction the plant will be built in Illinois and will produce a gas with a heating value of 950 Btu/SCF. Coal feed will be 2200 tons/day and this will produce 17.7 million SCF/day of pipeline gas and 2400 barrels/day of synthetic crude oil. Total cost of the project is $293 million, of which the government will fund $158 million. Construction is expected to be completed at the end of FY 1982 and shakedown and operation by the end of FY 1986.

The third high-Btu gas plant which is now under conceptual design would use the Hygas process developed by the Institute of Gas Technology. Cost sharing arrangements have not been worked out, but funds have been appropriated for a conceptual design. Gas with a heating value over 900 Btu/SCF will be produced with a thermal efficiency of approximately 70%. The first stage of development would consist of a demonstration plant producing 80 million SCF/day, using a single large reactor. Once the process had been demonstrated the plant would be expanded to a commercial size, producing 250 SCF/day from 20,500 tons/day of coal. If the process is ultimately funded, construction would be completed by mid-February 1983 and shakedown and operation be completed by the end of FY 1984.

C. Coal Liquefaction Programs

Coal liquefaction demonstration programs are not as far advanced as those for coal hydrogenation. At present only the Lurgi Ruhrgas low-pressure pyrolysis process, Fischer–Tropsch and methanol processes have been operated on a commercial scale. The H-coal process is to be installed at a site in Kentucky and will have a 600 tons/day coal capacity, small compared to the coal gasification demonstration plants under study. The comparative lack of activity in coal liquefaction compared to gasification is the result of the factors described in Section VII,C. In the long run, however, coal liquefaction could be of greater importance than coal gasification because it appears that liquids will be needed for the transportation sector for long periods, and for many uses oil and gas are substitutable.

One project to produce clean liquid fuels and high-Btu synthetic gas was proposed to ERDA in January 1975 by the Chemical Construction Company and Union Carbide, who formed a joint company called Coalcon. The proposal was to construct a plant which would convert 2600 tons/day of high sulfur coal to 3900 barrels of clean liquids and 22 million SCF/day of pipeline quality gas. However, after preliminary conceptual studies the project was dropped. Consideration of this proposal over two years also tended to slow down other liquid fuel demonstration projects.

IX. THE POSSIBLE GROWTH RATE OF COMMERCIAL COAL CONVERSION PLANTS

It is clear from the previous discussions that under normal development conditions there are a large number of institutional, legal, regulatory, financial, and social constraints on the growth rate of a coal conversion industry. However, many of these could be modified or even eliminated if the acceleration of such developments appeared to be in the national interest.

If one assumes that a "business as usual" philosophy will prevail, the number of large coal conversion plants that will be constructed by 1990 will be very small. This conclusion is reached because of the economic and technical status of coal conversion technologies and product costs using existing processes and the long lead times necessary to bring large, high capital intensive plants into operation.

In early 1978 no coal conversion plants could be constructed that could compete with oil and gas at current prices. Because the difference between the prices of synthetics produced using current technology and existing prices for oil and gas are so great and the economics of the more advanced technology processes under study appear to be only marginally more attractive, this situation is not likely to change for five to ten years. Under these circumstances the contribution to energy supply by the year 2000, could be less than 1 Q (approximately 10 plants each with a capacity of 250 million SCF/day) without government intervention. Even this share could be smaller if a guaranteed rate of return on investment and "rolled in" pricing for natural gas substitutes from coal are not allowed.

If, however, a commercially attractive technology now existed for coal conversion to liquids or high-Btu gas, several large plants could be designed, approved, and constructed by about 1985. Even if there were no unexpected technical, economic, or other barriers uncovered during this period, a large number of first-generation plants using the technology would not be operational before the 1990–1995 time period and it is unlikely that more than four to eight plants, at a maximum, using four different technologies (two coal to gas, two coal to oil) would be constructed. By 2000 under the most optimistic business as usual assumptions no more than 2 or 3 Q of energy (approximately 20–30 plants producing

250 million SCF of high-Btu gas) would be available from coal synthetics and possibly much less.

Another method of estimating what could probably be accomplished under business as usual conditions is to review the experience with the development of nuclear energy—a new fuel form that emerged in the early 1960s. The first commercial plant for generating electricity from nuclear fuel started operation in 1963. In the 15 years between the beginning of 1963 and the end of 1977 about 50,000 MW of nuclear generating capacity was installed at a cost of about $25 billion, expressed in 1972 dollars. No special priorities were given to nuclear plant construction so one could expect, given a commercially viable coal conversion technology, that the heavy construction industy and the architect engineers would be able to expand their capacity by at least $15 to $20 billion over 15 years (some heavy construction at electric generating plants using fossil fuels would have occurred in the absence of nuclear generating facilities). Using this growth rate as a measure of what could be attained, if the first plants were started now, by the end of 1992, approximately 10–15 large coal conversion plants could be in operation with a capacity of 1.0–1.5 Q/year.

On the other hand, if for policy or security reasons it was decided that the Federal government would support efforts to maximize synthetic fuels production from coal, the amount of synthetic fuels from coal could be much greater. To estimate the maximum amount that could be produced, it would have to be assumed that the only constraints on growth rate in conversion capacity would be the result of physical limitations—not regulatory, institutional, financial, or other factors. These physical constraints would include engineering manpower for plant design and operating construction and operating skills for the plants, the ability to increase coal production and to transport it to where it was required, and constraints on the industrial infrastructure to manufacture and erect the high-pressure high-temperature vessels that are needed, as well as the special instruments and other equipment used for plants that have not previously been manufactured in the United States.

Several attempts have been made to estimate the maximum productive capacity of coal conversion facilities that could be in operation if a decision was made to maximize production as rapidly as possible. One estimate (Senate Committee on Interior and Insular Affairs, 1973) was based on the constraints that would result in an all-out effort for commercializing coal conversion plants based on the size of the architect–engineer (AE) industry and the expansion rate in annual heavy industry construction expenditures that appeared to be possible.

In this study, the main bottleneck is how rapidly the industrial infrastructure could be mobilized to construct the plants. A number of knowledgeable people in the coal conversion field and AE firms estimated that about 36 plants (without regard to the economics of the process) could be constructed in about 13 years and these could produce about 3 Q of energy from coal in liquid or gaseous form by 1990.

If plants were standardized (venders could then gear up to mass produce the equipment needed) and the plants were considered prototypes for a large new industry, then the AE firms interviewed believed that up to as many as 70 plants could be operational in 13 years. This assumes that there would be no shortage of engineers with the required skills, a highly optimistic assumption since a large number of contractors would be competing for very scarce specialized engineering talent. Moreover, estimates of construction times for these plants (at the time of the study) were $4\frac{1}{2}$ years and it is more likely that, with the more complex installations now envisaged, that construction time will be closer to 6 years.

There are also some serious reservations about whether the specialized equipment needed for coal conversion could be provided at the rate necessary for even 36 plants to be constructed in 13 years. Shop fabrication facilities to construct large pressure vessels to operate at high temperatures are limited and the coal conversion plants would have to compete with the nuclear industry for these kinds of vessels. Moreover, because of the historical cyclic nature of the heavy construction industry there is generally a reluctance by construction firms to invest new capital to expand production facilities unless a long-term market is visible.

Other types of equipment constraints could also develop. A single unit sized gasification plant (2.5×10^{11} Btu/day) would require an oxygen plant capacity nearly equal to the entire United States capacity to build oxygen plants. Another equipment limitation, at least initially, would be the ability to obtain the multiple number of large compressors required for conversion plants. There are only a limited number of manufacturers of this type of equipment and, even if they decided to expand, there is not the foundry or forging capacity to supply their needs without this secondary type of facility being expanded. Given assurances that markets would exist long enough to amortize investments, these equipment limitations would disappear, but the rates at which the first conversion plants could be constructed would still be constrained.

Experience during World War II seems to indicate that estimates of how rapidly industrial expansion can occur, given a compelling national need, are generally too low (Krug, 1945). Production of planes increased from 25,228 in the 18 months from July 1940 to December 1941 to 96,400 per year in 1944; tanks from 4300 during that 18-month period to 29,500 in 1943; and synthetic rubber production from 1500 tons in the first quarter of 1941 to 227,865 tons in the first quarter of 1945. Thus, given the proper incentives, 3–5 Q/year of coal conversion capacity could be easily installed by 1990.

Without incentives and the elimination or reduction in the multitude of constraints discussed elsewhere it is very likely that there will be no commercial coal liquefaction production by 1990 or possibly even by 2000. Up to three or four coal to substitute natural gas plants may be in operation by 1990 (if "rolled in" prices and a guaranteed rate of return on the plant investment are permitted) and at least 20–30 could be constructed by the year 2000.

In another more recent study (National Research Council, 1977a) estimates were made of the rate at which 50,000 barrels per day coal liquefaction plants could be constructed. It was assumed that it would require four years to build each plant and that technology, economics, and engineering data required to construct such plants would permit the first plant to be started in 1980. Under these circumstances 3.7 million barrels of oil per operating day could be produced by 1994 or about 8 Q/yr. In the opinion of the NAS ad hoc panel, "this represents approximately the maximum rate of plant construction that is believed possible, even with considerable expansion of activity in the many industrial areas that would be involved."

A more likely set of assumptions is that construction on the first plant will be started in 1985 and at least the early plants will take 6 years to build so that only 1.7 million barrels per operating day would be produced, or 3.7 Q/yr. Since liquefaction plants would be competing for the same engineering skills, similar types of construction workers and plant equipment as coal gasification plants, if this number of liquefaction plants were constructed few if any gasification plants would be able to be built.

In another study (Dickson et al., 1976) an attempt was made to estimate the maximum rate at which a synthetic fuels industry could be deployed. The basic assumption in estimating the amount of synthetics that would

TABLE XX

Maximum Credible Synthetic Fuels Implementation Scenario (10^6 barrels/day)

	1980	1985	1990	1995	2000
Oil from shale	0.1	0.5	1.5	2.0	2.0
Methanol from coal[a]	0.05	0.3	1.0	2.5	4.0
Syncrude from coal	0.0	0.09	0.5	1.5	4.0
	0.15	0.89	3.0	6.0	10.0

[a] Oil equivalent energy.

be produced was that synthetic fuels could be produced at a profit and sold at prices competitive with imported oil. The estimates made in this study are shown in Table XX.

This estimate obviously overstates what could be physically accomplished by 1980, neglecting for the moment the question of economics. Shifting the estimates by 5 years gives a more realistic estimate of what might be done, *if* one of the conversion technologies were cost competitive with imported petroleum. Under these circumstances, by 1995 the maximum credible amount of synthetics that could be produced would be 3 million barrels/day, or 6.0 Q/yr. If, as is more likely, plants without government assistance or incentives do not become competitive before 1990 then only 0.89 million barrels/day, or about 2 Q, would be produced in 1995.

A study to investigate options for Phase II of a National Energy Policy (Hudson Institute, 1977b) attempted to estimate synthetic fuel production that would be in operation by 1995. On a business as usual assumption a large industry would not be possible until the twenty-first century. To accelerate the development of this industry, the United States would have to provide incentives to private industry or create government corporations to construct plants. A combination of price guarantees and other incentives and creation of a government corporation may be needed to obtain any large amount of productive capacity at an earlier date.

The estimates of the number of plants possible under "forced draft" conditions assume that construction of "first-generation" plants would start in 1978 while research and development on more advanced technology was being accelerated. Construction of "second-generation" plants are then assumed to start in either 1983 or 1984. On this basis it is estimated that 50 plant starts could be made by 1987, with a total capacity of 2.5 million barrels/day, and that 200 plant starts could be made by 1993, with a capacity of 10 million barrels/day. Actual production, even assuming 4-year construction time for each plant, would be 0.8 million barrels/day (1.6 Q/yr) in 1987 and 4.5 million barrels/day (8 Q/yr) by 1994.

This estimated amount of synthetic fuels production that could be in operation must be viewed as being very optimistic since it is highly unprobable that the first plant will be started as early as 1978, the initial plants can be constructed in as little as four years and "second generation" technology, even with government incentives, will be ready to be constructed as early as 1983 or 1984.

The report concludes that, "This study does not attempt to define specific bottlenecks such as steel plates, pumps or boilers, and the various possibilities for preproducing them, or specific institutional and regulatory difficulties. However, prior involvement in the study of mobilization bases for the Department of Defense has led us to conclude that very often, with relatively modest expenditures, great leverage can be obtained and impressive improvements made possible in the basic capability to procure equipment or to construct facilities rapidly.''

To summarize these various estimates: If coal conversion plants must produce a product that is competitive with petroleum and natural gas at or near current prices and no incentives or other forms of assistance are given to industry, there could be a few high-Btu plants in operation by 1990 and as many as 20 or 30 by 2000. For even this to occur, unless there is an unexpected technologic breakthrough, the natural gas plants would have to be given permission to ''roll in'' gas prices and to obtain a utility rate of return on the plant investment.

By 2000 there may be several coal liquefaction plants in operation but how many will depend on new technological developments in coal conversion as well as world oil prices.

If a national decision is made to offer incentives or other forms of assistance to stimulate plant development estimates of the installed capacity are for about 3 Q/year of coal synthetics by 1995 and about 6 Q/year by 2000 if the decision to go forward were made within the next two or three years. If, in addition, an all-out effort to construct plants were made in the next few years under war-time forced draft conditions 3–5 Q/yr could be produced by as early as 1990 and as many as 10 Q or more per year by 2000.

APPENDIX

Bibliography for Section II on Coal Conversion Technology

American Chemical Society (1976). *Symp. Commer. Synth. Fuels,* Colorado Springs, Colo.

Arthur D. Little, Inc. (1971). ''A Current Appraisal of Underground Coal Gasification,'' Rep. to U.S. Bur. Mines, C-73671. Cambridge, Massachusetts.

Atomic Energy Commission, Division of Applied Technology (1974). ''Synthetic Fuels From Coal,'' Rep. to AEC Senior Manage. Comm. Washington, D.C.

Energy Research and Development Administration (1975). ''Trails of American Coals in a Lurgi Gasifier at Westfield Scotland.'' Natl. Tech. Inf. Serv., Springfield, Virginia.

Energy Research and Development Administration (1976). ''Fossil Fuel Program Report 1975–1976, Vol. 2, Coal Gasification.'' Natl. Tech. Inf. Serv., Springfield, Virginia.

Energy Research and Development Administration (1976). ''Proposed Synthetic Fuels Commercial Demonstration Program: Fact Book.'' U.S. Gov. Print. Off., Washington, D.C.

Energy Research and Development Administration (1976). ''Synthetic Liquid Fuels Development: Assessment of Critical Factors, Vol. 1, Summary.'' U.S. Gov. Print. Off., Washington, D.C.

Federal Energy Administration (1974). "Report to Project Independence Blueprint," Interagency Task Force Synth. Fuels Coal. U.S. Gov. Print. Off., Washington, D.C.
General Accounting Office (1976). "Status and Obstacles to Commercialization of Coal Liquefaction and Gasification," Nat. Fuels Energy Policy Study, Senate Comm. Inter. Insular Aff. U.S. Gov. Print. Off., Washington, D.C.
Hudson Institute, Inc. (1977). "Issues Relative to the Development and Commercialization of a Coal-Derived Synthetic Liquids Industry." Natl. Tech. Inf. Serv., Springfield, Virginia.
Long, H. H., ed. (1963). "Chemistry of Coal Utilization." Wiley, New York.
Massey, L. G., ed. (1974). "Coal Gasification." Am. Chem. Soc., Washington, D.C.
National Research Council (1977). "Assessment of Low and Intermediate Btu Gasification of Coal." Natl. Acad. Sci., Washington, D.C.
National Research Council (1977). "Assessment of Technology for Liquefaction of Coal." Natl. Acad. Sci., Washington, D.C.
Tennessee Valley Authority (1975). "Evaluation of Fixed Bed Low Btu Coal Gasification Systems for Retrofitting Power Plants," Electr. Power Res. Inst., Palo Alto, California.

REFERENCES

American Chemical Society (1976). *Symp. Commer. Synth. Fuels,* Colorado Springs, Colo.
American Gas Association *et al.* (1975). *High Pressure Gasif. Slagging Conditions, Annu. Synth. Pipeline Gas Symp., 7th,* Chicago, Ill.
American Gas Association *et al.* (1976). "Reserves of Crude Oil, Natural Gas Liquids, and Natural Gas in the United States and Canada as of December 31, 1976." Am. Gas Assoc., Am. Pet. Inst., and Can. Pet. Assoc., New York.
American Gas Association *et al.* (1977). *Gas Supply Rev.* **3**(10/11), 7–9.
Ashworth, R. A. (1977). Coal gasification economics—on site vs. central plant and retrofitting to low and intermediate Btu gas. *Proc. Annu. Conf. Coal Conversion, 4th, Univ. Pittsburgh.*
Auldridge, L. (1976). *Oil Gas J.* **74**(52), 105.
Bardos, R. (1977). Gasifiers in industry. *Proc. Annu. Conf. Coal Conversion, 4th, Univ. Pittsburgh.*
Central Intelligence Agency (1977). "The International Energy Situation: Outlook to 1985." Lib. Congr., Washington, D.C.
Detman, R. (1976). "Factored Estimates for Western Coal Commercial Concepts," FE-2240-5, Contract E (49-18).
Dickson, E. M., *et al.* (1976). "Synthetic Liquid Fuels Development: Assessment of Critical Factors, Vol. 1, Summary." U.S. Gov. Print. Off., Washington, D.C.
Elliott, M. A., and Von Fredersdorff, C. G. (1963). Chemistry of coal utilization. *In* "Coal Gasification" (H. H. Lowry, ed.), pp. 829–1022. Wiley, New York.
Energy Research and Development Administration (1975). "Trails of American Coals in a Coal Gasifier at Westfield Scotland." Natl. Tech. Inf. Serv., Springfield, Virginia.
Energy Research and Development Administration (1976a). "Proposed Synthetic Fuels Commercial Demonstration Program: Fact Book." Energy Res. Dev. Adm., Washington, D.C.
Energy Research and Development Administration (1976b). "A National Plan for Energy

Research and Development, and Demonstration: Creating Energy Choices for the Future," Vol. 1, pp. 107, 110. U.S. Gov. Print. Off., Washington, D.C.

Enos, J. L. (1958). *J. Ind. Econ.* **6.**

Fisher, H. W. (1968). Innovations in a large company. *In* "The Process of Technological Innovation." Natl. Acad. Eng., Washington, D.C.

Gallagher, J. P. (1976). Political and economic justification for immediate realization of a synthetic fuels industry. *Proc. Annu. Conf. Coal Conversion, 3rd, Univ. Pittsburgh.*

Gardner, F. J. (1973). *Oil Gas J.* **71**(53), 86.

General Accounting Office (1976). "Status and Obstacles to Commercialization of Coal Liquefaction and Gasification," Table 5, p. 27. Natl. Fuels Energy Policy Study, Senate Comm. Inter. Insular Aff., U.S. Gov. Print. Off., Washington, D.C.

Goodman, M., and Bailey, E. *et al.* (1977). Synthetic medium Btu gas via Winkler process. *Proc. Annu. Conf. Coal Conversion, 4th, Univ. Pittsburgh.*

Hirsch, W. Z. (1952). *Rev. Econ. Stat.* **34,** 143.

Hoertz, C. D. (1976). Coal liquefication. *Symp. Commer. Synth. Fuels, Am. Chem. Soc., Colorado Springs, Colo.*

Hubbert, M. K. (1974). "U.S. Energy Resources, A Review as of 1972," Serial No. 93–40. Senate Comm. Inter. Insular Aff. U.S. Gov. Print. Off., Washington, D.C.

Hudson Institute, Inc. (1977a). "Issues Relative to the Development and Commercialization of a Coal-Derived Synthetics Liquids Industry: Manpower Requirements," Vol. III-5. Natl. Tech. Inf. Serv., Springfield, Virginia.

Hudson Institute, Inc. (1977b). "Suggestions for a Phase II National Energy Policy," FE-2660-1, Croton-on-Hudson, New York.

Hudson Institute, Inc. (1977c). "Issues Relative to the Development and Commercialization of a Coal-Derived Synthetic Liquids Industry: Management Issues," Vol. III-3. Natl. Tech. Inf. Serv., NTIS, Springfield, Virginia.

Jensen, G. W. (1970). "Energy: Global Prospects 1985–2000." Cambridge Univ. Press, London and New York.

Kimmel, S., Neben, G. W., and Pack, G. E. (1976). "Economics of Current and Advanced Gasification for Fuel Gas Production," EPRI AF-244, Project 239. U.S. Gov. Print. Off., Washington, D.C.

Krug, J. A. (1945). "Production—Wartime Achievements and the Reconversion Outlook," WPB Doc. No. 334, pp. 30, 32, 99 War Production Board Washington, D.C.

Linden, H. R. (1974). Review of world energy supplies. *In* "U.S. Energy Resources: A Review as of 1972," Comm. Print 93-75. Senate Comm. Inter. Insular Aff. U.S. Gov. Print. Off., Washington, D.C.

Mitzak, D. M., Bumbugh, J. W., and Cannon, J. F. (1976). Koppers–Totzek economics and inflation. *Proc. Annu. Conf. Coal Conversion, 3rd, Univ. Pittsburgh.*

Murray, R. H., and Chamberlin, P. E. (1977). Industrial fuel gas—how it is used and produced—costs and problems. *Proc. Annu. Conf. Coal Conversion, 4th, Univ. Pittsburgh.*

National Academy of Sciences, Committee on Mineral Resources and the Environment (1975). "Mineral Resources and the Environment," p. 98. Washington, D.C.

National Research Council (1977a). "Assessment of Technology for Liquefaction of Coal." Natl. Acad. Sci., Washington, D.C.

National Research Council (1977b). "Assessment of Low and Intermediate Btu Gasification of Coal." Natl. Acad. Sci., Washington, D.C.

Nuclear Energy Policy Study Group (1977). "Nuclear Power Issues and Choice," Mitre Corp. for Ford Found., Chs. 2 and 3. Bandinger, Cambridge, Massachusetts.

Nuclear Regulatory Commission (1976). "Final Generic Environmental Statement on the

Use of Recycle Plutonium Mixed Oxide Fuel in Light Water Cooled Reactors," Vol. 4, NUREG.002. Natl. Tech. Inf. Serv., Springfield, Virginia.

Organization for Economic Development and Cooperation (1974). "Energy Prospects to 1985," Vol. 1. Organ. Econ. Dev. Cooper, Paris.

Ridker, R. G., and Watson, W. D. (1978). "To Choose a Future," Resources for the Future, Washington, D.C. (in press).

Senate Committee on Interior and Insular Affairs (1973). "Energy Research and Development: Problems and Prospects," Serial No. 93–21 (96-56), pp. 117–121. U.S. Gov. Print. Off., Washington, D.C.

U.S. Bureau of Mines, Process Evaluation Group (1976a). "Preliminary Economic Analysis of Coal Gasification Process," IGT Hygas—ERDA 76-47, FE 2083.3. U.S. Gov. Print. Off., Washington, D.C.

U.S. Bureau of Mines, Process Evaluation Group (1976b). "Preliminary Economic Analysis of Coal Gasification Process," BCR BIGAS ERDA 76-48, FE-2083-2. U.S. Gov. Print. Off., Washington, D.C.

U.S. Bureau of Mines, Process Evaluation Group (1976c). "Preliminary Economic Analysis of Coal Gasification Process," Lurgi Plant ERDA 76-57, FE 2083-9. U.S. Gov. Print. Off., Washington, D.C.

U.S. Bureau of Mines, Process Evaluation Group (1976d). "Preliminary Economic Analysis of Coal Gasification Process," CO_2 Acceptor Process, ERDA 76-58, FE 2083-7. U.S. Gov. Print. Off., Washington, D.C.

U.S. Bureau of Mines, Process Evaluation Group (1976e). "Preliminary Economic Analysis of Coal Gasification Process," Synthane Plant—ERDA 76-59, FE 2083-10. U.S. Gov. Print. Off., Washington, D.C.

Wahrhaftig, L. (1977). Bureau of Natural Gas, Federal Power Commission. Personal communication.

Workshop on Alternative Energy Strategies (1977). "Energy: Global Prospects 1985–2000." McGraw-Hill, New York.

World Energy Conference (1974). "World Energy Conference Survey of Energy Resources," Tables III-12, IV-1—IV-3, VII-1, IX-2. U.S. Natl. Comm. World Energy Conf., New York, New York.

World Power Conference (1968). "Survey of Energy Resources" (periodic). World Power Conf., London.

District Heating with Combined Heat and Electric Power Generation

Richard H. Tourin

Stone & Webster Engineering Corporation
New York, New York

ISBN 1-12-014901-X

I. INTRODUCTION

Government leaders and the public have recently become aware that more than half of the fuel energy consumed in American public utility electric generating plants is dissipated in water-cooled steam condensers. Construction of new plants could result in doubling the amount of energy now wasted in this way within 25 years. As a result of this awareness, the word ''cogeneration'' has become popular. It refers to systems in which both electricity and heat are supplied for productive use, by the same generating unit, and the discharge of waste heat is reduced or eliminated.

Cogeneration has been applied successfully in American industrial plants, but its use by all but a few public utilities, to supply district steam, is negligible, and continues to decline. By contrast, combined heat and power (CHP) generation by public utilities is a large, successful, expanding enterprise in many European countries. It supplies district heat (DH) service by means of circulating hot water, along with electric service.

Development of feasible CHP/DH technologies for the United States is highly worthwhile. Fuel utilization efficiencies up to 80% are achievable, compared to 39% for electric generation alone, and thermal discharges into waterways are reduced or eliminated. In addition, the substitution of central stations for many small, inefficient individual heating plants with low chimneys adds to fuel economy and minimizes air pollution. Other advantages are: conversion to coal is facilitated by the reduced emissions from low chimneys, truck delivery of fuel oil is reduced, space is saved in buildings, maintenance and operating costs of buildings and fire hazards are reduced, service reliability is improved, and heat storage may be used to reduce utility peaks. The need for new transmission lines from remote sites is reduced, helping to preserve the countryside.

Steam-electric plants in the United States operate on a system utilizing condensing turbines. These are designed to maximize conversion of fuel energy to electricity. At the temperatures used in modern steam generation, the theoretical thermodynamic efficiency achievable is about 40%. A few existing units operate at nearly 39% efficiency, but the typical efficiency of American fossil-fueled plants is about 33%, corresponding to a heat rate of 10,000 Btu/kW h.

Some energy is unavoidably lost in the stack gases, in unburned fuel, and in other parts of the system, but most of the unused heat is rejected to the condenser cooling water in the low-temperature part of the steam cycle. For every unit of electrical energy generated, about $1\frac{1}{2}$ units of heat energy are rejected through the condenser. This energy is rejected at a temperature only slightly above ambient water temperature, in order to maximize thermodynamic efficiency of conversion to electricity.

Thus, modern electric generating stations reject heat energy amounting to over 60% of the total energy that their boilers produce. Projected increases in consumption of electricity would double this heat rejection by the year 2000 if current practices continue. Moreover, many utilities project nearly as much oil-fired capacity in 1992 as today and higher oil consumption (Empire State Electric Energy Research Corp., 1977). It seems imperative to reduce the waste of heat energy in the condensers of electric generating plants, especially for areas with high oil consumption and high energy costs.

Steam electric plants with fuel utilization efficiency over 70% can be built, by rejecting heat from the turbine cycle at a temperature sufficiently high to supply heat for use by customers, reducing the use of the condensers. This practice of combined heat and electric generation by a steam turbine entails some sacrifice of efficiency in conversion of steam energy to electricity, but the great increase in fuel efficiency makes it possible to balance the combined load at an economic return superior to the system that supplies electricity only.

Steam-plant reject heat constitutes a vast resource which might be utilized by a large heat load such as district heating. It is worthwhile to note that the energy requirements of commercial and residential buildings in the United States are roughly two units of heat energy for each unit of electrical energy consumed. This has direct correlation with the steam-electric plant, where one and one-half to two units of heat are rejected for each unit of electricity produced. Thus, district heating is an excellent application for the reject heat.

By combining heat and electric power generation, the fuel cost of electricity is reduced to about 40% of the fuel cost of electric production in a conventional plant, which rejects most of the fuel energy as waste heat to the condenser. In a condensing plant the fuel cost is fully charged against the electric output, including the cost of the fuel that produces only waste heat. In a combined plant the same quantity of fuel yields saleable heat in addition to the electricity, and the fuel cost is shared between the electric and heat sales.

The combination of electric power production with thermal energy supply systems has been called a "heat-drop" (McConnell and Elmenius, 1975), analogous to a waterfall that can be used to drive hydro turbines.

Figure 1 demonstrates some of the opportunities we have in the United States for heat-drop utilization (Berg, 1974). The four categories are not directly comparable without allowances for the means by which these functions are conveniently accomplished. For instance, air conditioning consumption is for the most part through motor-driven compressors and is thus electrical consumption, whereas space heating and industrial process

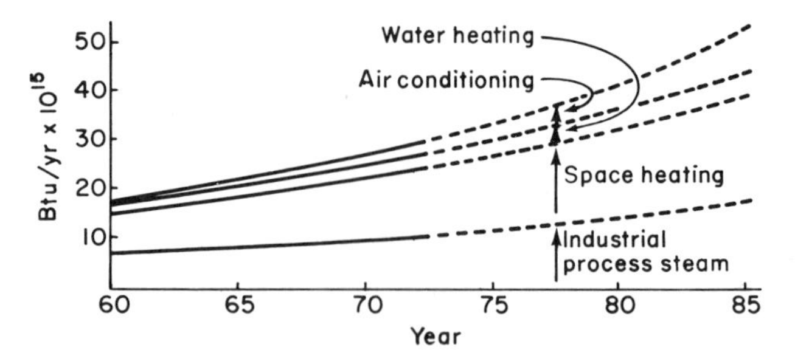

Fig. 1. Energy consumption in United States in categories having potential for by-product power generation.

steam are mostly energy consumed by direct firing, and contain all the inefficiencies associated with delivering the needed energy to its end use. By combining electric power production with thermal energy supply, many of these inefficiencies could be eliminated, thus substantially reducing the energy consumption for the four categories.

Unfortunately, considerably less than 10% of these potential heat drops are being tapped today.

Should 50% of the full potential be tapped by 1985, a conservatively estimated by-product power rate of 50 kW h/MBtu would result in an annual generation of by-product power amounting to 1.35×10^{12} kW h. Projected total electric power production by 1985 in the United States is 3.6×10^{12} kW h (Olmstead, 1974).

Inasmuch as the technology of combined heat and power (CHP) generation with district heating (DH) is well known and widely used in Europe, it might appear that the problems to be solved in the United States are purely institutional. If so, they would be amenable to such measures as financial and tax incentives, changes in regulations and codes, etc., such as are planned to encourage conservation of energy in transportation, buildings, and other fields. In a sense, the chief barriers to utilization of European CHP/DH practice in the United States are "institutional," but these barriers are essentially immovable. Therefore, new adaptations of the technology are required that can work within these barriers.

We distinguish combined heat and power generation in public utility systems from other types of combined heat and electric generation. In public utility CHP/DH heat carried by circulating hot water is distributed and sold to customers at large, in the same manner as electricity and gas. Such systems are popular, successful, and widely used in Europe, but nonexistent in the United States. The problems and possibilities of developing CHP systems for American public utilities are the subject of this chapter.

II. TYPES OF COMBINED HEAT AND ELECTRIC GENERATION

A. Definition of District Heating

Combined heat and power with district heating (CHP/DH), as the term is used here, refers to systems for producing electricity and heat in public utility stations and distributing both to subscriber networks. Such systems include as essential elements: (1) a conventional electric distribution system, (2) a "district heating" system for distributing heat as either steam or circulating hot water, and (3) a dual purpose generating plant, typically a back-pressure steam turbine generator. CHP/DH systems are of increasing interest in this era of rising fuel costs because they make possible overall efficiencies of fuel utilization in excess of 70%, as contrasted with 35% for the production of power alone.

Outside the scope of this definition are industrial power plants built for the combined production of electricity and process steam. This application is known as *in-plant generation* or *by-product power*. The primary product is process steam. Such systems are well established and are particularly advantageous in the process industries where load factors are usually high for both power and steam. There are some situations where interconnections between industrial dual purpose plants and CHP/DH systems are advantageous, but for the most part the industrial systems will prosper on their own merits.

Also beyond the scope of our definition are the small on-site "total energy systems" or "integrated utility systems" designed to serve a housing complex, university, or other institution. Many hundreds of such systems have been installed. Results have been mixed and highly dependent on the load factor and load balance and on the quality of design, construction, and operation. Where applicable, CHP/DH systems encompassing large areas seem likely to be preferred and for somewhat the same reasons that have led to the development of large interconnected electric systems. The large size permits economies in investment and operating costs. Interconnection increases reliability and/or decreases costs for backup facilities.

The future of on-site combined systems, both "in-plant" and "total energy," depends on economic and policy considerations, rather than technology. Their contributions to increased fuel utilization efficiency are welcome, but small compared to the need to utilize the heat from current and prospective electric utility capacity more efficiently.

B. Advantages of CHP/DH Systems

Advantages obtainable through the use of a well-designed CHP/DH system include the following:

1. The dual system for generating electric power and heat (in the form of steam or hot water) makes possible overall energy utilization efficiencies in excess of 70%, as contrasted with about 35% for the generation of electricity alone. A portion of the heat that would otherwise be lost to the power plant condenser is used to supply a district heating system.

2. Central station combustion facilities are substituted for the many small individual heating plants. The latter are inherently less efficient because of their small size, intermittent operation, and the necessity to design for safe automatic reliable unattended operation. Thus the district heating concept contributes to fuel economy over and above that resulting from its use as part of a CHP/DH system.

3. Air pollution is reduced. This results from the decrease in the amount of fuel burned (because of higher efficiency), the better combustion and emission control obtainable at central stations, and the elimination of low-level chimneys. Figure 2, taken from a paper by Unden (1975), illustrates a very strong correlation between SO_2 concentration and the amount of district heating used in some Swedish towns. Västerås (population 100,000) had converted about 95% to district heating and had an average SO_2 concentration in the air of about 0.5 parts per hundred million (pphm). Sundsvall (population 60,000) had very little district heating and an SO_2 concentration in the air of about 7.4 pphm. Although data for other pollutants are not available, substantial reductions are expected.

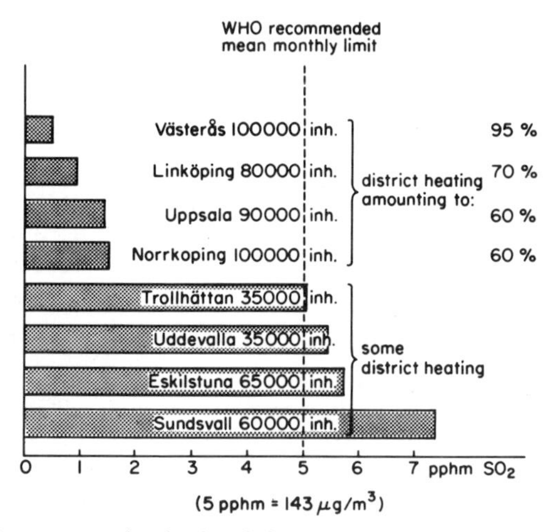

Fig. 2. SO_2 concentration in the air in some Swedish towns, February 1971.

4. Central stations have a wider choice of fuels. Residual oils (at a lower price) may be burned rather than the distillate oils required for domestic furnaces. A change to coal as the primary fuel is facilitated.

5. The delivery of fuel oil by truck to many locations is eliminated, with a savings in cost and traffic congestion.

6. Space required in buildings for heating equipment is reduced or in some cases eliminated.

7. Maintenance and operating costs are reduced.

8. Reliability of service is improved.

9. Fire hazards associated with the storage, handling, and combustion of fuel are eliminated. Insurance rates may be decreased.

10. If a hot water distribution system (or combination system) is used, storage tanks may be provided and used to reduce utility system peaks. This can be advantageous both as a means of reducing total generating capacity required and also as a means of reducing the amount of high-cost turbine fuel required for generating power at low efficiency in peaking units.

C. Differences between European and United States Conditions

All European countries except England, Ireland, and Norway make extensive use of CHP/DH systems. Perhaps the most extensive application and certainly the greatest sophistication in engineering and equipment design have been realized in Sweden. The first district heating plant in Sweden began operation in 1948 and by 1973 the total connected heat load exceeded 7000 MW. About 25% of the total building space in Sweden is now served by district heating systems. Dual purpose generating plants having capacities in excess of 200 MW electrical and 300 MW thermal have been built. Plans are being made for further expansion to cover all of the densely populated areas of the country, excluding only single-family residences and isolated buildings. It seems to be enthusiastically and almost universally accepted that CHP/DH is the preferred system for heating buildings. By contrast, district heating in the United States has been declining.

Historical factors appear to have been important contributors to the differences between Sweden and the United States in the development of, and attitude towards, district heating. The first district steam system in the United States was installed in 1877. The United States steam-electric generating industry developed substantially during the first quarter of this century. There was a prior history of isolated industrial power plants with steam engines that exhausted to building heating systems or process steam systems. The use of steam for district heating systems in large cities seems to be a logical extension of this early technology. The advantages of water were probably not apparent until much later, after the need for transport over longer distances became apparent and after simpler, cheaper, and more reliable pump and piping systems became available.

In contrast, the Swedish steam-electric generating industry has been developed since World War II. Prior to then abundant quantities of low-cost hydroelectric power were available from generation within Sweden and also by purchase from Norway. The importance of long-distance heat transmission and the superiority of water over steam for that purpose were recognized. There was even the hope and expectation during much of the development period that nuclear power, to be produced very cheaply and in very large plants, would predominate and that heat would be available at very low cost if it could be transmitted long distances and accepted into the users' equipment. The postwar period was also one of intensive housing construction in Sweden, and many large developments were built with a district system as the only source of heat.

Before going into the details of district heating, we digress now to discuss briefly the other two applications of cogeneration: industrial in-plant generation and total energy systems for small communities.

D. Industrial Cogeneration (In-Plant Generation)

Industrial plants that require process steam can find it advantageous to generate electricity as a by-product of steam generation. This practice is especially applicable in the pulp and paper, chemical, and petroleum refining industries. The technology is a well-developed commercial practice.

Substantial costs—both first costs and operating costs—are entailed for such systems. These costs are generally bearable only by installations with large steam demands and continuous loads—typically over one million pounds of steam per hour and over 8000 hours per year of plant operation (Coles, 1977). The economics of in-plant generaton has been covered in a number of recent papers (Wilson and Kovacik, 1976; McConnell, 1976; McConnell and Elmenius, 1976).

Attitudes in industry toward cogeneration vary widely. Historically, American industrial companies have considered the production of steam, for process or to heat buildings, as part of normal business operations. Electric generation, by contrast, has often been thought of as foreign to normal industrial activities. Electricity was a product to be purchased from a company specialized in supplying that commodity: an electric utility.

Expansion of cogeneration in industry, to save additional energy, is mainly a matter of providing economic incentives and removing institutional barriers. Formerly these systems were usually developed by their users independently of the local public utility, and the users paid the utility for standby capacity. More recently, efforts have been made to integrate such systems into the utility grid, which would result in overall more efficient use of capital and fuel. Public Service of New Jersey and Gulf States Utilities were early pioneers in such efforts. Their task was facilitated by having large steam-using chemical plants in their service territories.

Some industrial users have favored operating their own turbine generators because this gives them greater assurance they will have heat and power when needed and in the quantity needed. Today, many have real concern about the ability of the utility to serve the industrials' electric power load in all geographical locations in the years ahead. Some of this concern comes from people in areas where there is a high incidence of lightning that can affect lines feeding power from the outside.

By contrast, a common reaction to in-plant generation is "We are in the business to make salable products (chemicals, steel, lumber, etc.) and will leave power generation to the electric utility."

In many instances, this has been the economic thing to do. In other instances, this attitude has eliminated the consideration of alternatives that would have been a profitable investment for the user and conserved our United States energy resources.

Environmental cleanup has consumed much of the capital funds that would normally have been available for industries to invest in cogeneration.

Power costs continue to rise steeply and this has made in-plant generation attractive to many companies. By contrast, some believe that regulatory agencies will hold purchased power costs down but fuel prices will continue to rise steeply. Fixed charges on investments for industrial electric plants ordered today will probably be $0.03/kW h, which is higher than total purchased power costs that many industrials are paying today.

Many industrials feel that the regulatory agencies will tend to let large industrial electric rates rise to keep residential rates low. Thus, they believe in-plant generation (IPG) will surely be a good investment.

It is not always possible for a company to work out a reasonable arrangement with the utility if the company generates a part of its power and still wants to buy some from the utility. Utility charges for service should bear a relation to the cost to serve, and a reasonable charge for demand and other elements of the power rate should be expected. The economics of IPG must include an assessment of the relative reliability/availability of

IPG, purchased power, or some combination of the two. Economics will dictate whether an extra turbine generator, a utility stand-by charge, or a curtailment in production are the preferred way to permit a turbine shutdown for normal maintenance or handle an unscheduled outage of a turbine or a utility tie. Some industrials feel utilities may impose undue restrictions that discourage the generation of fuel-conserving by-product power in the industrial plant.

Many industrial firms are confused by uncertainties in future costs, availability of fuels, and environmental regulations. They are inclined to let the utility worry about securing fuel for power generation and to buy power without regard to economic analysis. The future costs and availability of fuels are legitimate concerns. But the large process plant with a significant need for process heat will have to face the problems of fuel supply and any associated environmental regulations. In most energy-conserving IPG applications, the additional fuel that must be secured and burned will be a relatively small increment (20–30%). So the basic problems of securing, handling, burning, and environmental cleanup must be faced regardless of whether IPG is considered.

Some industries feel the electric utility should finance and operate energy-conserving IPG in their plants. Joint industry–utility ventures can be advantageous. The technical application factors could be the same regardless of ownership. The regulated utility will usually accept a longer payout on an investment than would be considered acceptable to the industrial. Regardless of ownership, the energy-conserving features could be the same. A utility might consider a large plant, co-located in a cluster of industrials, with the necessary steam distribution system to supply steam to the industrial complex. In this case, the scale factor should result in some reduction in the per unit investment costs for boilers, turbines, and other equipment. These savings would then have to be compared with the extra costs for piping systems, rights-of-way, etc. to distribute the steam, chilled water, or other services supplied to the various customers. The success of Public Service and Gulf States Utilities in this approach has already been mentioned.

Even though industry has installed and continues to install energy-conserving cogeneration systems, they usually demand a high rate of return, equivalent to a discounted rate of return on the investment of perhaps 15% minimum and ranging up to 25–30%. The criteria for approval of a project to reduce operating costs will vary widely and depend on the number and type of other projects being considered and the general business climate at the time.

Some of the incentives that might turn marginal IPG energy-conserving projects into projects approved for installation are as follows:

1. Continue to help decision makers understand the potential for an acceptable rate of return on an investment for energy-conserving IPG.

2. Consider various government incentives to increase the profitability of projects that can conserve energy, such as increasing the investment tax credit, reducing the depreciation period, provide a tax incentive for the user who installs a high-pressure 400–1500 psi (gauge) coal-fired boiler feeding IPG, instead of a boiler designed for low-pressure process steam.

3. Each user must try to realistically appraise the favorable effect of future increases in energy costs (fuel and purchased power) on the profitability of an investment for IPG today. Many feel that the regulatory agencies have tended to hold the "lid" on the price of purchased electric power. The cost of fuels in the United States, influenced by the sharp rise in world market prices and the increased percentage of imported oil, has risen sharply in the last three or four years. This has upset the historic relation between fuel cost and purchased power costs. In addition, new utility generating and transmission capacity must be added and few predict that future utility system base load capacity will cost less than $800–$1000/kW. With 18% fixed charges and a plant factor of 68% (6000 h/yr at rating), the fixed charges on this new generation capacity at $100/kW would be $0.03/kW h. So fixed charges on this new generating capacity are more than the total being paid for purchased electric power by most large industrials in the United States today. People have little basis to project the cost, the quality, and availability of fuels that will be so dependent on government energy policies, environmental, and other regulations. But if the large industrial company and the utility burn fuels having the same per unit costs, then future increases in fuel costs will increase IPG generation costs less than one-half the increase for a normal utility cycle.

4. Seek opportunities where a large system might be installed to supply heat and power to one or more industrial or other type customers. This system might be owned and operated by the electric utility, the industrial user, or a combination of both.

It should be mentioned that many process plants are recovering low-level heat, e.g., through heat exchangers or bottoming cycles, where it was not economic to do so when energy costs were lower. This recovery of heat in process reduces the potential for IPG.

E. On-Site Nonindustrial Cogeneration

This is most commonly called *total energy*, also called *integrated utilities* and *community energy systems*. The basic characteristics are (1) fuel is delivered to the site and converted there to heat and/or electricity,[1] (2) the fuel is commonly petroleum distillate or natural gas, (3) electric generating capacity, if any, must be backed up by duplicate capacity, either on-site or from a public utility, available through a stand-by connection.

In the vanished era of cheap, plentiful oil and gas, such plants had advantages to their owners, but current conditions favor large-scale (district heating) systems over on-site systems. The United States Department of Energy has initiated a program to study integration of on-site systems with the public utility grid. This is a step in the right direction.

On-site systems have a place in the current energy picture, but it is a small one. The total electric generating capacity of such systems in the United States is comparable to the capacity of one or two modern public utility plants. The general case against proliferation of on-site systems has been well stated by Muir (1973). The present status has been summarized by Mullen (1977). Current trends seem unfavorable to major reliance on such systems, for the following reasons:

1. They require oil (usually distillate) or gas to operate diesel engines, gas turbines, or low-pressure boilers. This is contrary to the national policy of shifting to coal.
2. Efficient load balancing on site is usually not feasible, because the heat and electric production capacities of such plants do not match the respective on-site heat and electric loads. Consequently there is a surplus of electricity to export, mainly during periods when it is least in demand elsewhere.
3. The construction of such plants reduces the potential heat load for prospective district heating systems. It thereby makes it harder to use public utility power plant heat, which will otherwise be rejected to the environment.
4. The efficiency of converting fuel energy to useful heat and to electricity is much lower for such plants than for central stations.
5. Air quality is adversely affected by having a multitude of small plants with short stacks instead of a few central stations equipped with taller stacks and modern pollution control facilities.
6. On-site plants are specified, designed, and constructed by a large number of small organizations, without the rigid standards and supervision characteristic of the regulated public utility business. This factor has contributed to the poor efficiency, poor design, and unfavorable operating experience found so often in on-site systems.

Total energy systems have been in operation in the United States since

[1] Many on-site plants supply central heat and air conditioning to a military base, housing complex, etc. without generating electricity.

the late 1950s with only limited success. In the last few years, retirements have greatly exceeded new installations, and government policy to curtail gas and oil usage has virtually halted private sector applications. The federal government is, however, deeply involved in investigating the potential of this technology in terms of economics and fuel efficiency. This work, rather than the efforts of commercial interests with a stake in the success of total energy technology, will form the basis for any further applications in this field.

The possible economic advantage of total energy systems stems from fuel savings brought about by a plant that is designed to utilize fuel at approximately twice the efficiency of a conventional utility installation designed for electric generation only. The small scale of such systems makes this a slight advantage at best. For a majority of such systems this economic picture, coupled with some inherent problem—technical, institutional, environmental, or fuel related—has caused many owners to abandon their substantial investments in total energy systems.

Several agencies, led by the United States Department of Housing and Urban Development, are involved in the modular integrated utility systems (MIUS) project, which proposes to mass produce modular total energy systems providing all utility services. This is a complex, highly sophisticated system—the "Cadillac" of total energy systems. Its application is not a near-term solution. More immediate results can be expected from the integrated community energy systems (ICES) program undertaken by U.S. ERDA. This program deals primarily with institutional problems involved in total energy applications, and emphasizes integration with the public utility electric grid. Four projects are being funded to demonstrate a wide spectrum of institutional arrangement, service community size, and corresponding system configuration and components. The results of this project will be available within the next few years to exemplify the features required to achieve a successful total energy technology application.

The on-site total energy system (TES) meets the heating and cooling demands of a single building, or group of buildings, using the waste heat of on-site electric generation. In this way the definition includes systems that supply electric power to the utility grid, along with the more familiar systems that attempt to supply the complete electric demands of the service community.

TES today can be divided into three groups on the basis of institutional arrangement. The first group is the privately owned, normally isolated, system which was first demonstrated in the late 1950s, reached the peak of its popularity in the late 1960s, and finds only limited application today. These systems are generally poor economic ventures, due to poor systems design, bad management, and dependence on scarce fuels (distillates and gas). The second group is the MIUS, supported by the combined efforts of several federal agencies, particularly HUD and NASA. Designed as a complete utility module (including waste and sewage treatment), these units will be "plugged in" to a residential load. The last group is the ICES, funded by ERDA.

Conventional total energy systems, first demonstrated in 1958, reached a peak of approximately 600 installed units in 1967 and have declined from that point. Early predictions were for 25,000 units by the mid-1970s; however, very few new units are being started or considered today and fewer than 600 are still in operation.

These systems were promoted by an expanding gas utility industry through the Group for the Advancement of Total Energy (GATE), which no longer exists. These systems offered the owner 60–70% fuel efficiency for communities with a 0.5–2 MW electrical load. They offered isolation from the utility (especially popular after the 1965 blackout) and offered landlords another source of income from tenants. The widest application has been in the commercial sector, mainly in shopping centers and malls. Hospitals and schools are also good loads, especially where a hospital requires back-up power. Some apartment complexes are also involved, although they never achieved more than 6% of the market. These same institutions constitute the ICES and MIUS demonstration communities.

The total capacity of the 600 units installed was about 0.2% of the installed electrical generation in the country. These systems depend primarily on natural gas or high-quality distillates; about 3.5% are dual fuel units. Reciprocating engines account for 70% of the installations, and gas turbines for about 15%. One to three prime movers are generally installed with waste heat recovery from circulating cooling fluid. Absorption air conditioning is used almost exclusively for all cooling requirements; stand-by and peaking boilers are usually employed to guarantee thermal supply. Recent studies show that more than one-half of these TES have been retired (McClure, 1977; Spielvogel, 1977).

There are several technical reasons for failures and retirements, all of which translate to economic penalties. Important and generally applicable conditions leading to failure are given below.

1. Systems are often designed to meet the worst case peak load plus some safety margin. The result is equipment normally running at much less than full load and much less than full efficiency.

2. Electrical and thermal load seldom coincide with system design. Thus, with no storage of either energy form, both electricity and heat are often wasted.

3. With no electrical standby power available, except for utility con-

nection and its high demand charge, isolated systems require high reliability components and investment in little-used stand-by equipment. This means high initial cost and quite often decreased system efficiency.

4. Absorption air conditioners are used almost exclusively to serve the cooling load. Installed units have shown very low coefficients of performance (down to 0.4 or 0.3), especially at partial load. At full load these units can require the waste heat of electrical generation plus supplemental heating from the standby boilers. These heat rates translate to very poor fuel efficiencies and eliminate the combined heat and power advantage of these systems.

5. TES have generally been designed and installed on an individual basis, each is a custom system. Little feedback from installed operation is used in the design of new systems. Operation and maintenance personnel normally keep no records and, thus, no input is available to allow for subsequent design improvements.

6. Since each scheme is an individual design, no "champion" of TES has emerged. Correspondingly, no equipment has been designed or manufactured specifically for TES operation. Thus, designers often apply peaking equipment for the constant operation required in a TES, resulting in frequent breakdown, low equipment life, and generally poor reliability.

7. Systems are often designed by qualified technical firms that have no further contact with a system after start-up. Thus, there is no technical input after this point and no satisfactory solution to operational problems.

8. Maintenance is generally specified by the outfit which will perform the maintenance. Unnecessary and expensive operations have been carried out at the expense of the owner.

9. Preventive maintenance is seldom performed, resulting in many unnecessary breakdowns.

10. Many states require a licensed engineer to be on duty at all times. Thus, owners must pay for this service with little return.

11. Operational personnel seldom handle system economics and thus, pay little attention to waste and inefficiency.

12. Improperly designed systems can be noisy and can emit foul odors which reduce their desirability (especially in residential applications), or require longer distribution lines in order to isolate the plant. TES emit more pollutants than a conventional heat plant, due to their fuel consumption for electrical production, thus, ambient air quality standards have limited their application.

This is a quick summary of the more important technical problems leading to failures of conventional TES. The interested reader will find more detailed discussions in McClure (1977) and Spielvogel (1977), which include a historical survey of the development of conventional TES, and a survey of a random sample of 40 TES, with a good analysis of their track records.

The technical problems just listed can all be translated into economic penalties. It is important to note that theoretically the economic advantage to TES is a result of the increased fuel efficiency offered by combined heat and power generation. With the small scale of these systems, relative to utility generation plants, this becomes a marginal advantage that is often not realized in practice.

In most cases, the institutional arrangement (which, like design, is different for each installation) has a great impact on system economics. Hidden penalties and subsidies characterize most of these arrangements, clouding the true economics of this technology, and blocking any generalizations concerning TES economics. The most important institutional arrangement is between the owner and the local utility. The questions of back-up electric supply, demand charge, and use of surplus TES electrical power are normally the deciding factors in system economics. Other institutional arrangements that can play a lesser role are government subsidy, arrangements with local utility rate commissions, relationship between plant owner and service community owner, local environment protection agency interactions, and maintenance companies, among others. Again, each of these agreements can be translated to some economic penalty or benefit, with essential concern lying in the owner-utility agreement. To get a true picture of TES economics, an in-depth study of a system under various management, ownership arrangements is required; for these private systems, this has never been done.

III. COMBINED HEAT AND POWER TECHNOLOGY

A striking feature of the subject of cogeneration in utilities is the contrast between Europe and the United States. In many European countries, CHP/DH is highly successful, profitable, and popular, and is expanding rapidly. In the United States, CHP and district heating are relatively rare, and utilities have not been seeking to expand the business. When we consider the numbers involved, the contrast is staggering. Sweden, with a population of 8 million, currently has a connected district heat load of 7000 MW (t) (t, thermal), of which 30% is supplied by CHP/DH systems; the Soviet Union, with a population slightly larger than the United States, has a connected heat load of 450,000 MW (t), of which 60% is provided by CHP/DH systems. United States public utilities currently have zero connected heat load on district (hot water) heating, and only 15,000 MW (t) of steam sendout capacity in district steam systems.

TABLE I

District Heating Loads, 1975

	Population	Connected Heat Load (MW)[a]
Sweden	8,200,000	7,000
USSR	246,300,000	450,000
West Germany	61,900,000	23,454
East Germany	17,000,000	7,700
Denmark	5,000,000	9,885
Poland	34,500,000	35,400
		Peak steam sendout
United States	210,000,000	16,000[b]

[a] Winkens *et al.* (1976).
[b] *Proc. Int. Dist. Heat. Assoc.* (1974).

Table I shows district heating statistics for various countries. Figures for the United States (steam systems) are given in terms of steam sendout capacity, while European figures are in terms of connected heat load. About one-third of total United States capacity is in the Con Edison steam system, which supplies parts of Manhattan.

People with a prejudice against utility companies might imagine that United States public utilities are indifferent to waste, as the result of many decades of operation with cheap fuel. It is then an easy step to the erroneous belief that we have only to influence the utilities into mending their wasteful ways and adopting superior European methods, and we can then utilize the power plant heat now rejected to the condensers. The actual situation is not so simple. It is true that the historical development of electric generation practice in the United States was profoundly influenced by the ready availability of cheap fuel, and by the higher value of energy in the form of electricity than in the form of fuel. This provided a strong incentive to strive for higher efficiency of converting fuel energy to electricity, and simultaneously made it uneconomical for a utility to sell heat to its electric customers, who could buy cheap fuel and burn it themselves.

The development of ever more efficient steam electric systems has been proceeding steadily for over 50 years, on the premise that fuel would remain cheap and abundant. This development has culminated in the widespread use in the United States of very large generating units, of 500–1300 MW (e) (e, electrical) capacity, with steam at supercritical pressures and with superheat and reheat above 1000°F. Efficiencies close to 39% are achievable, and many currently operating units achieve 36% in practice. The need for energy-consuming auxiliary equipment for pollution control, such as scrubbers and cooling towers, makes it even more imperative for utilities to strive for higher efficiency. Such large units and advanced steam conditions are unknown in the European countries most advanced in CHP/DH.

The crux of the problem, and the major obstacle to adoption of European district heating practice in the United States, is simply that generating units and loads on United States public utility systems are incompatible with European CHP/DH practice. With 500,000 MW (e) of capacity on line or in advanced planning, this "institutional" obstacle will be with us for many years. New technological approaches are needed to develop CHP/DH systems compatible with the reality of United States generating plants and loads. Elaboration of the problem and presentation of possible solutions are our concerns.

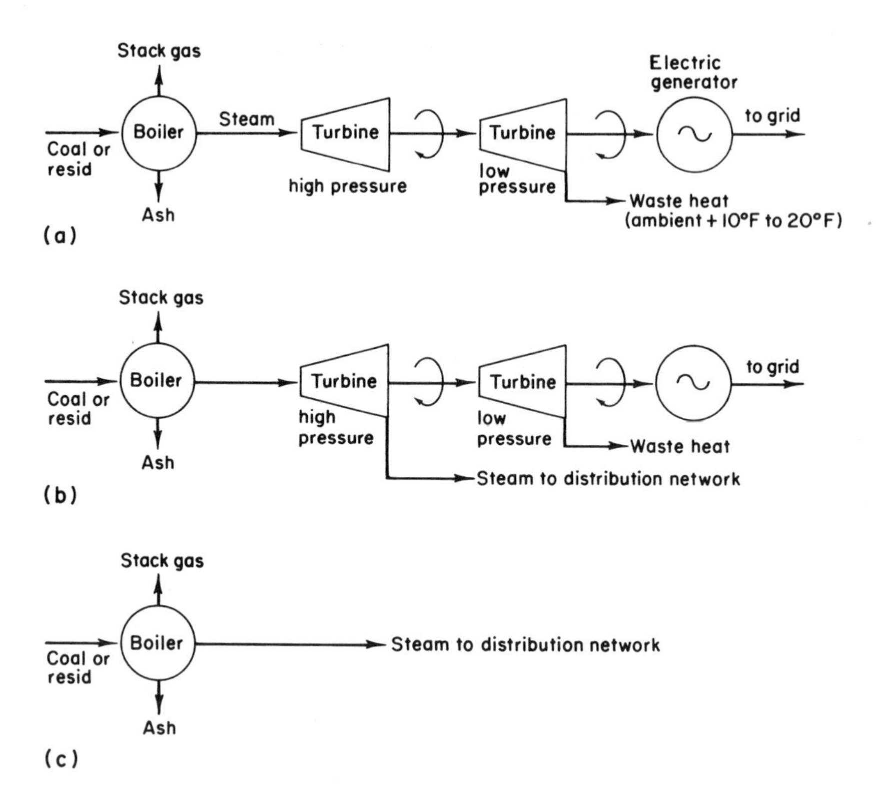

Fig. 3. Heat power technology in United States public utilities. (a) Electric generation. (b) District steam with electric generation. (c) District steam alone.

A. Heat Power Technology in the United States and Europe

Figures 3, 4, and 5, respectively, depict United States public utility practice, nonutility practice, and European practice for public utilities, in highly simplified form. Figure 3(a) illustrates the typical setup of a large modern electric generating plant in the United States. The plant could be oil or gas fired, but coal is the preferred fuel. Fuel is burned to produce steam, with about 90% efficiency, the balance of the energy being lost mainly in the stack gases. Steam from the exhaust of the high pressure turbine is reheated in the boiler, adding to the steam energy in the low-pressure stage, and the spent steam is discharged to the water-cooled

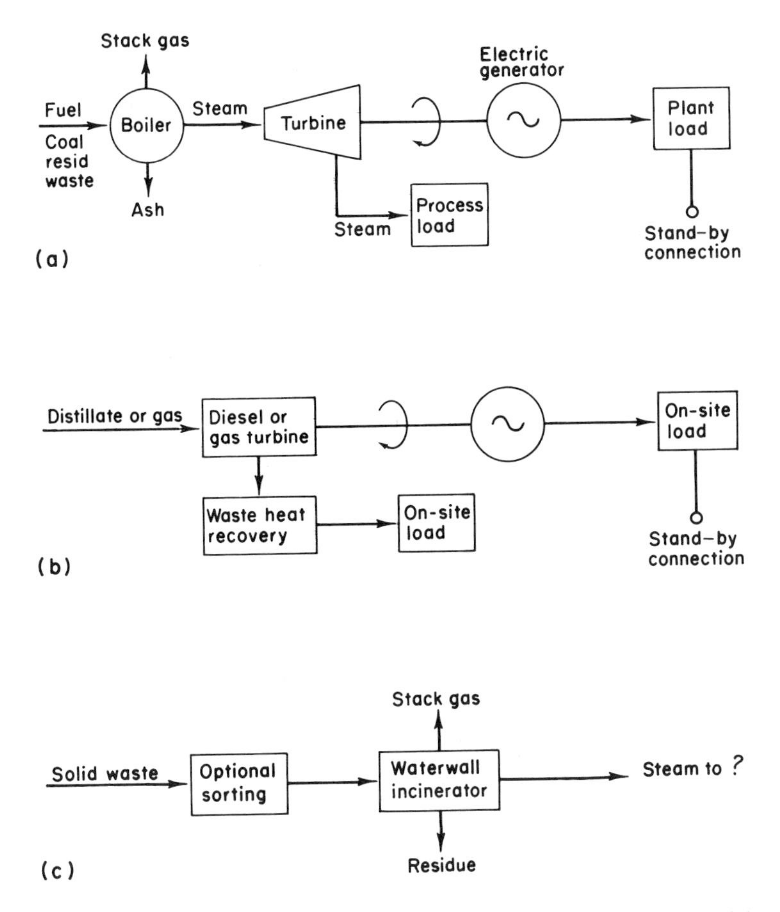

Fig. 4. Heat power technology in the United States (nonutility). (a) Industrial process steam and by-product electric generation. (b) On-site "total-energy" system. Variants: oil-fired boilers, chilled water circulation, etc. (c) Incinerator with heat recovery.

condenser. In the most efficient modern units, about 39% of the fuel energy is converted to electricity; most of the remainder is dissipated in the cooling water, which is discharged from the condenser at a temperature 10–20°F above ambient water temperature.

Figure 3(b) illustrates a typical arrangement of a United States district steam system for a public utility cogenerating electricity and steam. This scheme achieves a higher fuel efficiency, although it has found limited application in United States district steam systems. In the 1975 International District Heating Association report, only 6 of the 44 district steam systems listed employed any cogeneration (*Proc. Int. Dist. Heat. Assoc.*, 1976). Instead, the trend over the last 20 yr has been to utilize direct fired package boilers, as shown in Fig. 3(c). There is evidence, however, that the higher cost of fuels has started a reversal of this trend, and greater use of topping cycle steam generation.

Figure 4 shows three typical arrangements for electric and heat production outside public utilities. Figure 4(a) shows the setup used in process industries. It is similar to Fig. 3(b), except that a large fraction of the steam is extracted for process use, and the electricity is produced as a by-product. The steam is essential for plant operation. Cogeneration is often particularly atttractive because of high load factor and because steam and electric loads tend to rise and fall together. This contrasts with the utility case, where the steam is being sold for space heating and has to compete with the customer's own boilers at prevailing fuel prices, taxes, etc. In some industrial cogeneration plants, the fuel consists largely of plant wastes. This is especially true in the pulp and paper industry, where cogeneration is applied. The industrial plant must have on-site stand-by facilities or a stand-by connection to the utility grid, as shown in Fig. 4(a).

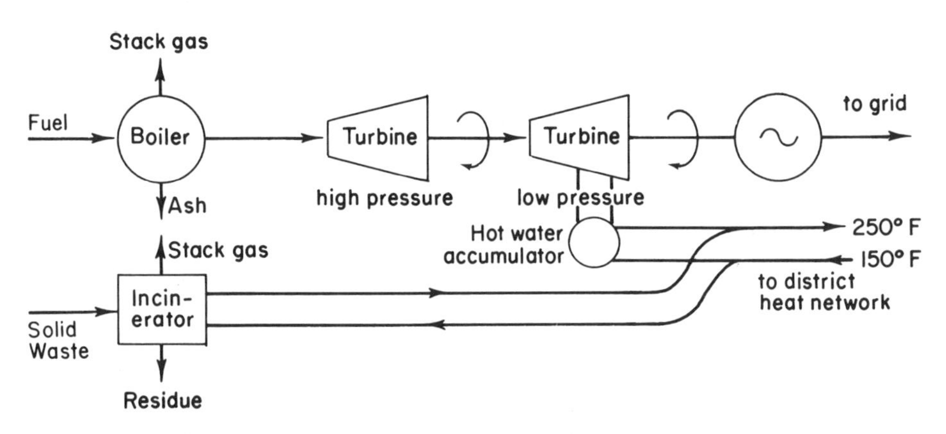

Fig. 5. European CHP/DH generation.

A more progressive approach might be to have the plant owned and operated by the utility and integrated with the grid.

Figure 4(b) illustrates the typical on-site "total energy" system driven by diesel or gas turbine power. Heat is recovered from the exhaust gases, and produces steam in a waste heat boiler for heating on the site. Either on-site backup or a stand-by grid connection is required. These systems are small compared to utility units, ranging up to a few megawatts.

Many variations and elaborations of these systems are used, including oil-fired on-site boilers to produce steam for heating and chilled water for cooling through absorption systems or steam-driven compressors, and steam-electric generation on site. The latter is not common, because production of electricity up to the thermodynamic capability of such a system usually requires an arrangement to sell excess electricity to a utility.

Figure 4(c) shows another system for producing steam that is independent of public utility operations. This is the heat-recovery incinerator, which extracts energy from solid waste in the form of steam. In principle, the steam can be used to make electricity, but the efficiency is low and the sizes are small, so that utilities have limited interest in such systems.

Figure 5 illustrates the typical district heating system with combined heat and power generation employed in Europe. It integrates electric production and distribution; heat production, distribution, and storage; and solid waste disposal, all on the public utility system.

The discussion here deals with fossil-fuel systems, preferably coal fired. It is technically possible to apply cogeneration and district heating with nuclear-powered units, and this has been the subject of extensive study (Margen, 1974). The ratio of heat to electric output is very high for nuclear plants, and they are remote from population centers, so that transmission mains for heat many miles long are required. There are other special problems—social and political—associated with use of nuclear power, that make it an unlikely source for district heating in the United States.

B. Steam "District Heating" Systems

Outside the United States, "district heating" is virtually synonymous with distribution of heat by means of circulating hot water, although distribution by steam is found in a few older systems in Europe. The reasons for the ascendancy of hot water have been well presented by Muir (1975), and will be considered in Section III,E.

Steam district heating in the utility sector of the United States has declined in the last few years. Several factors have made it economical for building owners to operate individual, on-site boilers rather than utilize the district steam line that may pass through their property.

Present steam district heating systems suffer from poor economy for several reasons:

1. U.S. systems have not made extensive use of cogeneration cycles in producing the steam for district heating. Thus they have lost the good economy of electric generation possible from topping turbines.

2. Increasingly stringent emission control standards have forced the use of low-sulfur fuels or large capital expenditures in control equipment.

3. Steam sales are in most cases taxed much more than fuel sales, making the operation of on-site boilers more economical. This has artificially reduced the economic advantage of the more efficient and cleaner district steam system.

As a consequence of these factors, it is not surprising that district heating steam systems are declining in the United States. Increased appli-

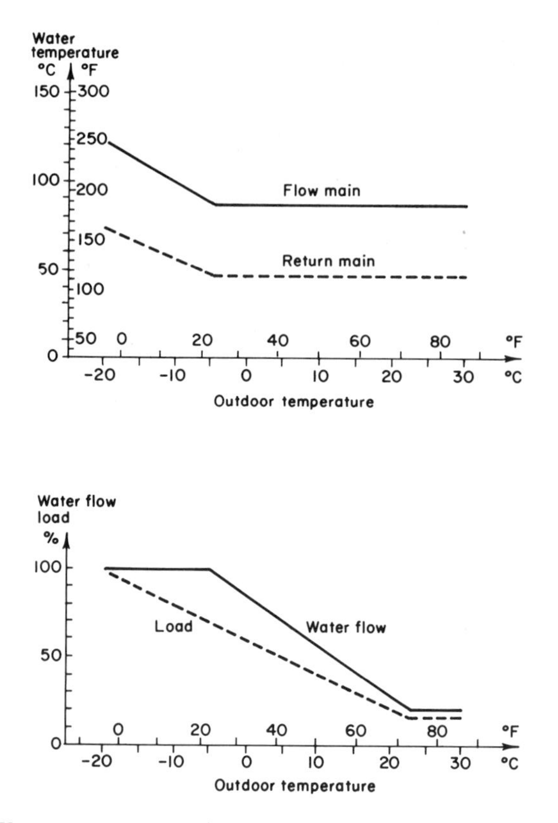

Fig. 6. Water temperature and water flow in relation to outdoor temperature.

cation of district heating in the United States has important fuel conservation and emission control advantages which should be exploited. However, the European hot water systems demonstrate great advantage over the United States steam systems, as we shall see, and thus hot water systems must be demonstrated in the United States.

C. Some Characteristics of District Heating Hot Water Systems

Figure 6 illustrates the temperatures and water flow used in district heating hot water systems, as a function of outdoor temperature. Figure 7 is a typical load curve. The combined heat and power plant with back-pressure turbines is sized to carry up to 50% of the maximum heat load. To cover the winter peaks, heat is supplied directly from boiler plants, often fired with municipal waste. This is the most economical arrangement. The dashed line in Fig. 7 indicates a possible additional load of absorption air conditioning.

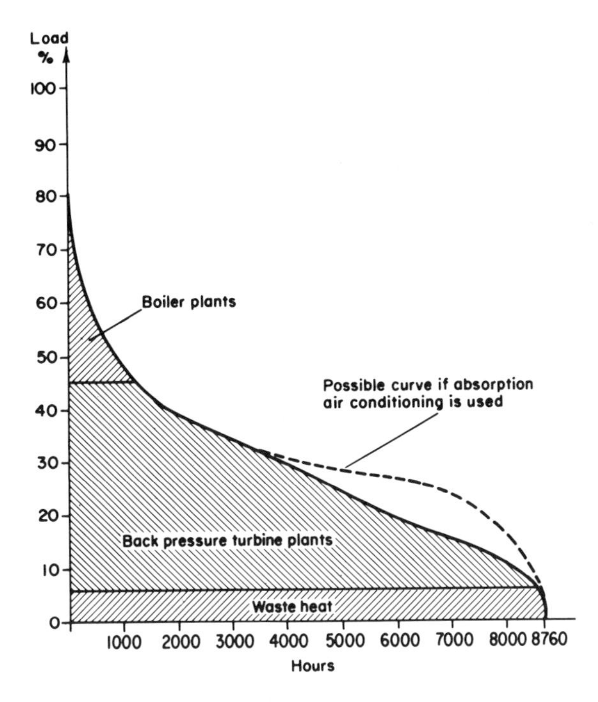

Fig. 7. Annual load diagram. Heat consumption in a small town in the middle of Sweden. The equivalent full load operating time is about 2200 h/yr. Power factor depending on the number of consumers connected to the heat producer is 0.8.

When using back-pressure turbines, all steam is condensed in the hot water condenser for district heating. and there is no need for cooling water, because the generated electricity is produced according to the hot water demand. If there is a demand for electricity without a corresponding heat demand, either the heat is stored or the necessary steam for power production can be condensed in a recooler chilled with cold water. This subject has been discussed by Muir (1973). A brief treatment, adapted from Muir's paper, is given below.

Electricity is the more valuable product in a combined heat and power system for district heating. The ratio of electricity to heat is called the α-ratio, and varies from 0.45 up to 0.70. It depends on the steam inlet data, the number of heating condensers, and the temperature of the district heating water. Once a turbine is built, the first two parameters are fixed, and it is only the third factor which varies. It is desirable to maximize the α-ratio to achieve greatest economy.

The α-ratio can be improved by heating the circulating water in a number of stages. Two stages are normally used. These stages, or hot water condensers, are connected in series on the water side and this gives about 4–5% greater electrical output than would heating in one stage. Apart from the change brought about by seasonal weather variations, the α ratio is fixed, which means that full electrical output is not available if the heat demand is less than the maximum heat output capacity of the turbine. Some means must therefore be provided so that the output of electrical energy can be varied independently of the heat output. There are two ways in which this can be done.

The first way of increasing the electrical output is by artificially increasing the heat load. This is done by means of a recooler, which is a large heat exchanger cooled either by air or by water and connected into the hot water system, as shown in Fig. 8. When the load is low, pumps A and B are running fairly slowly or have their guide vanes set for low flow. But the output of pump C can be increased, thus circulating more or less of the water through the recooler. More water passes through the turbine condenser, which means that more steam can be passed through the turbine and thus more electricity generated.

Hot water accumulators are used to smooth daily variations in heat load. These are large insulated vessels connected into the hot water circuit as shown in Fig. 8. Suppose that the maximum electrical output is required at a time when the heat load is not at a maximum. Full steam flow is passed through the turbine and to the hot water condensers. The flow through the condensers is increased by means of pump C, and some of the water is taken off and passed into the top of the accumulator. The flows are adjusted so that the right amount of water is left to be pumped out into

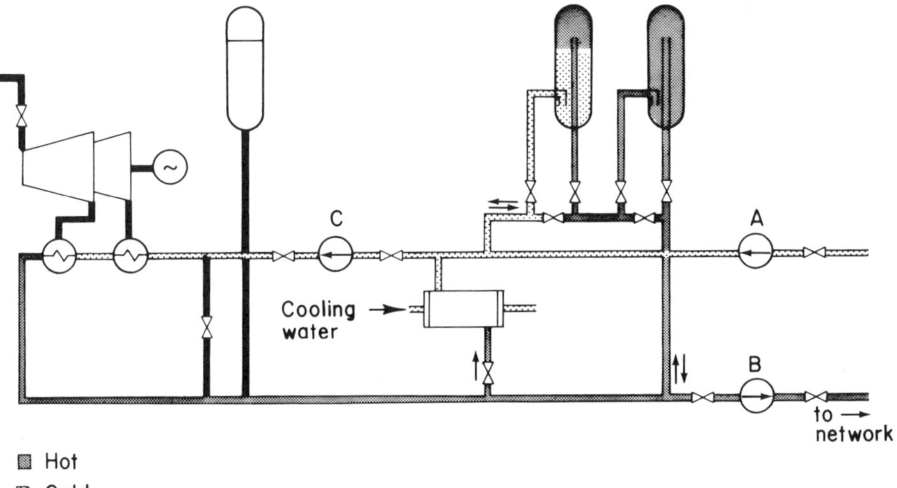

Fig. 8. District heating turbine with recooler and accumulators.

the distribution system. The colder water in the accumulator is displaced and pushed out of the bottom and back to the hot water condensers. Thus the heat load on the turbine has been effectively increased: the rate of flow of water has been increased and colder water is being supplied for reheating. More electricity can be obtained from the turbine, and the extra heat which has been transferred to the hot water is stored.

The accumulators are big enough for it to take some hours before they are full of hot water. At some other time during the day, if the electrical demand drops while the heat load rises, or if perhaps the heat load rises to more than the turbine can supply, the accumulators are discharged by changing over the valves so that the hot water which was heated earlier is supplied to the distribution system.

The turbine can also be constructed with a stage for extraction of steam to heat exchangers for district heating and a cold condensing "tail." In the last case there will be need of cooling water, but the amount required is considerably smaller than in a pure condensing turbine, and the heat rejected as waste is also much smaller. The production of hot water district heat from turbine extractions gives a considerable heat saving, as only one fifth as much heat needs to be fed into the turbine in the form of admission steam as is extracted. This appears in the turbine schematics in Fig. 9. The output of 1000 MW (e) is shown only to illustrate the calculation. Units of this type now in operation are below 300 MW (e) in size.

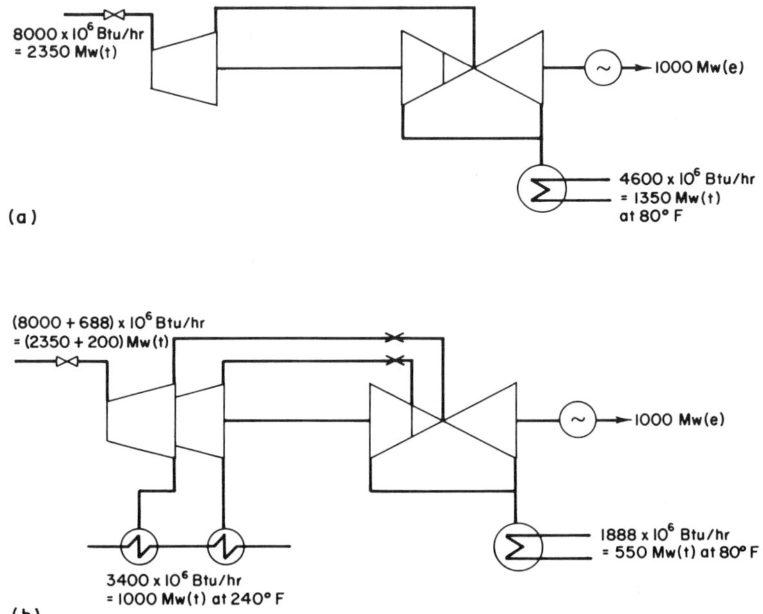

Fig. 9. (a) Pure condensing turbine. (b) Combined condensing and extraction turbine arranged to obtain two different exhaust pressures.

D. Distribution Systems

District heating mains exist in a variety of forms, for street excavations, earth cuts, and rock tunnels. These have been developed to a high degree of reliability (Muir, 1973, 1975). Annual water losses generally do not exceed the volume of a system. In some European cities, district heating mains run through the cellars of buildings, rather than in the street. The absence of conventional boiler plants makes space available for this purpose.

Individual buildings are connected to the district heating network through heat exchangers. The advantages of heat exchangers are:

1. No risk of water hammering and leakage in the radiators.
2. Lower static pressure in the local system, compared to the high pressure of 16 bars (230 psi) in the primary network.
3. The secondary system is independent of the network when the water must be tapped off.
4. Dirt in the consumer's side cannot get into the mains. The only disadvantage is the higher price of the heat exchanger than the pressure-regulating valves that would be used without a heat exchanger.

District heating also heats the tap water at a lowest temperature of 60°C (140°F). There is no need for hot water storage for tap water: the units are quick acting with very low thermal inertia.

Normally load changes are met by adjusting both the flow and the temperature of the outgoing water. A certain minimum outgoing temperature is always maintained in order to provide tap water at a minimum temperature of 60°C (140°F). With falling outdoor temperature the water flow is first increased at constant temperature until the maximum flow rate is reached. Thereafter the temperature is raised at constant flow. The load is altered by the consumer's heat exchangers opening and closing their valves and so causing changes in the mains differential pressure and/or flow. These changes in differential pressure are sensed by transducers out in the distribution system and transmitted to the power station.

Water returning to the plant for reheating is at a temperature of between 50 and 90°C (120 and 195°F). It can be used for street heating, where water at about 25°C (77°F) is pumped from heat exchangers through secondary circuits in plastic pipes buried in the asphalt or in sand beneath pavements. This eliminates the need for snow-clearing equipment.

E. Hot Water versus Steam District Heating Systems

At this point it is important to discuss the advantage of hot water district heating used in Europe over the present United States steam systems. There are two major advantages to a hot water district heating system: improved system control (including load leveling) and increased fuel efficiencies (for combined heat and power generation).

Control of the quantity of heat which reaches the consumer in a hot water system is achieved by the control of both flow rate and temperature. These two parameters are routinely monitored and controlled at the central heat plant in response to the consumer's heat requirement (based on ambient temperature) and electrical loading. The hot water system allows for greater flexibility in matching electrical load to generating capacity. The heat energy can be stored in hot water tanks during periods of high electrical demand, and sent out during periods of low demand. A system with these capabilities can level the system heat demand, while continuously following the electrical load (Muir, 1973, 1975).

To illustrate the fuel savings obtainable by using a hot water distribution system rather than a steam system, calculations have been made for the systems shown in Figs. 10 and 11. The hot water combined heat and power generation system is represented schematically in Fig. 10. Figure 11 represents our model of a steam combined heat and power generation system. Calculated data for the two models are given in Table II. Each

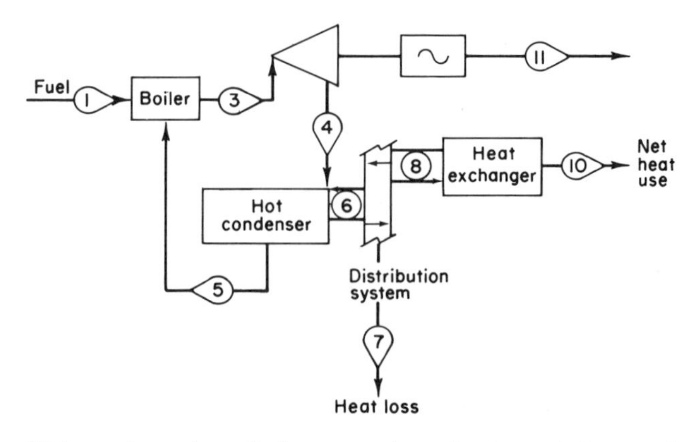

Fig. 10. Water system schematic for comparison of water and steam systems (see Fig. 11).

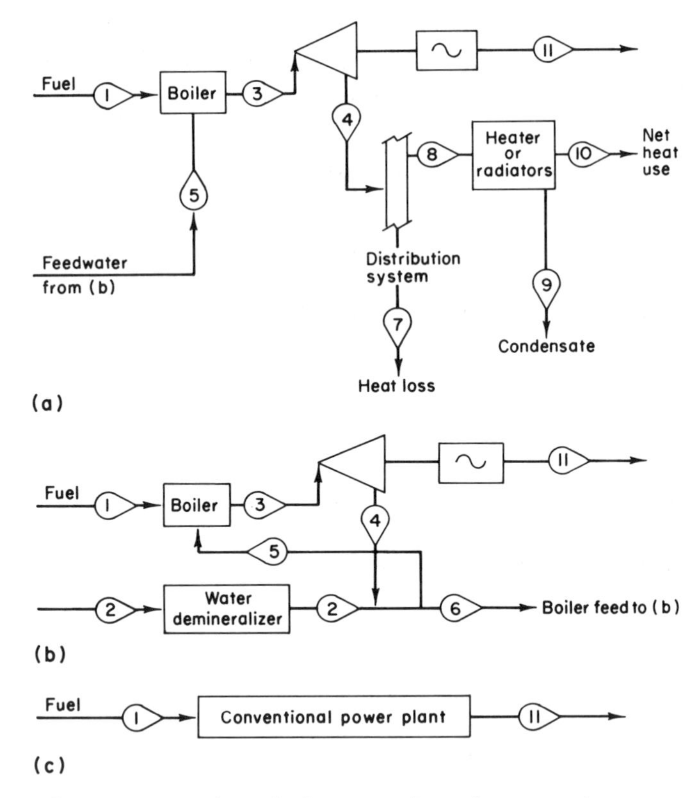

Fig. 11. Steam system schematic for comparison of water and steam systems. (a) Back-pressure steam system. (b) Feedwater heating for (a). (c) Conventional unit to supplement (a) and (b).

TABLE II. Comparison of Water and Steam Systems

Location	1	2	3	4	5	6	7	8	9	10	11
I. Water system											
a. Substance[a]	F	—	S	S	W	H	H	—	—	H	E
b. Quantity (lb)	—	—	1080	1080	1080	—	—	—	—	—	—
c. Pressure psi (absolute)	—	—	1500	17.2	17.2	—	—	—	—	—	—
d. Temperature (°F)	—	—	1000	220	220	—	—	—	—	—	—
e. Enthalpy (Btu/lb)	—	—	1490	1142	188	—	—	—	—	—	—
f. Heat content (K Btu)	1654	—	1609	1234	203	1031	31	1000	—	1000	—
g. Electricity (kW h)	—	—	—	—	—	—	—	—	—	—	110.02
IIA. Steam system											
a. Substance[a]	F	—	S	S	W	—	S	S	W	H	E
b. Quantity (lb)	—	—	905	905	905	—	90	815	815	—	—
c. Pressure [psia (absolute)]	—	—	1500	200	—	—	200	200	—	—	—
d. Temperature (°F)	—	—	1000	549	220	—	549	549	100	—	—
e. Enthalpy (Btu/lb)	—	—	1490	1295	188	—	1295	1295	68	—	—
f. Heat content (K Btu)	1387	—	1349	1172	170	—	117	1055	55	1000	—
g. Electricity (kW h)	—	—	—	—	—	—	—	—	—	—	51.66
IIB. Feed heating for IIA											
a. Substance[a]	F	W	S	S	W	W	—	—	—	—	E
b. Quantity (lb)	—	905	161	161	161	905	—	—	—	—	—
c. Pressure [psi (absolute)]	—	—	1500	17.2	—	—	—	—	—	—	—
d. Temperature (°F)	—	50	1000	220	220	220	—	—	—	—	—
e. Enthalpy (Btu/lb)	—	18	1490	1142	188	188	—	—	—	—	—
f. Heat content (K Btu)	247	16	240	184	30	170	—	—	—	—	—
g. Electricity (kW h)	—	—	—	—	—	—	—	—	—	—	16.40
IIC. Conventional unit to supplement IIA and IIB											
a. Substance[a]	F	—	—	—	—	—	—	—	—	—	—
f. Heat content (K Btu)	420	—	—	—	—	—	—	—	—	—	—
g. Electricity	—	—	—	—	—	—	—	—	—	—	41.96
Total for IIA, IIB, and IIC	2054	—	—	—	—	—	—	—	—	1000	110.02

[a] F = Fuel, W = water, S = steam, E = electricity, H = heat.

system supplies 1,000,000 Btu of useful heat energy and 110.02 KW h of electric energy. Column 1 in Table II illustrates that 24% more fuel is required by the steam system to supply the same power requirements as the hot water system.

The hot water system model used here is a straightforward application of combined heat and power technology to a back-pressure turbine scheme (see columns 1–11 which refer to Fig. 10 under *Water system,* Table II). The steam system model, however, is complicated by the need for feedwater heating (no condensate is returned to the steam combined heat and power generation plant), and by the need for additional electric generation to match the output of the hot water system. (This power output difference is made up by a conventional condensing-type electric generating plant.) Thus, Fig. 11 consists of the following:

(a) A straightforward back-pressure steam system which supplies 1,000,000 Btu of useful heat to the distribution system while generating electric power (but substantially less power than the hot water system supplying the same quantity of heat) (IIA in Table II).

(b) A back-pressure steam system which supplies heat for boiler feedwater while producing a small quantity of electricity. (At the steam combined heat and power generation plant, this steam and electric generation would be integrated with the system of (a). It is separated here to demonstrate the fuel requirements for feedwater heating which are associated with the steam system, which does not incorporate condensate return) (IIB in Table II).

(c) A conventional condensing power plant which supplies the electric power required to match steam system output to the higher electric output of the hot water system (IIC in Table II).

The data associated with each component of this steam system model are displayed in Table II under the appropriate column numbers (which refer to the schematics of Fig. 11). To summarize the numbering scheme of Table II, we have:

1. Fuel input to boiler.
2. Water input to boiler feedwater makeup
3. Steam from boiler output.
4. Steam exhausted from back pressure turbine.
5. Boiler feedwater.
6. Boiler feedwater makeup.
7. Heat lost in district heating distribution system.
8. Heat input to customer's heating system.
9. Condensate discharged to sewer.

10. Heat used in customer's heat system.
11. Electric energy generated.

In the calculations for Table II a boiler efficiency of 85% has been used. Turbine efficiency to electric power has been taken as 80% of the maximum theoretically obtainable by expansion to the pressure indicated, with the balance of the heat appearing in the exhaust steam. For simplicity in calculation only single-stage feedwater heating has been assumed, hot water heating in system I has been assumed to be single stage, and auxiliary power requirements have been ignored.

It may be noted that in the above example the water and steam systems are brought to equal outputs, for comparative purposes, by adding the production of electric power in a conventional system (IIC). This implies that the heat load is limited in size and that the electric load will never be limited. This will always be the case when there is a connection to the grid of adequate capacity. For isolated systems (no grid connection) the higher fuel used by the steam system is only about 10%.

The major reasons for the higher fuel consumption for the steam system are (1) the need to generate more electric power in conventional units, (2) higher distribution losses (10% versus 3%), and (3) loss of heat by the customer in the condensate.

There are other less dramatic advantages to the hot water system including the following:

1. Hot water has economically been distributed, at a constant pressure, as far as 60 km (37 miles) with pumping power requirements of only 0.5% to 3% of the system's thermal power capacity. This permits greater flexibility in scheduling heat delivery from the most economical stations at times of low load. By contrast, steam distribution is practicable only up to distances a mile or two from the steam plant.
2. The simplicity of a low-pressure integrated hot water system affords great system reliability.
3. Steam metering is much more difficult than hot water metering, causing greater amounts of unaccounted-for steam.

These are the factors that have caused the longevity and continuing growth rates for the European hot water systems. The greater fuel efficiencies of hot water systems can reverse the unfavorable economics for district heating in the United States, considering the escalating cost of fuel. The interested reader will find a more detailed comparison of steam and hot water district heating systems in Muir (1975).

IV. DEVELOPING DISTRICT HEATING SYSTEMS

A. The European Step-by-Step Approach

The definition of district heating in Sweden and other countries in Europe is that of an economical combination of electric power and heat generation from hot water condensers and the distribution of that hot water in a large distribution network.

To get a profitable system, district heating has to be built up in at least three stages. The first stage (Fig. 12) in this system starts with district heating in new building projects situated not far from the town center or in the rebuilding of large areas. These buildings are centrally heated and will be supplied with heat from one or several hot water boiler stations or transportable boilers. This is called group heating schemes.

In due course a point is reached when it is time to link up the isolated group heating schemes and the old small boiler stations in the city and supply them with heat from one source. This is where the municipality or power company may step in and take over the heat supply, which is stage two.

Some of the smaller boiler stations will then be shut down and scrapped, and the transportable boilers will be moved to another area, but the larger ones will still be required for providing stand-by capacity and for meeting peak loads during cold weather.

During these first two stages, the electricity demand is provided for in the conventional way. When the heat demand has grown so much that it becomes economical to generate electricity and heat in combination, a back-pressure turbine power plant will be installed. This is stage three. In Sweden many years may pass between the first group-heating scheme and this combined power and heat generating plant.

To realize this system it is necessary to have a strong organization represented by some authority or promoter which can give the directions for feasibility studies, the investigations, the planning, and the realization of the whole program. The authority or promoter must also decide on political, economic and technical questions and coordinate different utilities. For these purposes consultants or other technical experts can give technical advice and do the investigations and the project work. The promoter is usually the local power company, typically owned by the municipality. An early individual promoter was John Sintorn in Västerås, who has given an interesting and often amusing account of his pioneering effort (Sintorn, 1974).

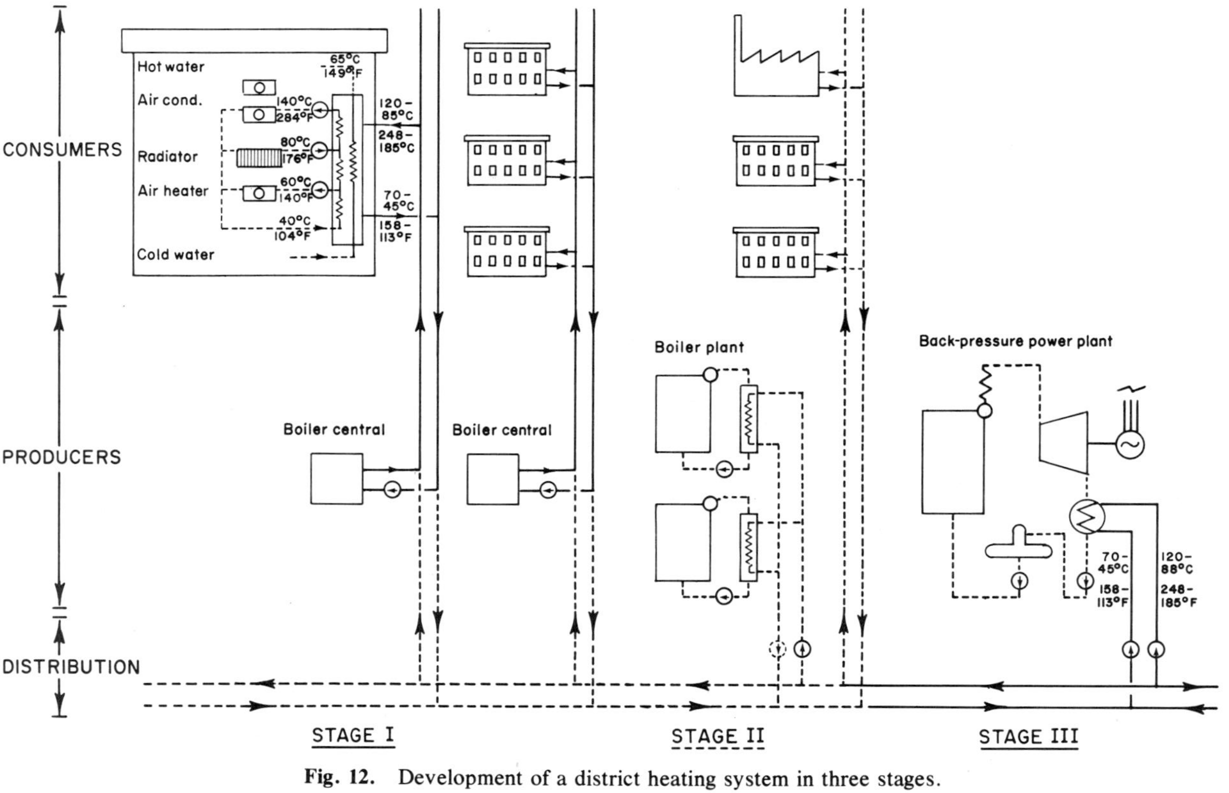

Fig. 12. Development of a district heating system in three stages.

B. Requirements for Combined Heat and Power Generation in American Public Utilities

The following barriers prevent utilization in the United States of established European technology for CHP/DH:

1. We have already pointed out that United States public utility heat power practice is incompatible with European CHP practice. This is illustrated in Table III. The large United States units, with advanced steam conditions, do not lend themselves to the typical European approach of obtaining hot water for the district heating distribution system by back pressure operation of the turbine.

2. Peak loads in most United States utility systems are produced by air conditioning, in the summer. Moreover, many systems have long slack periods for both heating and cooling in spring and fall. This is an institutional problem, in that the public cannot be induced to give up air conditioning. The air conditioning load now requires the utility to maintain large extra capacity that is little used. Addition of a district heating system of the European type would constitute another large investment in capacity that would not be used during the summer. These load conditions are much different from what is found in Europe.

3. Steam and electric heating systems in buildings are incompatible with modern district heating technology. A large proportion of existing buildings in American cities are heated by these systems. In the course of 40 years or so, replacement of old buildings with new buildings with modern, hot water heating systems may take care of this problem. This leads to barrier No. 4.

4. The step-by-step approach is the fundamental basis for the success of the much admired Swedish model of CHP/DH. It has been described above. The step-by-step approach consists essentially in building up a load of buildings heated by modern hot water systems until it becomes economical to construct new central CHP/DH stations. The DH network is therefore already in place in advance of the CHP station. By contrast, the immediate need in the United States is for retrofit. It would take too

TABLE III

Characteristics of United States and Swedish Steam Turbines for Electric Generation

	United States	Sweden
Sizes[a]	500–1300 MW	50–300 MW
Throttle pressure	2400–3500 psi	≤2400 psi
Mode	Condensing	Back pressure

[a] Estimate.

long and cost too much to convert existing steam and electric heating systems in buildings to hot water in sufficient numbers prior to building a CHP station. The Swedes estimate 50,000–100,000 urban population is the smallest feasible DH load required to justify construction of a new CHP station. This corresponds to about 60 Mw (e) to 120 Mw (e) electric generating capacity. For the United States, the minimum population required is likely to be over 300,000, because of larger generating units and other differences.

5. Public utility systems take 5–10 yr to build and must operate for 35–40 yr. District heating mains last even longer: 50–60 yr. Therefore, changes appear as increments to the existing system, and spread slowly over a period of many years. To have any significant impact in the near term (5–15 yr) during which our oil and gas supplies will be vanishing rapidly, approaches to CHP/DH are required for the United States that do not have to wait for radical changes in existing public utility systems and loads.

C. Institutional Barriers

Swedish and Finnish engineers pioneered the development of the system of CHP/DH used in Europe. The basic design concept of back-pressure steam turbine combined with hot water distribution satisfied the load conditions and utility practices found in Europe very well. Engineers then worked as a team with planners and architects to overcome the institutional barriers to implementing the technology.

In Western Europe, the direct institutional influence on the development of district heating has been negligible. District heating developed in a free market. Customers hook up to the system because it is to their advantage. Only recently have governments intervened to encourage district heating as a conservation measure. However, European promoters of district heating had the inherent advantage that *all* utilities in many European cities are municipally controlled. Thus taxes on the generating and distribution systems are often low or nonexistent, and the utility need not earn a profit. Nevertheless, in general many institutional barriers had to be overcome. Some of these barriers were:

Tariff structure
Tax disincentives
Building codes and practices
Public utility charters: electricity and gas only
Relative return on investment
Relative payback times
Capital shortage
Cheap fuel favors waste

Cheap hydropower
Restrictive labor practices
Public apathy

Sintorn's (1974) article illustrates very well how Swedish engineers overcame these institutional barriers.

It should be noted that very little European development involved retrofit. Sweden had little steam electric capacity in 1948, and the other leading countries in CHP/DH had to rebuild after being devastated by World War II.

It appears that a major institutional factor inhibiting the growth of district heating and CHP in the United States has been the absence of a "champion" or "promoter." In Sweden the role is filled by city or town administrative bodies charged with the responsibility for delivering all utility services to the community. Connection to the district heating system has been on a voluntary basis and the systems have not been directly subsidized, but property taxes are not assessed against the generating plants or distribution facilities and they are operated on a nonprofit basis. These amount to substantial financial benefits, as compared to the situation faced by American investor-owned utilities, but it appears that an even more important factor is the provision in Sweden of dedicated and informed attention to planning and implementing the needed projects. In the United States those few public utilities that still operate steam distribution systems seem to have no incentive to expand or modernize them nor to devise and promote new and improved systems.

If a viable, economical, and advantageous CHP/DH industry is to be developed in the United States it will need, for each location, some "agency" or "authority" or "company" or other group to be responsible for planning, financing, building, and operating the system. The identity of such an entity is not now clear and indeed it seems likely there will be no need for it until such time as the merits of CHP/DH are firmly established.

V. APPROACHES TO CHP/DH SYSTEMS FOR THE UNITED STATES

Hypothetical examples of directly applying the Swedish system in the United States have been worked out by Swedish experts in designing CHP/DH systems. These were necessary and important exercises, contain much useful material, and may be a basis for future developments, but they do not provide a basis for rapid action within the constraints we have discussed.

As first requirements, we must accept the immovable institutional barriers as working constraints. For new cities, we must develop a workable CHP/DH system that includes air conditioning. This alone is a formidable task; but the era of rapid expansion of cities and suburbs is over. Our main need is for retrofit designs, to serve existing loads while reducing fuel usage. We will consider several possible approaches, suitable for application in American cities.

In developing approaches to district heating for the United States we must consider in more detail the differences between Europe and the United States that have already been mentioned. These differences relate to: (1) weather, (2) custom or historical development, (3) technology, and (4) institutional factors.

Figure 13 is a plot of average monthly temperatures for three areas: New York City, upstate New York, and Stockholm. The shaded area designated New York City includes average monthly temperature readings taken at Central Park, LaGuardia Airport, and Kennedy Airport (National Oceanic and Atmospheric Administration, 1974, pp. 271–272). The shaded area designated upstate New York includes average monthly readings taken at the airports that serve Albany, Binghamton, Buffalo, Rochester and Syracuse (National Oceanic and Atmospheric Administration, 1974, pp. 270–271, 273). Data for Stockholm were obtained from a Swedish publication (Anderson, n.d.).

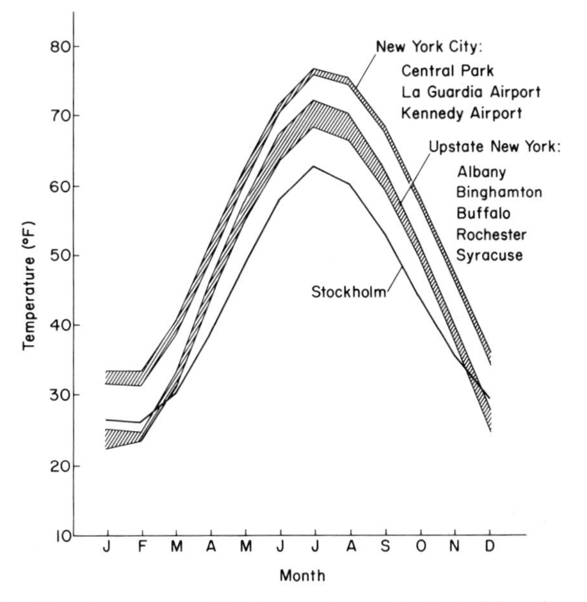

Fig. 13. Average monthly temperature at indicated locations.

Examination of Fig. 13 indicates that Stockholm does not have substantially different temperatures from New York during the winter months but that it does have cooler summers. There is little need for air conditioning in Stockholm and its use is limited primarily to a small number of modern office buildings, etc. The summer air conditioning load is one of the most important differences between Stockholm and New York. Summer electric peaks have exceeded winter peaks in most portions of this country for many years. Maximum hourly peaks in megawatts in New York have been as follows:

	Winter	Summer
1973	5898	8220
1974	5898	7973
1975	6056	8051

Planning for increased electric generating capacity in the United States must give major consideration to helping meet the summer electric peak load. A back-pressure steam turbine would have limited value, and for any large steam unit only a configuration that permits full condensing should be given serious consideration. This conclusion is reinforced by other factors, such as: (1) large condensing units conform to established design practices, whereas large backpressure units do not, and (2) even in Sweden there seems to be a provision for full condensing operation for units that exceed about 100 MW.

The prosaic conclusion that steam electric units should be designed for full condensing summer operation does not mean that there is no place here for combined heat and power production. Rather, it means that the American configuration should be different than the Swedish configuration to adapt to the different requirements. One adaptation that has been suggested is the widespread use of absorption refrigeration air conditioning systems.

Absorption systems use heat in the form of steam, hot water, or a gas flame to provide refrigeration for use in air conditioning systems. They operate without moving parts and therefore require a minimum of maintenance and operating attention. They will continue to be favored where these attributes are more important than energy consumption, but they are inherently low-efficiency units and will not be considered seriously for new large district heating and cooling systems. In terms of coefficient of performance (COP), which is defined as the ratio of energy removed to energy input (at refrigeration temperature), a common value for an electric driven mechanical refrigeration system is 4.0 (Holmes, 1951) compared with 0.6 for a one-stage absorption system or 0.9 for a two-stage system (Trane Company, 1976). Thus even when allowance is made for low conversion efficiencies from low-pressure steam to electric power, the objective of minimum fuel oil consumption to serve a given load will be achieved through the use of electric driven (or occasionally direct steam driven) mechanical refrigeration systems, rather than absorption systems. The energy savings may not be sufficient to justify the replacement of existing absorption units, but there does not seem to be any place for new absorption units for use with a district heating and cooling system.

The major technical factor distinguishing existing American from Swedish district heating systems is the use of steam rather than hot water as a distribution medium. Replacement of steam with water seems completely impractical for an entire system, but not necessarily so for limited portions of existing systems and for extensions thereof. Another technical factor is the very large size and extreme sophistication of American steam–electric generating plants. These do not lend themselves readily to adaptation for the dual supply of electricity and heat.

A. Alternative Approaches to the Development of a Viable CHP/DH System

Alternatives that have been considered for the development of CHP/DH systems in the United States are (1) coordination with new city construction, (2) the step-by-step approach, (3) retrofit of an old condensing turbine unit, and (4) integration of CHP/DH with existing major electric generating stations.

1. *Coordination with New City Construction*

When a new town or a new major complex of buildings is planned there is an opportunity to provide economies by designing for CHP from the start. The cost of separate furnaces and separate fuel handling facilities can be avoided. This approach has been advantageous in Sweden where CHP/DH is accepted and it is known that the district heating supply system is available. Similar advantages in the United States may be expected in the future after CHP/DH is accepted.

2. *The Step-by-Step Approach*

This approach is described in Section IV,A. It has been followed in Sweden with obvious success. Throughout the development of a system in this manner there are continual extensions to the heat distribution system and continual opportunities for the connection of existing individual heating systems to the network.

The success of this method of development in Sweden is beyond question and it may some day be applicable in this country. For the present, however, this does not seem to be a promising approach here for the following reasons:

1. Viability and future existence of CHP systems have not been established and accepted.
2. Required standardization of pressure, temperature, and system designs cannot be enforced.
3. Motivation to proceed is inadequate.
4. Even if a subsidized and intensive development program were undertaken in a selected area, a great many years would be required before the load could be developed to a size that would justify a new back-pressure steam turbine plant. Additional time would be required for siting studies, approvals, etc.

3. *Retrofit of an Old Condensing Steam-Turbine Plant*

This approach was the subject of a paper by Muir (1977). It describes the conversion of old, small condensing turbine units to back-pressure service by removing the last few rows of turbine blades and making other appropriate changes. The approach seems to have substantial merit, particularly for consideration for a community of moderate size where no district heating system has previously been installed. Drawbacks, from a development standpoint, are as follows:

1. It is not widely applicable, but is limited in application to those areas near a suitable old power plant.
2. It seems to avoid, rather than face up to, the problems of high building density areas where there is the greatest long-term potential for energy savings.
3. It accepts the low level of steam temperatures and pressures common to old units, rather than devising an arrangement that permits maximum power-to-heat ratios obtainable with modern steam temperatures and pressures.

B. CHP/DH Integration with an Existing Major Generating Station

The planning, authorization, and construction of a new power plant is a lengthy and uncertain process. Probably 10 yr would be required under the most favorable circumstances, and perhaps longer where innovation is involved or if it is planned for an area of high population density. It would also be difficult to choose between a smaller plant, with higher unit costs, and a large plant, with the requirement for a larger district heating load. Some of these problems can be avoided by utilizing an existing modern electric generating plant as a source of heat for a CHP/DH system. This might make possible earlier unsubsidized operation of a modern CHP/DH system and provide the basis for future growth.

The recent poor experience of American public utilities with district heating is largely due to their exclusive use of steam as the distribution medium. But this poor experience has undermined confidence in district heating *per se* as a profitable business venture. In order to demonstrate that hot water distribution can make a major difference, it is desirable that hot water and steam systems be observed in close proximity. Moreover, the best chance for a successful demonstration in the United States would seem to be in a densely populated area (to provide a high ratio of load to capital cost), and on a system of a utility with experience in district heating, of which there are 44 in the United States. In order to illustrate the type of system that might be developed, we focus our attention on one of these systems. The service area is shown on Fig. 14. Reasons for choosing this area are as follows:

(1) The area includes buildings having various occupancies, some now served by the existing district steam system and some that have separate fired boilers. It therefore may illustrate benefits of both revision and expansion.

(2) The age and character of the buildings is such that it may be expected that hot water heat distribution systems predominate within the buildings. These are more readily adaptable to the proposed technology; the conversion of a steam radiator system to hot water would be very costly.

(3) An existing 20-in. steam line can supply the necessary connection from a modern steam electric plant.

(4) The power station connected to the steam line is the most modern one accessible to the system.

(5) The steam supply to the area now comes from oil-fired boilers located at the power station supplying the 20-in. steam line, and topping turbines at three older plants.

Figure 15 represents a modern steam electric generating unit, and Fig. 16 shows a means by which the unit may be used to supply heat to a hot water district heating system. A portion of the steam would be diverted from the reheat system, desuperheated to the extent necessary and passed through the existing steam line to feed a new back-pressure steam turbine-generator. Turbine exhaust would heat water in a new district heating piping system connected to a group of buildings in the area.

The hot water distribution system would follow the same design principles as are common in Sweden but would be used for hot water only during the winter months. A separate small domestic hot water distribution system would be provided.

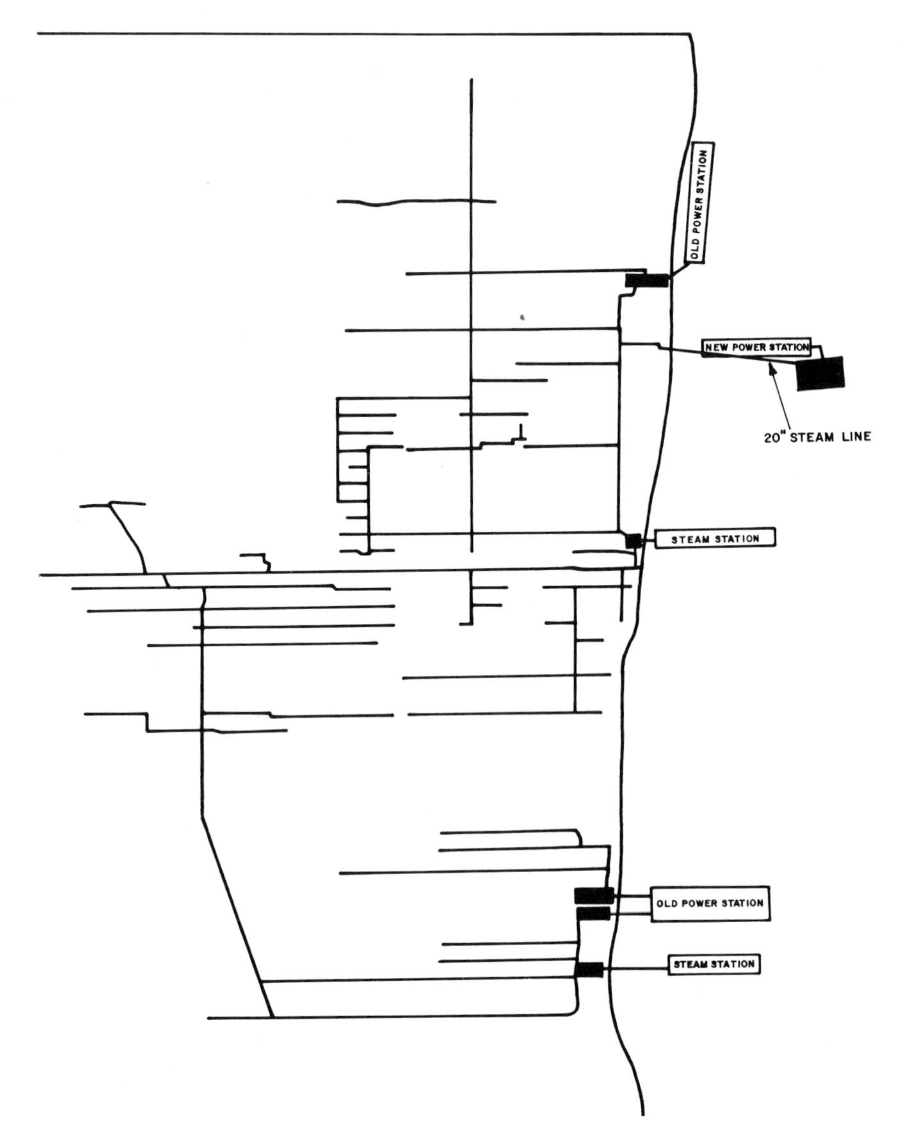

Fig. 14. Existing district steam system.

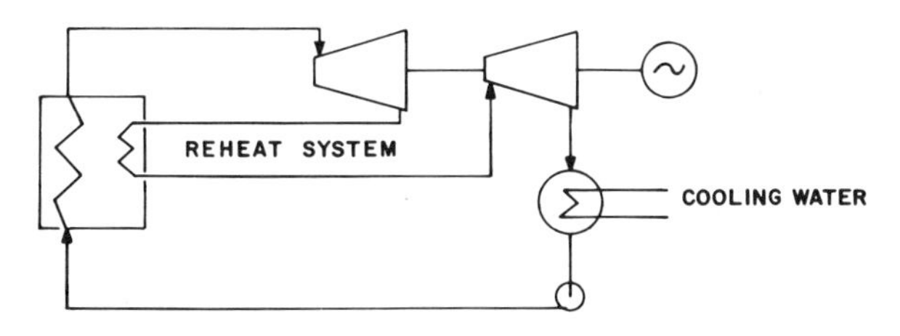

Fig. 15. Conventional turbine setup.

During the summer months the distribution system would be used for chilled water service, supplying the air conditioning needs. Existing electric driven (and possible some other) refrigeration machines would be used to supply the cooling. This would be, in essence, the interconnecting of existing chilled water systems to obtain the advantages, by selection, of optimum machine loading and reduced operating and maintenance costs.

Substantial storage capacity for hot water in winter and chilled water in summer would be used to reduce peak loads and obtain higher utilization of night capacity. In winter this would apply to the steam system and would permit decreased use of direct steam supply and increased use of CHP/DH. In summer, this would apply to the electric system, and would decrease the requirement for peaking capacity and also decrease the requirement for supply of electricity from the higher cost sources, such as purchase or from gas turbines.

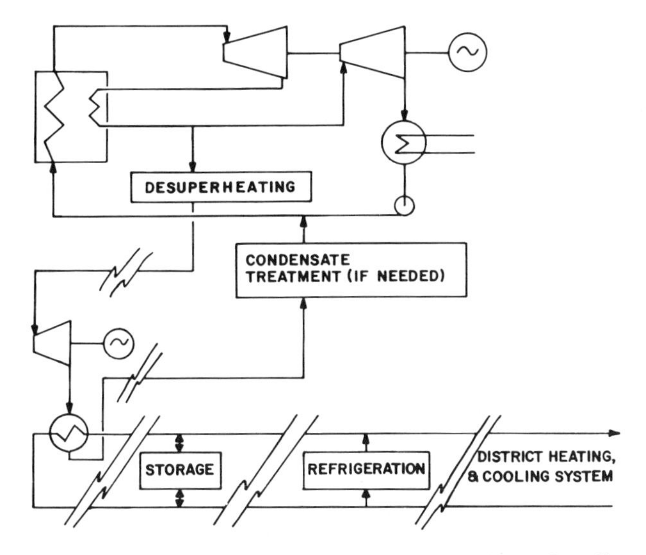

Fig. 16. Turbine setup modified for district heating and cooling.

Advantages may be summarized as follows:

1. Reduced winter fuel oil consumption, resulting from the use of a high efficiency high power-to-heat ratio, combined heat and power system as a partial replacement for:
   (a) small separate building heating furnaces
   (b) steam from packaged boilers now used
   (c) steam from existing old, low power-to-heat ratio CHP/DH units
2. Further reduced winter fuel oil consumption resulting from higher load factors made possible by hot water storage
3. Reduced summer fuel oil consumption resulting from:
   (a) preferential use of higher efficiency refrigeration units
   (b) off-peak use of higher efficiency generating units, made possible by chilled water storage
4. Reduced air pollution
5. Reduced maintenance and operating costs
6. Greater dependability

This system seems to have promise as both a prospective demonstration project and as a nucleus for expansion. Implementation could be achieved at a much earlier date than would be possible if the construction of a new steam plant were required. The analysis itself could provide valuable information as to the practical problems and costs to be encountered in planning for retrofitting CHP/DH systems to existing buildings.

VI. CONCLUSION

Swedish success in combined heat and power/district heating is a source of inspiration and much technology, but not a model for the United States because of the wide differences we have discussed. For this reason, American engineers with experience in CHP/DH plant design are much less euphoric about the ready adoption of CHP/DH by American utilities than popular accounts of the subject would lead one to expect.

On the other hand, the extremely pessimistic attitude about CHP/DH often displayed by American utilities is the result of poor experience with steam systems, and is not justified in relation to modern developments. Application of such developments continues to have very low priority in the plans of utilities struggling to satisfy electric service demands.

The retrofit approaches suggested in this chapter appear advantageous as first steps to breaking this apparent deadlock. They make good use of existing plant capacity, and require minimum investment in new facilities to achieve a demonstration of appreciable size. They can be implemented quickly, since they are applicable to serve existing buildings. They would enable ready comparison with other types of heating system in the same area.

The next step required is to verify the premises of the proposed retrofit design, and to do preliminary calculations of engineering and economic feasibility. If such a preliminary analysis confirms the apparent merit of the design, a full-scale engineering analysis would follow, and produce a prospectus for gaining approval and financing of the demonstration project.

The Swedes developed a successful case history, Västerås, which showed the possibilities. Public demand for district heating enabled breaching the movable institutional barriers and rapid expansion followed. Government was apathetic at first. Enterprising engineers and managers led the way, because of economic benefits of more efficient fuel utilization.

The United States with its urban concentrations and escalating energy costs, needs combined heat and power/district heating to conserve scarce, expensive oil supplies. The driving force is here: the need to save some of the fuel energy now rejected in electric generation. What is needed is a successful case history: a workable retrofit design that will be the American equivalent of Västerås. After such a design has been demonstrated to work, planners and builders can be called in to develop its use on a large scale.

ACKNOWLEDGMENTS

The assistance of R. D. Glenn and T. E. Mullen was particularly helpful in the preparation of this chapter. The concept of a hybrid system, extracting turbine steam to drive a back pressure district heating turbine, was originated by R. D. Glenn. J. E. McConnell and W. B. Wilson contributed time and much valuable material. G. Berg and E. Wahlman helped by working out a detailed example of district heating for the United States, and through many stimulating discussions. Thanks are due to N. Muir for several discussions and papers, and for making a special trip from Stockholm to New York to aid the author's studies of district heating. Others whose cooperation was most helpful include H. Bremer, D. Romaine, and J. E. Ryman. Lastly, we are indebted to A. Squires and M. Gluckman for arousing our interest in district heating several years ago.

BIBLIOGRAPHY

A great deal has been written during recent years on subjects related to district heating and the combined production of heat and power. Most of it is not published in the usual technical journals, but rather as articles contained in the proceedings of conferences and privately published articles from industrial firms and government agencies.

This bibliography is a brief listing with comments, of some of the work we have found most useful, in addition to the references given in the text. We have made no attempt to cover all of the literature and have omitted many articles solely because the information they contained has been covered in some other article that we happened to have come across earlier. In short, the attempt here is to be helpful to those who wish to extend their knowledge of the subject.

1. *Proceedings of the International District Heating Association.*

This association, with offices at 5940 Baum Square, Pittsburgh, Pa. was organized in 1909 and has held an annual meeting each year except 1918. Proceedings are published and contain extensive statistics for the industry and a series of technical papers. They are essentially concerned with *steam* distribution systems.

2. *District Heating.*

This publication is a magazine printed quarterly by the International District Heating Association. It is also primarily concerned with steam distribution systems.

3. *Proceedings of the First International Total Energy Congress,* sponsored by Energy International, Copenhagen, October 4–8, 1976 (E. Jeffs, ed.). Freeman, San Francisco, California, 1977.

About 35 articles were presented. Those of particular interest and within the scope of our interests in CHP are as follows:

3.1 Aamot, H. W., and Phetteplace, G., Long distance heat transmission with steam and hot water, p. 517.

3.2 Larson, K., Computer optimization of district heating schemes, p. 550

3.3 Farkas, G.S., and Quaresi, S., District heating and cooling in Canada—the energy profile of cooling, p. 718 h

3.4 Muir, N., Conversion of existing turbines for combined power and heat production.

4. *Energy Conservation Through District Heating—A Technical Study Trip to Sweden, April 11–19, 1975.*

This trip was organized by the Swedish Export Council and the Swedish Trade Commissioner for a Canadian delegation. The Swedish Trade Commission maintains its United States office and staff in Chicago (333 N. Michigan Ave., Chicago, Ill. 60601) Contributions of particular interest are:

4.1 Unden, B., Production of heat and electric power, separate and combined.

4.2 Muir, N., Turbinology—a heat power and fuel balancing problem.

4.3 Wahlman, E., Systems and hardware for local and district distribution. Water versus steam, pros and cons.

4.4 Ryman, J.-E, District heating—an introduction to the Stockholm system.

4.5 Sintorn, J., How we started in Västerås.

5. *District Heating, Avenue to Energy Conservation—Swedish Experience and Technology.*

This volume is a compilation of articles used in a series of symposia arranged by the Swedish Trade Commission and presented at several locations in Canada in the fall of 1976. Articles of particular interest are as follows:

5.1 Wahlman, E., District heating—a step by step approach.

5.2 Muir, N., Combined district heating and electricity production in Sweden. Stal-Laval Tech. Inf. 1/76.

6. Harboe, H., "Importance of Coal for Heat and Power Generation." Stal-Laval, Finspong, Sweden, 1974.

This work deals with CHP, fluidized-bed combustion of coal, air storage, etc., and their interrelationships.

7. *Proceedings of the Dual Energy Use Systems Workshop.* Electric Power Research Institute, Palo Alto, California, 1978.

This volume is a compilation of articles presented at a conference held in September 1977. It covers all forms of combined heat and electric generation.

REFERENCES

Anderson, (n.d.). "Swedish Temperature and Precipitation Records Since the Middle of the 19th Century," p. 108. Natl. Swed. Inst. Build. Res., Stockholm.

Berg, C. A. (1974). *Mech. Eng.* May 1974.

Coles, W. (1977). Costs of adapting boilers to coal. *Proc. Coal Conversion Conf.*, New York State Legislative Comm. Energy Syst., Albany.

Empire State Electric Energy Research Corp. (1977). ESEERCO Rep. No. 149b, Vol. 1, pp. 369, 391. New York.

Holmes, R. E. (1951). "Air Conditioning in Summer and Winter," 2nd ed., p. 210. McGraw-Hill, New York.

McClure, C. J. R. (1977). Total energy plant concept failures. *ASHRAE Energy Conv.*, Feb. 1977.

McConnell, J. E. (1976). "A Coordinated Approach To Industrial Energy Systems Design." Allm. Sven. Elek. Ab., White Plains, New York.

McConnell, J. E., and Elmenius, L. (1975). Improved evaluation techniques for dual-purpose power plants. *Am. Power Conf., 37th,* April, 1975.

McConnell, J. E., and Elmenius, L. (1976). A systematic approach to the economic selection of design parameters for an integrated industrial power plant. *Annu. Am. Power Conf., 38th, Chicago, Ill.*

Margen, P. (1974). *In* "Combined District Heating and Power Generation," p. 38 Swed. Export Counc., Stockholm.

Muir, N. (1973). "District Heating in Sweden, 1973," (together with two exercises on load forecasts and costs associated with the introduction of district heating in Dublin), Stal-Laval Tech. Inf. 6/73. Stal Laval, Finspong, Sweden. (Also available from Allm. Sven. Elek. Ab., White Plains, New York.)

Muir, N. (1975). "District Heating—The Water Approach," Lecture to the Canadian Boiler Society. Stal-Laval Turbin AB, Finspong, Sweden.

Muir, N. (1977). Conversion of existing turbines for combined power and heat production. *Proc. Int. Total Energy Cong., 1st, Copenhagen, 1976* p. 767.

Mullen, T. E. (1977). "Total Energy System Application in the U.S." New York State Energy Res. Dev. Auth., New York.

National Oceanic and Atmospheric Administration (1974). "Climates of the United States," Vols. 1 and 2. U.S. Dep. Commer., Water Inf. Cent., Port Washington, New York, pp. 271–272.

Olmstead, L. M. (1974). *Electr. World* Sept. 15, 1974.

Proc. Int. Dist. Heat. Assoc. (1974). p. 20.

Proc. Int. Dist. Heat. Assoc. (1975). p. 22.

Sintorn, J. (1974). How we started in Västerås. *In* "Combined District Heating and Power Generation," p. 10. Swed. Export Counc., Stockholm.

Spielvogel, L. G. (1977). "Total Energy." Lawrence G. Spielvogel, New York.
Trane Company (1976). "2-Stage Absorption Cold Generator, Tech. Bull. TS-ABS2, p. 2. La Crosse, Wisconsin.
Unden, B. (1975). Production of heat and electric power separately and combined. *In* "Energy Conservation Through District Heating," Ch. 4. Swed. Export Counc., Stockholm. (Also available from Swed. Trade Comm., Chicago, Illinois.)
Wilson, W. B., and Kovacik, J. M. (1976). Selection of turbine systems to reduce industrial energy costs. *Proc. Mid-Year Meet., 41st Am. Pet. Inst.,* Reprint No. 18-76.
Winkens, H. P., Molter, F. J., and Neuffer, H. (1976). *Fernwarme Int. FWI* **5,** No. 5, p. 134.

Subject Index

T

A
B 8
C 9
D 0
E 1
F 2
G 3
H 4
I 5
J 6